Basic Quadratic Forms

Basic Quadratic Forms

Larry J. Gerstein

Graduate Studies in Mathematics

Volume 90

American Mathematical Society
Providence, Rhode Island

2000 *Mathematics Subject Classification.* Primary 11Exx, 12Exx, 15–XX.

For additional information and updates on this book, visit
www.ams.org/bookpages/gsm-90

Library of Congress Cataloging-in-Publication Data

Gerstein, Larry J.
Basic quadratic forms / Larry J. Gerstein.
p. cm. — (Graduate studies in mathematics, ISSN 1065-7339 ; v. 90)
Includes bibliographical references and index.
ISBN 978-0-8218-4465-6 (alk. paper)
1. Forms, Quadratic. 2. Equations, Quadratic. 3. Number theory. I. Title.

QA243.G47 2008
512.7′4–dc22 2007062041

∞ The paper used in this book is acid-free and falls within the guidelines
established to ensure permanence and durability.
Visit the AMS home page at http://www.ams.org/

10 9 8 7 6 5 4 3 2 1 13 12 11 10 09 08

To

SUE, DAVID, and BENJY

Contents

Preface

The theory of quadratic forms has a long and glorious history: launched in ancient Babylonia between 1900 and 1600 BC, taken up again by Brahmagupta in the Seventh Century, and then—another thousand years later—by the great genius Fermat, followed by a succession of extraordinary mathematicians, including Euler, Lagrange, and Gauss, who brought the subject closer to its modern form. The work of Minkowski in the late Nineteenth Century, coupled with the extension of his work by Hasse in the early Twentieth Century, led to a great broadening and deepening of the theory that has served as the foundation for an enormous amount of research that continues today.

Though the roots of the subject are in number theory of the purest sort, the last third of the Twentieth Century brought with it new links of quadratic forms to group theory, topology, and—most recently—to cryptography and coding theory. So there are now many members of the mathematical community who are not fundamentally number theorists but who find themselves needing to learn about quadratic forms, especially over the integers. There is thus a need for an accessible introductory book on quadratic forms that can lead readers into the subject without demanding a heavy background in algebraic number theory or previous exposure to a lot of sophisticated algebraic machinery. My hope is that this is such a book.

One of the special attributes of number theory that distinguishes it from most other areas of mathematics is that soon after a subject is introduced and objects are defined, questions arise that can be understood even by a newcomer to the subject, although the answers may have eluded the experts for centuries. Even though this is an introductory book, it contains a substantial amount of material that has not yet appeared in book form, and the reader will be exposed to topics of current research interest. I will be happy if the readers find themselves wanting to pursue some aspects of the subject in more detail than this book can provide; accordingly, I will offer some references to the literature and recommendations for further study.

Before 1937, quadratic forms were treated primarily as homogeneous polynomials of degree 2 acted on by transformations that could change a given quadratic form into certain other ones. (And a fundamental question was: into *which* other ones?) But a pioneering paper by Witt in 1937 brought a more geometric flavor to the subject, putting it on the border of linear algebra and number theory—roughly speaking, a theory of generalized inner products on modules. Our coefficient ring of interest will most often be the ring $\mathbb{Z}$ of rational integers, though we will also give special attention to the polynomial rings $\mathbb{F}_q[x]$. (Here $\mathbb{F}_q$ denotes a finite field with q elements.) We will see that before we can effectively explore quadratic forms over a given domain R, we may need to extend R, perhaps in many ways, to larger rings. The extended domains (specifically, the p-adic number fields, their rings of integers, and their function-field analogues) may possess complications of their own that require clarification before we can consider quadratic forms over them; but once we have achieved that clarification, we may find that quadratic forms over those extensions are far more tractable than over R. When that happens, the trick is to then bring that information down to R and apply it to the original forms.

This book has evolved from lecture notes for introductory graduate courses on quadratic forms I have taught many times at the University of California, Santa Barbara, and once at Dartmouth College. Typically these courses have been populated by second-year graduate students who have already had a basic course in algebraic structures, and this is the primary audience I have had in mind during the writing process. But in fact the book should be readable by anyone with a strong undergraduate background in linear and abstract algebra who has also seen the construction of the real numbers from the rationals.

Naturally the contents of this book have been shaped by my own interests, experience, and tastes, and I have no doubt that some mathematicians will lament the absence of one or more of their favorite topics in the theory of quadratic forms. But I hope that their concerns will be eased by seeing in these pages some new perspectives—and occasionally something completely new—and that where the material is familiar they will experience the joy of revisiting old friends.

I thank Miklós Ajtai, Mark Gaulter, Arnold Johnson, Timothy O'Meara, Martin Scharlemann, Thomas Shemanske, and the anonymous referees for their helpful comments, and I especially thank Melissa Flora for her detailed reading—and numerous corrections—of nearly the entire manuscript. Of course any errors that remain are my own doing. I have appreciated TeXnical rescues from Caroline Johnson, Barbara Beeton, and Richard Spjut. Natalya Pluzhnikov's perceptive and thorough copy editing helped me put the manuscript in final form. And I thank editor Ina Mette of the AMS for her patience and encouragement, and for her thoughtful selection of excellent referees.

Finally, I thank my dear family for the inspiration, love, and encouragement that have sustained me throughout my work on this project.

Chapter 1

A Brief Classical Introduction

A development in graph theory or topology from last July may already be viewed as "ancient history," but in the theory of quadratic forms, and indeed throughout number theory, "ancient" really means ANCIENT. For example, Babylonian tablets dating back to 1900–1600 BC suggest that it was then known that there are infinitely many **primitive Pythagorean triples**: solutions $(a, b, c) \in \mathbb{Z}^3$ of the equation $x_1^2 + x_2^2 - x_3^2 = 0$, with $\gcd(a, b, c) = 1$; and it was known how to produce them. These results could be said to constitute the first theorem in quadratic forms; and historians suggest that this is the first instance on record in which mathematics was clearly being done for fun, not for commercial purposes or to determine property boundaries.

Historical Remark. Pythagoras actually lived during the period 580–500 BC, long after the above discovery to which we apply his name. See Boyer [**Bo**] for a thorough treatment of this early history.

We begin with the classical definition of quadratic forms. Later on we will see that it can be advantageous to view a quadratic form as a mapping on a module bearing an inner product.

1.1. Quadratic Forms as Polynomials

Definition 1.1. Let R be an integral domain of characteristic not 2, and suppose F is a field containing R. An **n-ary quadratic form over F** is a polynomial of the form

$$q(x_1,\ldots,x_n) = \sum_{i,j=1}^{n} a_{ij}x_ix_j \in F[x_1,\ldots,x_n].$$

Because multiplication in F is commutative, in considering $q(x_1,\ldots,x_n)$ as a function $R^n \longrightarrow F$ without loss of generality we can (and will, from now on) suppose that the coefficients of q satisfy the condition $a_{ij} = a_{ji}$; just replace each of these coefficients by $\dfrac{a_{ij}+a_{ji}}{2}$. (Note that here the underlying assumption that char $F \neq 2$ is needed.) With this convention in force, q is now given by the symmetric matrix $A = (a_{ij}) \in M_n(F)$. This symmetric matrix is the **Gram matrix** of q.

Let $\alpha \in F$. We say the form q **represents** α **over** R, denoted $\alpha \underset{R}{\longrightarrow} q$, if there are $\lambda_1,\ldots,\lambda_n \in R$ such that $q(\lambda_1,\ldots,\lambda_n) = \alpha$. More briefly: viewing q as a function $R^n \longrightarrow F$, the statement $\alpha \underset{R}{\longrightarrow} q$ means $\alpha \in$ range q. (A more general notion of representation will come in Definition 2.8.)

An instance of the **representation problem** is this: Given a form q, for which $\alpha \in F$ is it true that $\alpha \underset{R}{\longrightarrow} q$?

Example 1.2. Pythagorean triples, mentioned earlier, are precisely the solutions for the representation $0 \underset{\mathbb{Z}}{\longrightarrow} x_1^2 + x_2^2 - x_3^2$.

Example 1.3. Brahmagupta (598–668) could generate Pythagorean triples, and he studied representations by other quadratic forms as well. For instance, from a given representation $1 \underset{\mathbb{Z}}{\longrightarrow} x^2 - ay^2$ (with $a \in \mathbb{N}$) he could produce infinitely many.

Example 1.4. (i) Fermat (1601–1665) studied sums of integer squares, and here is one of his typical results, stated by him in 1640 in a private communication—along with hints on his method of proof—and with a formal proof published by Euler in 1754: If $n \in \mathbb{N}$ has prime factorization $n = p_1^{\alpha_1}\cdots p_r^{\alpha_r}$, then $n \underset{\mathbb{Z}}{\longrightarrow} x^2 + y^2$ if and only if $\alpha_i \equiv 0 \pmod 2$ for all i such that $p_i \equiv 3 \pmod 4$.

(ii) In 1967 Leahey [**Le**] returned to Fermat's result, but replacing $\mathbb{Z}$ with a polynomial ring. Let $h(x) \in \mathbb{F}_q[x]$. (Here $\mathbb{F}_q$ denotes a finite field with q elements, with q odd.) Say $h(x) = \varepsilon p_1^{\alpha_1}(x)\cdots p_r^{\alpha_r}(x)$, with the $p_i(x)$ prime (monic and irreducible) polynomials and $\varepsilon \in \mathbb{F}_q^*$. Then $h(x)$ is a sum of two polynomial squares—that is, $h(x) \underset{\mathbb{F}_q[x]}{\longrightarrow} X^2 + Y^2$—if and only if $\alpha_i \equiv 0 \pmod 2$ for all i such that $\partial p_i(x) \equiv 1 \pmod 2$. (Here we use "$\partial$" for the degree function.)

Example 1.5. Euler (1707–1783) considered representation of primes by quadratic forms of the form $x^2 \pm Ny^2$ over $\mathbb{Z}$. A typical result: If p is prime and satisfies $p \equiv 1 \pmod{20}$, then $p \underset{\mathbb{Z}}{\longrightarrow} x^2 + 5y^2$.

Example 1.6. Lagrange (1736–1813) showed that every nonnegative integer n can be expressed as a sum of at most four squares; that is,

$$n \underset{\mathbb{Z}}{\longrightarrow} x_1^2 + x_2^2 + x_3^2 + x_4^2.$$

Lagrange also substantially developed the theory of binary (two-variable) quadratic forms over the integers.

Example 1.7. Let p and q be odd primes, with $p \leq q$, and consider representations of the product pq by the quadratic form $Q(x, y) = x^2 - y^2$. Over the field $\mathbb{Q}$ of rational numbers there are infinitely many representations $pq \underset{\mathbb{Q}}{\longrightarrow} Q$. In fact if β is any rational number, then for all $\lambda \in \mathbb{Q}^*$ we have

$$\beta = \left(\frac{\lambda^2 + \beta}{2\lambda}\right)^2 - \left(\frac{\lambda^2 - \beta}{2\lambda}\right)^2.$$

On the other hand, a straightforward arithmetic argument (Try it!) shows that there are just two nonnegative *integral* representations $pq \underset{\mathbb{Z}}{\longrightarrow} Q$, given by the pairs

$$(x, y) = \left(\frac{p+q}{2}, \frac{q-p}{2}\right) \quad \text{and} \quad (x, y) = \left(\frac{pq+1}{2}, \frac{pq-1}{2}\right).$$

Now we will move in the opposite direction and *start* with an integer value for x, say $x = n \geq 2$, and ask this question: Must there exist an integer k satisfying $0 \leq k \leq n-2$ and primes p, q such that $n^2 - k^2 = pq$? If the answer is "yes," then (if $p \leq q$, say) we must have $p = n-k$ and $q = n+k$ by the Fundamental Theorem of Arithmetic, and hence $2n = p+q$. In other words, an affirmative answer to this question on integral representations by the binary quadratic form $Q(x, y) = x^2 - y^2$ would affirmatively settle the celebrated **Goldbach Conjecture: Every even integer ≥ 4 is the sum of two primes.**

Conversely, if the Goldbach Conjecture is true and for each integer $n \geq 2$ there are primes $p \leq q$ such that $2n = p + q$, then $p = n - k$ and $q = n + k$ for some integer k satisfying $0 \leq k \leq n - 2$ and hence $n^2 - k^2 = pq$. So the Goldbach Conjecture and our quadratic forms conjecture are actually equivalent. Note that a supplementary conjecture on $Q(x, y) = x^2 - y^2$ that there are infinitely many values of n for which $k = 1$ "works" (i.e., $n^2 - 1 = pq$ for some primes p and q) is equivalent to the **Twin Prime Conjecture: There are infinitely many primes p such that $p + 2$ is also prime.**

1.2. Representation and Equivalence; Matrix Connections; Discriminants

The theory of quadratic forms underwent a dramatic change of style and perspective in the Twentieth Century; but before getting on with that (in the next chapter), in keeping with the tenor of this brief look at history we will first consider—without formal proof—an example involving binary quadratic forms in the style of Lagrange. Lagrange observed that if each variable in a given quadratic form is replaced by a linear combination of variables, then the result is again a quadratic form. It may turn out that representations by the transformed form are easier to determine than those of the original form; and then one can carry those representations back and produce representations by the original form. With this general outline, our goal here is to determine what integers n are represented over $\mathbb{Z}$ by the form

$$f = 17x^2 + 94xy + 130y^2.$$

In particular, can we find an explicit representation $5 \underset{\mathbb{Z}}{\longrightarrow} f$?

Consider the Gram matrix $A = \begin{pmatrix} 17 & 47 \\ 47 & 130 \end{pmatrix}$ for f. First observe that a sequence of elementary row and column operations for $\mathbb{Z}$-matrices (each time following a row operation with the corresponding column operation to get a symmetric resulting matrix) reduces A to a matrix in which the off-diagonal entry is smaller than the diagonal entries:

$$\begin{pmatrix} 17 & 47 \\ 47 & 130 \end{pmatrix} \mapsto \begin{pmatrix} 17 & 47 \\ 13 & 36 \end{pmatrix} \mapsto \begin{pmatrix} 17 & 13 \\ 13 & 10 \end{pmatrix} \mapsto \begin{pmatrix} 4 & 3 \\ 13 & 10 \end{pmatrix}$$
$$\mapsto \begin{pmatrix} 1 & 3 \\ 3 & 10 \end{pmatrix} \mapsto \begin{pmatrix} 1 & 3 \\ 0 & 1 \end{pmatrix} \mapsto \begin{pmatrix} 1 & 0 \\ 0 & 1 \end{pmatrix}.$$

Here $T = \begin{pmatrix} 3 & -11 \\ -1 & 4 \end{pmatrix}$ is the product of the elementary matrices—listed from left to right in their order of use—corresponding to the column operations, and we have observed that ${}^tTAT = \begin{pmatrix} 1 & 0 \\ 0 & 1 \end{pmatrix}$. Now use the matrix T to dictate a change of variables: in our given form, make the substitutions

$$x \mapsto 3u - 11v \qquad \text{and} \qquad y \mapsto -u + 4v.$$

Then upon carrying out all the arithmetic we get

$$f(x, y) = g(u, v) = u^2 + v^2.$$

(In other words, we have gotten a new quadratic form with Gram matrix ${}^tTAT$.) But we know from Fermat's result cited earlier exactly which integers are represented by g. And given explicit values for u and v that solve the equation $n = u^2 + v^2$, we can use the equations relating x and y to

u and v to get a representation $n \underset{\mathbb{Z}}{\longrightarrow} f$. For instance, the representation $5 = 1^2 + 2^2$ gives

$$\begin{pmatrix} x \\ y \end{pmatrix} = \begin{pmatrix} 3 & -11 \\ -1 & 4 \end{pmatrix} \begin{pmatrix} 1 \\ 2 \end{pmatrix} = \begin{pmatrix} -19 \\ 7 \end{pmatrix}$$

as a solution to $5 = 17x^2 + 94xy + 130y^2$. Also note that since $T \in GL_2(\mathbb{Z})$, we can use the inverse of this transforming matrix to express u, v as integral linear combinations of x, y. This sets up a bijection between the representations

$$n \underset{\mathbb{Z}}{\longrightarrow} f \qquad \text{and} \qquad n \underset{\mathbb{Z}}{\longrightarrow} g.$$

Quadratic forms f and g related in this way via invertible matrices over a ring R are said to be **equivalent** over R. As the example suggests, equivalent forms represent the same elements, and they represent them the same number of times.

Now we will describe the above process in greater generality. Given a quadratic form $q = \sum_{i,j=1}^{n} a_{ij}x_ix_j$, with $A = (a_{ij}) \in M_n(F)$; if each x_i is replaced by a linear combination of $x_1, \ldots, x_n$ with coefficients in R, then the result is a new quadratic form h. More explicitly, given a matrix $T = (t_{ij}) \in M_n(R)$, replacing (in q) each x_i by $\sum_{j=1}^{n} t_{ij}x_i$ yields a new quadratic form $h = h(x_1, \ldots, x_n)$ with associated Gram matrix ${}^tTAT$. We say h **is represented by** q **over** R, denoted $h \underset{R}{\longrightarrow} q$. It is useful to observe that q can be viewed as the result of a matrix product:

$$q(x_1, \ldots, x_n) = (x_1, \ldots, x_n)A \begin{pmatrix} x_1 \\ \vdots \\ x_n \end{pmatrix} = {}^txAx, \text{with } x = \begin{pmatrix} x_1 \\ \vdots \\ x_n \end{pmatrix}.$$

Thus

$$h(x) = {}^tx{}^tTATx = {}^t(Tx)A(Tx) = q(Tx).$$

It follows that if $\alpha \underset{R}{\longrightarrow} h$ and $h \underset{R}{\longrightarrow} q$ then also $\alpha \underset{R}{\longrightarrow} q$.

Definition 1.8. Let q, h be n-ary quadratic forms over F, with respective Gram matrices A, C. We say q and h are **equivalent** over R and are in the same **class** over R if there is a matrix $T \in GL_n(R)$ such that $h \underset{R}{\longrightarrow} q$ via T. That is, $h(x) = q(Tx)$ with T an **invertible** R-matrix. We denote equivalence of quadratic forms h and q over R by $h \underset{R}{\cong} q$. (If R is understood from the context, to avoid visual clutter we will usually denote equivalence by $h \cong q$ and representation by $h \to q$.)

Remark 1.9. A matrix $T \in GL_n(R)$ is called a **unimodular** R-matrix. It determines an automorphism of R^n via matrix multiplication. So from the equation $h(x) = q(Tx)$ we see that equivalent forms represent the same elements of F over R; moreover, for every $\alpha \in F$ represented by these forms, there is a bijection between the subsets $q^{-1}(\alpha)$ and $h^{-1}(\alpha)$ of R^n.

We now have two fundamental problems for quadratic forms over R: Determine a computationally effective set of necessary and sufficient conditions for two given forms to be equivalent or for one to represent the other. Incidentally, notice that the question of whether a given form represents a given element amounts to a special case of the problem of whether one form represents another: $\alpha \underset{R}{\longrightarrow} q(x_1, \ldots, x_n)$ if and only if $\alpha x_1^2 \underset{R}{\longrightarrow} q$.

The equivalence problem for quadratic forms over R can be stated as a matrix question: Given symmetric matrices $A, C \in M_n(F)$, is there a matrix $T \in GL_n(R)$ such that $C = {}^tTAT$? If the answer is affirmative we write $A \underset{R}{\cong} C$ and say that A and C are **congruent** over R. (As before, if R is understood we usually write "$A \cong C$".) The representation problem for quadratic forms has a similar matrix formulation, except that T no longer is required to be unimodular or square.

Example 1.10. Let $q = x_1^2 + x_2^2$ and $h = 2x_1^2 + 2x_2^2$. Then $h(x) = q(Tx)$, where $T = \begin{pmatrix} 1 & 1 \\ 1 & -1 \end{pmatrix}$. Since $T \in M_2(\mathbb{Z}) \cap GL_2(\mathbb{Q})$, it follows that $h \underset{\mathbb{Z}}{\longrightarrow} q$ and $h \underset{\mathbb{Q}}{\cong} q$. But it is NOT the case that $q \underset{\mathbb{Z}}{\longrightarrow} h$, since q represents 1 over $\mathbb{Z}$ while h does not, and hence $h \underset{\mathbb{Z}}{\ncong} q$. In terms of matrix congruence,

$$\begin{pmatrix} 1 & 0 \\ 0 & 1 \end{pmatrix} \underset{\mathbb{Q}}{\cong} \begin{pmatrix} 2 & 0 \\ 0 & 2 \end{pmatrix} \quad \text{via} \quad T = \begin{pmatrix} 1 & 1 \\ 1 & -1 \end{pmatrix}$$

while

$$\begin{pmatrix} 1 & 0 \\ 0 & 1 \end{pmatrix} \underset{\mathbb{Z}}{\ncong} \begin{pmatrix} 2 & 0 \\ 0 & 2 \end{pmatrix}.$$

Notice that if $q \underset{R}{\cong} h$, say $C = {}^tTAT$ with $T \in GL_n(R)$, then

$$|C| = |A| \cdot |T|^2 \in |A| \cdot U^2 \in F^*/U^2 \cup \{0\}.$$

Let us dissect this latter expression. Here $F^* = F - \{0\}$ is a multiplicative group, $U = R^*$ is the group of units of R, and U^2 is its subgroup of squares. So F^*/U^2 is a multiplicative group, and $F^*/U^2 \cup \{0\}$ then becomes a semigroup if we define $0 \cdot x = 0$ for all x. The **discriminant** of q is the element $dq = |A| \cdot U^2 \in F^*/U^2 \cup \{0\}$. It follows that if $q \underset{R}{\cong} h$ then $dq = dh$; that is, the discriminant is a **class invariant** of quadratic forms. (We will have more to say about discriminants—including some variations—later on. See 2.6 and 2.48.)

Example 1.11. If $R = \mathbb{Z}$ then $U = \{\pm 1\}$, hence $U^2 = \{1\}$ and so $|A|$ is a class invariant, often called the **determinant** of the form. In practice (even when R is not $\mathbb{Z}$) it is common to abuse terminology and refer to the determinant of the Gram matrix as the discriminant of the form, with the understanding that two discriminants are to be regarded as equal if one is the square of an R-unit times the other.

Exercises

(1) Describe the groups F^*/F^{*2} when F is each of the following: $\mathbb{C}$, $\mathbb{C}(x)$, $\mathbb{Q}$, $\mathbb{F}_q$, $\mathbb{F}_q(x)$. (Here $\mathbb{F}_q$ is a finite field of odd order q.)

(2) Consider the quadratic form $f(x,y) = 62x^2 + 176xy + 125y^2$. Is there a representation $4 \underset{\mathbb{Z}}{\longrightarrow} f$? Is there a representation $11 \underset{\mathbb{Z}}{\longrightarrow} f$? In each case, if there *is* such a representation, find them all.

(3) Is the quadratic form $f(x,y) = 3x^2 + 2xy + 2y^2$ equivalent over $\mathbb{Z}$ to a quadratic form $g(x,y) = ax^2 + by^2$ with $a, b \in \mathbb{Z}$?

(4) Show that over a field F of characteristic not 2 every binary (i.e., two-variable) quadratic form $f(x,y)$ is equivalent to a form $\alpha x^2 + \beta y^2$ with $\alpha, \beta \in F$. If F is an *ordered* field (such as $\mathbb{Q}$ or $\mathbb{R}$), deduce a necessary and sufficient condition on the discriminant df for f to represent both positive and negative elements of F.

1.3. A Brief Historical Sketch, and Some References to the Literature

Many of the greatest mathematicians have contributed to the theory of quadratic forms, and fortunately much of the early history has been well treated elsewhere, so we will not linger on it here. Particularly recommended are André Weil's book, *Number Theory: An Approach Through History. From Hammurapi to Legendre* [**We**]; and Winfried Scharlau and Hans Opolka's *From Fermat to Minkowski: Lectures on the Theory of Numbers and its Historical Development* [**SO**].

Much of the more recent work continues in the spirit of Gauss and the other masters of the past in that it seeks to understand issues of equivalence and representation of quadratic forms through an assortment of **reduction** theories, wherein one attempts to find a hopefully unique canonical form in the equivalence class of a given form. Details of the efforts in this direction—and the degree of success—depend heavily on the underlying coefficient ring of the form. For example, over an algebraically closed field we'll see that there is only one equivalence class of "nondegenerate" quadratic forms in a given number of variables. (A form is **nondegenerate** if it is not equivalent to a form with fewer variables.) So the classification and representation problems are trivial in this case. But as the coefficient ring becomes more restricted the theory becomes more interesting and the mathematics becomes more challenging.

Sylvester, in the mid-Nineteenth Century gave us his "law of inertia," which settled matters over $\mathbb{R}$. Hermite developed some ideas on reduction over $\mathbb{Z}$, hoping to extend the remarkable earlier work by Gauss and Legendre on reduction and classification of forms in two variables. In particular Hermite obtained a bound on the **minimum** of a positive definite quadratic form over $\mathbb{Z}$; that is, the smallest positive integer represented by a quadratic form with integer coefficients that represents only positive integers when its variables are replaced by integers. But Hermite's work just starts the more general reduction process. Toward the end of the Nineteenth Century Minkowski began what is now called the "local-global" approach, considering the classification of quadratic forms over $\mathbb{Z}$ through the classification over $\mathbb{Z} \pmod{p^k}$, as p ranges through the primes and k grows. In the early Twentieth Century **valuation theory** was developed by Hensel, allowing for the exploration of problems given over the field $\mathbb{Q}$ of rational numbers (and other fields of number-theoretic interest) through the discussion of the same problems over certain extension fields: the fields of **p-adic numbers**. Then Hasse (Hensel's student) was able to use valuation theory to reformulate and extend Minkowski's work to a wider number-theoretic context. Roughly speaking, the idea is as follows. A given equation over a given domain may be difficult to solve, and even the question of whether a solution *exists* may be difficult to determine. But if one extends the coefficient ring to a larger and more tractable domain—or to a family of such domains—the solvability "upstairs" may be easy (or at least easier) to determine. Obviously if no solution exists in an extension of the original ring R then there is no solution in R. On the other hand, if there *is* a solution in the extensions then perhaps one can use this information to produce a solution in R, or at least to show that such a solution exists. In order for the process to be successful, R needs to be densely embedded in the extensions, and there must be a kind of coherence or "reciprocity" among the extensions.

The local-global approach developed by Hasse had great success when the coefficient ring was $\mathbb{Q}$ or an algebraic number field. But over $\mathbb{Z}$ (and the other rings of algebraic integers) the success was much more limited, in the sense that without further special information one could not conclude that the existence of **local** solutions—i.e., solutions over the rings $\mathbb{Z}_p$ of **p-adic integers**—forced a **global** solution: a solution over $\mathbb{Z}$.

In the 1930s C. L. Siegel introduced the methods of complex analysis into the study of quadratic forms in a series of ground-breaking papers. Siegel used his approach to explore the gap—between local integral solutions with respect to every prime p and global solutions—through the notion of the **genus** of a quadratic form (where he expanded on ideas first developed by Gauss [**Ga**] in the binary case) and the concept of the **class number**: the number of equivalence classes in a genus.

Also important in the 1930s was a pioneering paper by Witt, who developed a more geometric approach to quadratic forms over fields, for instance transforming the process of equivalence of quadratic forms (via linear changes of variables) into the study of isometries of appropriate associated inner-product spaces of a more general sort than had been considered earlier. Witt's work not only provided a more workable and intuitively fruitful framework for the old questions, but led to what is now called the **algebraic theory of quadratic forms**, which studies the subject over arbitrary fields, not necessarily of a number-theoretic origin. Many of the algebraic issues raised by Witt lay dormant for many years, though a paper by Kaplansky in 1953 brought fresh attention to the subject, and then the remarkable work by Pfister in the 1960s led to an explosion of the algebraic theory. While we will briefly touch on some of that theory in this book, we refer the reader to T. Y. Lam's book, *Introduction to Quadratic Forms over Fields* [**Lam**], for a full treatment.

Interest in Witt's more geometric approach was heightened by the work of Emil Artin (as represented in his book *Geometric Algebra* [**A**]) in the 1940s and 1950s, particularly in connection with the developing theory of the **classical groups**. And so by the mid-'50s it was becoming more standard to view quadratic forms not as polynomials to be treated as abstract symbols, but as inner products on modules: **quadratic spaces** if the modules were vector spaces; and **lattices** for appropriate submodules of quadratic spaces. So at this point the old problem of determining whether two quadratic forms were equivalent had become the problem of determining whether two lattices on quadratic spaces were isometric.

The pursuit of the local-global approach to the theory over $\mathbb{Z}$ demanded the solution of the classification problem over $\mathbb{Z}_p$, and this was the work of Jones and Pall in the 1940s. Their work was extended to the classification of lattices (on quadratic spaces) over the rings of integers of arbitrary **local fields** in the 1950s by O'Meara, who also made major advances on the local **representation** problem: the determination of when one lattice contains an isometric copy of another. In the early 1960s further progress was made by Riehm [**Ri**] on the representation problem, and recently the local representation problem has apparently been solved in its full generality by Beli [**Be**].

From the 1950s through the 1970s, Kneser and O'Meara were leaders in exploring the linkage of local and global invariants of lattices, and the impact of that linkage on the study of the classification problem. Particularly important is Kneser's **neighbor lattice** construction [**K2**], which allowed for a step-by-step transformation of a lattice in one isometry class (in the so-called **unimodular** case) into a lattice in another class. O'Meara's book, *Introduction to Quadratic Forms* [**O'M1**], which appeared in 1963, is a wonderful introduction to the subject as of that time.

Since the late 1960s there have been important new links forged between the theory of quadratic forms and other areas of mathematics, and corresponding new demands on the subject. In 1968, Conway's discovery [**Con1**] of a new finite simple group through the study of the **orthogonal group** (the group of inner-product preserving automorphisms) of the **Leech lattice** brought the attention of the group theory community. In topology, work by Milnor, Freedman, and many others reduced certain problems on the classification of manifolds to problems on the structure of associated lattices over $\mathbb{Z}$. For example, if M is a closed, oriented, simply-connected 4-manifold, then the homology group $H_2(M)$ is a $\mathbb{Z}$-lattice, and the "intersection pairing" gives $H_2(M)$ a unimodular structure. It turns out that there is an orientation-preserving homotopy equivalence between two such 4-manifolds if and only if the corresponding $\mathbb{Z}$-lattices are isometric. See *Symmetric Bilinear Forms*, by J. Milnor and D. Husemoller [**MH**], for more details.

The extraordinary development and proliferation of computing power since the 1970s has of course made possible a broad range of algorithmic approaches to the subject that before were out of the question. And procedures developed earlier (such as Kneser's neighbor-lattice process) that were practical only in very low dimensions were now able to be extended to higher dimensions. In the 1980s came the so-called **LLL-** or **L^3-algorithm**, due to A. Lenstra, H. Lenstra, and L. Lovasz [**LLL**], giving a sort of reduced form for a lattice that, while not necessarily canonical, allows for much useful information to be deduced. For instance, L^3 can often be used to find the shortest nonzero vector in a lattice. Finding the length of such a vector is the lattice equivalent of the problem mentioned earlier (pursued by Hermite) of finding the minimum of a quadratic form, and while the inequality on the minimum guaranteed by L^3 is weaker than Hermite's in the worst case, in practice the L^3 inequality is usually stronger than Hermite's; and L^3 actually produces short vectors, whereas Hermite's result does not.

The development of the Internet and other sophisticated electronic means of communication led to the subject of algebraic coding theory and new forms of cryptography. Electronically transmitted messages are usually encoded as n-tuples of 0's and 1's; that is, as elements of $\mathbb{Z}_2^n$. Because no medium of communication is perfect, a 0 digit may be received as a 1, or vice versa. Coding theory deals with the practice of introducing redundancy into the transmitted message so that when occasional errors occur they can be recognized and corrected. The greater the likelihood of errors, the more substantial the redundancy needs to be in order to handle the errors. On the other hand, redundancy slows the communication process, so it is best to avoid more redundancy than is absolutely necessary to do the job. Algebraic coding theory deals with the associated problems. If a given code C is a **linear code**, meaning a subspace of $\mathbb{Z}_2^n$, then one can view C as a subset of $\mathbb{Z}^n$ and define an associated lattice L as the set of all vectors in $\mathbb{Z}^n$ congruent $\pmod 2$ in each coordinate to an element of C. In this way and other ways, problems in coding theory become recast as problems on lattices in $\mathbb{R}^n$ and related problems on **sphere packing**. We refer the reader to the massive *Sphere Packings, Lattices and Groups*, by Conway and Sloane [**CS**], and Ebeling's *Lattices and Codes* [**Eb**] for thorough introductions to lattices in coding theory. We note that the Conway–Sloane book has one of the most extensive bibliographies in the mathematical literature.

Cryptography is an ancient subject, but the classical encryption systems—many going back to antiquity—were so-called **private key** systems, in the sense that sender and receiver would agree in advance on an encryption method and key (the "key" being the essential information needed in order to understand the details of how the messages were being encrypted), and anyone knowing the key would be able to decrypt the message. In the 1970s the first of the **public key** cryptosystems (namely RSA encryption) was developed. In such a system, a party wishing to receive a message publically announces the method of encryption to be used—a method involving some kind of number-theoretic manipulation of the message—but someone intercepting the encrypted message will not be able to decrypt it within a reasonable time period without special additional information known only to the receiver. For example, RSA encryption involves use of a number m that is a product of two large primes; successful decryption requires knowledge of m's factorization, and factorization is a computationally complex problem. Not long after RSA came **knapsack cryptosystems**, which allowed for much faster encryption than RSA, but it wasn't long before these began to be successfully attacked by a variety of methods, including approaches through lattice reduction based on the L^3 algorithm. In the 1990s an assortment of encryption schemes was developed in which the fundamental difficulty of lattice problems was exploited in the encryption method. Here, though the setting is new, some of the issues go back to the same things that concerned Hermite: the difficulty of finding short vectors in a lattice: the **shortest vector problem (SVP)**; or the more general **closest vector**

problem (CVP): given a lattice L in $\mathbb{R}^n$ and a vector $v \in \mathbb{R}^n$, find a vector $x \in L$ closest to V. A lattice is usually specified by giving a basis for it, and it turns out that questions of the sort just described can be much easier to solve in one basis than another, and the computational complexity of these problems is the basis for some of these cryptosystems. We will have a brief look at lattices in cryptography in the final chapter. Good book sources for more complete treatments of much of this material: *Complexity of Lattice Problems: A Cryptographic Perspective*, by Micciancio and Goldwasser [**MG**], and *Cryptography and Lattices*, ed. by Silverman [**Si**].

Chapter 2

Quadratic Spaces and Lattices

Now we start over again, with a more contemporary view of quadratic forms. The connection with the classical formulation in Chapter 1 will soon be apparent.

2.1. Fundamental Definitions

Let R be an integral domain of characteristic not 2, let F be its quotient field, and let M be a free R-module of finite rank.

Definition 2.1. A mapping $B : M \times M \to F$ is a **bilinear form** on M if it is linear in each variable when the other is held fixed. More explicitly, given $v \in M$, linearity in the second variable means $B(v, x+y) = B(v, x) + B(v, y)$ and $B(v, \alpha x) = \alpha B(v, x)$ for all $x, y \in M$ and $\alpha \in R$; and linearity of $B(\cdot, v)$ in the first variable is similar. The bilinear form B is said to be **symmetric** if $B(x, y) = B(y, x)$ for all $x, y \in M$. Finally, if B is a symmetric bilinear form, the mapping $q : M \to F$ defined by $q(x) = B(x, x)$ is the **quadratic form** or **quadratic map** associated with B, and (M, B) (or (M, q)) is said to be a **quadratic R-module**. (For simplicity we usually just write M.) If R is a field we call M a **quadratic R-space**; if R is a principal ideal domain and not a field then M is a **quadratic R-lattice**.

Let B be a symmetric bilinear form on M with associated quadratic form q. Then the following properties are easily checked:

(1) $q(\alpha x) = \alpha^2 q(x)$ for all $\alpha \in R, x \in M$.

(2) $B(x,y) = \dfrac{q(x+y) - q(x) - q(y)}{2}$.

Therefore, given that a mapping q is a quadratic form and hence defined by *some* symmetric bilinear form, it is in fact defined by a *unique* symmetric bilinear form. [Remark: This intimate linkage between quadratic forms and symmetric bilinear forms is not the case in the theory of quadratic forms over domains of characteristic 2.]

Examples 2.2. (i) With R and M arbitrary, define $B(x,y) = 0$ for all $x, y \in M$, and hence $q(x) = 0$ for all x. This is a **trivial** quadratic module.

(ii) Let $M = \mathbb{R}^n$ and $R = \mathbb{R}$, and let B be the standard dot product. Then for $x = {}^t(x_1, \ldots, x_n)$ we have $q(x) = \sum x_i^2$, the **squared length** of x. In view of this example, the term "inner product" is often used instead of "symmetric bilinear form," even in settings dramatically different from this important example in Euclidean space.

(iii) Take $M = \mathbb{R}^4$, with symmetric bilinear form $B(x,y) = x_1y_1 + x_2y_2 + x_3y_3 - x_4y_4$. This is **Minkowski space**, useful in special relativity. Notice that in this space $q(1,0,0,1) = 0$, and this leads to the following definition.

Definition 2.3. In a quadratic module M, an **isotropic** vector is a nonzero vector v such that $q(v) = 0$; and a vector is **anisotropic** if it is not isotropic. We say M is isotropic if it contains an isotropic vector, and anisotropic otherwise. (In particular, the trivial space $\{0\}$ is anisotropic.) Finally, if $M \neq 0$ and and $q(v) = 0$ for all $v \in M$ then M is **totally isotropic**.

Incidentally, because of connections with the theory of relativity, physicists call the set of isotropic vectors (along with the 0 vector) in a quadratic space the **light cone** of the space. [In general, a subset of a vector space is a **cone** if it is closed under scalar multiplication.]

Now suppose the R-module M has basis $\mathbb{B} = \{v_1, \ldots, v_n\}$, and let B be a symmetric bilinear form on M. The symmetric matrix $A = (a_{ij}) \in M_n(F)$ defined by $a_{ij} = B(v_i, v_j)$ is the **Gram matrix of M with respect to** (or "**in**") $\mathbb{B}$; we write $M \cong A$ in $\mathbb{B}$. Conversely, given the module M, a basis $\mathbb{B}$, and a symmetric matrix $A = (a_{ij}) \in M_n(F)$, we can *define* a bilinear form B on M by setting $B(v_i, v_j) = a_{ij}$ and "extending bilinearly": $B(\sum_i \alpha_i v_i, \sum_j \beta_j v_j) = \sum_{i,j} \alpha_i \beta_j a_{ij}$ for all $\alpha_i, \beta_j \in R$. This makes M a quadratic R-module, with $M \cong A$ in $\mathbb{B}$.

Remark 2.4 (Notation)**.** Given a symmetric matrix A and a quadratic R-module M, the expression $M \cong A$ (with no basis mentioned) means that $M \cong A$ in *some* basis. The symbol $\langle A \rangle$ is used for an unspecified free quadratic R-module having A as its Gram matrix with respect to some basis. For $\alpha \in F$ we write $\langle \alpha \rangle$ instead of $\langle (\alpha) \rangle$.

Remark 2.5. Suppose $M \cong A$ in $\mathbb{B} = \{v_1, \ldots, v_n\}$, and $x \in M$, say $x = \sum x_i v_i$, with $x_i \in R$. Then $q(x) = B(x,x) = \sum_{i,j} a_{ij}x_i x_j = \sum_{i=1}^n a_{ii}x_i^2 + 2\sum_{i<j} a_{ij}x_i x_j$. By viewing the x_i as variables taking their values in R, the mapping q can be regarded as a quadratic polynomial form in the sense of Chapter 1. Conversely, one can identify a given quadratic form $q(x_1, \ldots, x_n) \in R[x_1, \ldots, x_n]$ with the quadratic map on a suitable R-lattice on a quadratic F-space of dimension n. For example, the form $q(x,y,z) = 3x^2 + 2xy - y^2 + 5xz \in \mathbb{Z}[x,y,z]$ corresponds to the form on the $\mathbb{Z}$-lattice $\left\langle \begin{pmatrix} 3 & 1 & \frac{5}{2} \\ 1 & -1 & 0 \\ \frac{5}{2} & 0 & 0 \end{pmatrix} \right\rangle$.

What happens to the Gram matrix of a free quadratic R-module under change of basis? Suppose $M \cong A$ in $\{v_1, \ldots, v_n\}$ and $M \cong A'$ in $\{v_1', \ldots, v_n'\}$, where $v_j' = \sum_i t_{ij}v_i$ with $T = (t_{ij}) \in GL_n(R)$. Then it is easily checked that $A' = {}^tTAT$; that is, $A' \underset{R}{\cong} A$. Thus $|A'| = |A| \cdot \varepsilon^2$ for some unit $\varepsilon \in R^*$.

Definition 2.6. If $M \cong A$, the **discriminant** of M is $dM = |A| \cdot R^{*2} \in F^*/R^{*2} \cup \{0\}$. Following the convention in Chapter 1, if $|A| = \alpha$, we usually write $dM = \alpha$ rather than the more complete $dM = \alpha R^{*2}$.

Note: in the literature (e.g., [**Lam**], [**Sz**]) sometimes dM is called the **determinant** of M. (Also see Remark 2.48.)

Example 2.7. Assume $2 \in R^*$, and suppose $M \cong \begin{pmatrix} 0 & 1 \\ 1 & 0 \end{pmatrix}$ in $\{v_1, v_2\}$. Define $v_1' = v_1 + \frac{1}{2}v_2$ and $v_2' = v_1 - \frac{1}{2}v_2$. So $T = \begin{pmatrix} 1 & 1 \\ \frac{1}{2} & -\frac{1}{2} \end{pmatrix}$. Then $\{v_1', v_2'\}$ is also a basis, since $T \in GL_2(R)$, and $M \cong \begin{pmatrix} 1 & 0 \\ 0 & -1 \end{pmatrix}$ in $\{v_1', v_2'\}$. (This can be checked in two ways: either evaluate B on the four relevant pairs (v_i', v_j') or compute ${}^tT \begin{pmatrix} 0 & 1 \\ 1 & 0 \end{pmatrix} T$.) Incidentally, a quadratic space $H \cong \begin{pmatrix} 0 & 1 \\ 1 & 0 \end{pmatrix}$ is called a **hyperbolic plane**, and the associated basis vectors (such as $\{v_1, v_2\}$ in this example) are said to be a **hyperbolic pair**. Notice that a hyperbolic plane H has discriminant $dH = -1$.

Here is a useful computational observation. If $M \cong A$ in the basis $\mathbb{B} = \{v_1, \ldots, v_n\}$ and x and y are the column vectors of coefficients of v and w with respect to $\mathbb{B}$, then $B(v,w) = {}^txAy$ and $q(v) = {}^txAx$.

Definition 2.8. Let L and M be quadratic R-modules. A homomorphism $\varphi : L \to M$ such that

$$B(\varphi x, \varphi y) = B(x,y)$$

for all $x, y \in L$ is a **representation of L into** (or **by**) M. Equivalently, $q(\varphi x) = q(x)$ for all $x \in L$. [Note: Strictly speaking, the B's in the definition are different functions and should perhaps be subscripted by L and M accordingly; similarly for the q's in the equation preceding this note. But to avoid notational clutter we follow custom and omit the subscripts when confusion is unlikely.] If there is such a representation we write $L \to M$. If there is a representation $\langle \alpha \rangle \to M$, we usually just write $\alpha \to M$ and say that M **represents** α. An **isometry** is an injective representation. If there is a surjective isometry $L \to M$ we say L and M are **isometric**, denoted $L \cong M$.

Now suppose L and M are free quadratic R-modules of respective ranks l, m and Gram matrices A, C with respect to bases $\mathbb{B} = \{v_1, \ldots, v_l\}$ and $\mathbb{B}' = \{w_1, \ldots, w_m\}$. Suppose further that $\varphi : L \to M$ is a representation, with $\varphi(v_j) = \sum_{i=1}^{m} t_{ij} w_i$. Then $T = (t_{ij})$ is an $m \times l$ R-matrix, and we write $\varphi \sim T$ with respect to $\mathbb{B}$ and $\mathbb{B}'$; or, more compactly, $(\varphi)_{\mathbb{B}}^{\mathbb{B}'} = T$. Since φ is a representation, we have

$$a_{ij} = B(v_i, v_j) = B\big(\varphi(v_i), \varphi(v_j)\big)$$

and after expressing $\varphi(v_i)$ and $\varphi(v_j)$ in terms of $\mathbb{B}'$, a straightforward computation (try it!) shows that $A = {}^tTCT$. Moreover, φ is actually an isometry of L onto M if and only if $T \in GL_l(R)$ and hence $A \underset{R}{\cong} C$. So the issues of representation and isometry of free quadratic R-modules are equivalent to the representation and equivalence of polynomial quadratic forms over R discussed in Chapter 1.

2.2. Orthogonal Splitting; Examples of Isometry and Non-isometry

Let M be a quadratic R-module, and let X be a submodule. Then the **orthogonal complement of X (in M)** is the submodule

$$X^\perp = \{v \in M \mid B(v, x) = 0 \text{ for all } x \in X\}.$$

Vectors $x, y \in M$ are **orthogonal** if $B(x, y) = 0$, and subsets X, Y of M are said to be orthogonal if $B(x, y) = 0$ for all $x \in X, y \in Y$. We write $M = \perp_i M_i$ (an **orthogonal sum**) if $M = \bigoplus_i M_i$ with the M_i pairwise orthogonal submodules. Such a decomposition is an **orthogonal splitting** of M, and each M_i is an **(orthogonal) component** of M and is said to **split** M.

Remark 2.9. If M is a free quadratic R-module and $A = (a_{ij})$ is a symmetric matrix, then the expressions $M \cong A$ and $M \cong \langle A \rangle$ provide exactly the same information about M, namely that it has a basis $\{v_i\}$ such that $B(v_i, v_j) = a_{ij}$. Therefore, to reduce notational clutter, we will feel free to drop the outer "$\langle\quad\rangle$" if no confusion is likely; that is, when there is no need to make a careful distinction between a matrix and a quadratic module having that matrix as its Gram matrix with respect to some basis. So, for example, if $M_1 \cong A_1$ and $M_2 \cong A_2$, with the M_i quadratic modules and the A_i symmetric matrices, we may write

$$M_1 \perp M_2 \cong A_1 \perp A_2 \quad \text{instead of} \quad M_1 \perp M_2 \cong \langle A_1 \rangle \perp \langle A_2 \rangle.$$

In both cases the meaning is that the lattice $M = M_1 \perp M_2$ has a Gram matrix with block decomposition

$$\begin{pmatrix} A_1 & 0 \\ 0 & A_2 \end{pmatrix}.$$

Definition 2.10. For a quadratic space V, the **radical** is the subspace

$$\operatorname{rad} V = V^{\perp} = \{\, v \in V \mid B(v,x) = 0 \text{ for all } x \in V\}.$$

Clearly $V = \operatorname{rad} V \perp V_1$ for some subspace V_1. (Take a basis for rad V and extend it to a basis for V.) Such an expresssion is called a **radical splitting** for V. We say V is **regular** if rad $V = 0$.

The proof of the following theorem imitates the **Gram–Schmidt** procedure for inner-product spaces over $\mathbb{R}$ as seen in an elementary linear algebra course, but here we allow for the possible presence of nonzero vectors v satisfying $q(v) = 0$. (We will continue to refer to this as the Gram–Schmidt procedure.) A piece of useful notation: if v is a vector in an R-module M, then Rv denotes the submodule $\{rv \mid r \in R\}$.

Theorem 2.11. *Let F be a field, and let V be an n-dimensional quadratic F-space. Then V has an orthogonal basis; that is, a basis $\{v_1, \dots, v_n\}$ such that $V = Fv_1 \perp \cdots \perp Fv_n$.*

Proof. If $B = 0$ the result is trivially true, since then *every* basis is an orthogonal basis. So we assume $B \neq 0$, and the argument will be by induction.

The result is clear if $n = 1$. So assume that $n \geq 2$ and that the result is true for spaces of lower dimension. Start with a basis $\{w_1, \dots, w_n\}$. If some $q(w_i) \neq 0$, call one such vector v_1; and if $q(w_i) = 0$ for all i, locate w_i and w_j such that $B(w_i, w_j) \neq 0$ (such vectors exist because $B \neq 0$), and put $v_1 = w_i + w_j$. Having fixed v_1, extend v_1 to a basis $\{v_1, z_2, \dots, z_n\}$ for V. Then for $2 \leq i \leq n$ define

$$z_i' = z_i - \frac{B(z_i, v_1)}{q(v_1)} v_1,$$

and check that $V = Fv_1 \perp (Fz_2' \oplus \cdots \oplus Fz_n')$. Now apply the induction hypothesis. $\square$

Corollary 2.12. *A quadratic space $V \neq 0$ is regular if and only if $dV \neq 0$.*

Proof. Let $\{v_1, \ldots, v_n\}$ be an orthogonal basis for V. Then V is regular if and only if $q(v_i) \neq 0$ for $1 \leq i \leq n$, from which the result follows. □

Remark 2.13. In the preceding theorem, were we to know V to be anisotropic, the construction of an orthogonal basis $\{v_1^*, \ldots, v_n^*\}$ from a given basis $\{v_1, \ldots, v_n\}$ could be said much more quickly: Define $v_1^* = v_1$; and having defined $v_1^*, \ldots, v_j^*$ for some $j < n$, define

$$v_{j+1}^* = v_{j+1} - \sum_{i=1}^{j} \frac{B(v_{j+1}, v_i^*)}{q(v_i^*)} v_i^*.$$

From now on we will usually write $\langle \alpha_1, \ldots, \alpha_n \rangle$ instead of the lengthier $\langle \alpha_1 \rangle \perp \cdots \perp \langle \alpha_n \rangle$. This is the standard notation for an unspecified space or lattice having an orthogonal basis.

We will see shortly (in fact, in the exercise set at the end of this section) that the guaranteed existence of an orthogonal basis for a space V is a luxury that might not exist for a lattice, and in trying to develop criteria for two quadratic F-spaces to be isometric we will take full advantage of these wonderful bases. But if the group of F^*/F^{*2} of square classes is large—for example, it is infinite if $F = \mathbb{Q}$—then there may be no straightforward way to canonically select the scalars appearing in the symbol $\langle \alpha_1, \ldots, \alpha_n \rangle$ that represents V: a new choice of orthogonal basis produces a new symbol $\langle \beta_1, \ldots, \beta_n \rangle$ that may look quite different. Imagining what does **not** change—that is, what remains invariant—when we switch from one orthogonal basis to another can be a major challenge. [All we know at this point is that the number n, being the dimension of V, is the same in both symbols, and

$$\prod_{i=1}^{n} \beta_i \in \left(\prod_{i=1}^{n} \alpha_i \right) F^{*2}$$

from our discussion of the discriminant dV.] We will return to this issue in Chapter 4 as we struggle to classify quadratic spaces over the p-adic number fields, a major step in the classification of spaces over $\mathbb{Q}$.

Example 2.14. Consider whether the following quadratic modules are isometric:

$$\langle 1, 1 \rangle, \qquad \langle 2, 2 \rangle, \qquad \langle 2, 3 \rangle.$$

(i) Over $\mathbb{R}$: Yes; in fact if $M \cong \langle \alpha_1, \ldots, \alpha_n \rangle$ in $\{v_1, \ldots, v_n\}$, with $\alpha_i > 0$ for all i, then $M \cong \langle 1, \ldots, 1 \rangle$ in $\left\{ \frac{v_1}{\sqrt{\alpha_1}}, \ldots, \frac{v_n}{\sqrt{\alpha_n}} \right\}$.

(ii) Over $\mathbb{Q}$: $\langle 1,1\rangle \cong \langle 2,2\rangle$ since if $M \cong \langle 1,1\rangle$ in $\{v_1, v_2\}$ then $M \cong \langle 2,2\rangle$ in $\{v_1+v_2, v_1-v_2\}$. Thus the transition matrix for the base change is $T = \begin{pmatrix} 1 & 1 \\ 1 & -1 \end{pmatrix} \in GL_2(\mathbb{Q})$.

On the other hand, $\langle 1,1\rangle \not\cong \langle 2,3\rangle$ since the discriminants are different: $1 \in \mathbb{Q}^{*2}$ but $6 \notin \mathbb{Q}^{*2}$.

(iii) Over $\mathbb{Z}$: The three modules are in different isometry classes, since they have different discriminants (1, 4, and 6 respectively). Now we will see it another way. If $M \cong \langle 2,2\rangle$ in $\{v_1, v_2\}$ then $q(\alpha v_1 + \beta v_2) = 2(\alpha^2+\beta^2) \in 2\mathbb{Z}$ so clearly $M \not\cong \langle 1,1\rangle$. If $J \cong \langle 2,3\rangle$ then clearly $q(v) \geq 2$ if $0 \neq v \in J$, so $J \not\cong \langle 1,1\rangle$; and $M \not\cong J$ because M represents 2 four times while J represents 2 only twice. (An isometry from M onto J would restrict to a bijection

$$\text{from} \quad \{\, v \in M \mid q(v) = \alpha \,\} \quad \text{to} \quad \{\, w \in J \mid q(w) = \alpha \,\}$$

for every scalar α.)

Example 2.15. Consider quadratic $\mathbb{Z}$-lattices

$$L \cong \begin{pmatrix} 2 & 1 & 0 \\ 1 & 2 & 1 \\ 0 & 1 & 5 \end{pmatrix} \text{ in } \{v_1, v_2, v_3\} \quad \text{and} \quad M \cong \begin{pmatrix} 2 & 1 & 0 \\ 1 & 3 & 1 \\ 0 & 1 & 3 \end{pmatrix} \text{ in } \{w_1, w_2, w_3\}.$$

Are these lattices isometric? They both have discriminant 13, so isometry cannot be ruled out on that account; we must look further.

Our method involves expressing the given quadratic forms on L and M as sums (in the usual sense of function sum) of others. Define a new quadratic form q_0 on L by setting

$$(L, q_0) \cong \begin{pmatrix} 2 & 1 & 0 \\ 1 & 2 & 1 \\ 0 & 1 & 2 \end{pmatrix}$$

in the given basis $\{v_1, v_2, v_3\}$. In fact such a lattice L can be constructed in $\mathbb{R}^3$, with the standard dot product serving as the bilinear form, by taking

$$v_1 = \left(\sqrt{2}, 0, 0\right), \qquad v_2 = \left(\frac{\sqrt{2}}{2}, \frac{\sqrt{6}}{2}, 0\right), \qquad v_3 = \left(0, \frac{\sqrt{6}}{3}, \frac{2\sqrt{3}}{3}\right)$$

or in $\mathbb{R}^4$ using the basis

$$v_1 = (1,1,0,0), \qquad v_2 = (0,1,1,0), \qquad v_3 = (0,0,1,1).$$

[The quadratic $\mathbb{Z}$-lattice (L, q_0) (or any lattice isometric to it) is usually denoted by A_3.] Hence $q_0(v) > 0$ for all nonzero $v \in L$; and, since

$$q_0\Big(\sum_i x_i v_i\Big) = 2(x_1^2 + x_2^2 + x_3^2 + x_1x_2 + x_2x_3),$$

it follows that $q_0(v)$ is an even integer ≥ 2. The original quadratic form (call it q) on L now satisfies

$$q\Big(\sum_i x_i v_i\Big) = q_0\Big(\sum_i x_i v_i\Big) + 3x_3^2.$$

Therefore $q(v) \geq 2$ for all $v \in L - \{0\}$, and $q(v) \geq 5$ if $x_3 \neq 0$. In particular, all representations of 2 by (L, q) are in fact representations by the sublattice $A_2 = \mathbb{Z}v_1 + \mathbb{Z}v_2 \cong \begin{pmatrix} 2 & 1 \\ 1 & 2 \end{pmatrix}$. Since

$$q(\alpha v_1 + \beta v_2) = (\alpha + \beta)^2 + \alpha^2 + \beta^2,$$

it is easy to check that A_2 (and hence L) represents 2 exactly six times. On the other hand, if in a similar way we view the given form on M as $q_0 + x_2^2 + x_3^2$, we see that in M the only vectors representing 2 are $\pm w_1$, and from this it follows that L and M are not isometric.

Remark 2.16. The lattice A_2 in the preceding example has the property that the lattice vectors closest to any given lattice vector v are the vertices of a hexagon centered at v, hence A_2 is called the **hexagonal lattice**. It is familiar to any child who has arranged pennies on a flat surface in what appears to be the most efficient way, in the sense that there is the least amount of "wasted space." In fact, what appears to be most efficient really is: A_2 consists precisely of the centers of nonoverlapping disks of equal radius when packed in the plane in the densest possible way. Though this may seem obvious, even to the child arranging pennies, the proof—sketched by Axel Thue in 1892, and proved by him in more detail in 1910—is not easy. See Rogers [**Rog**] for precise definitions and details.

Exercises

(1) Express the idea of regularity in the language of quadratic polynomial forms as in Chapter 1.

(2) In a radical splitting $V = \text{rad } V \perp V_1$, show that V_1 is regular.

(3) Show that if $V = \perp_i V_i$ then $\text{rad } V = \perp_i \text{rad } V_i$. Deduce that V is regular if and only if each V_i is regular.

(4) Suppose $\mathbb{B} = \{v_1, \ldots, v_n\}$ is an orthogonal basis for the quadratic F-space V, and suppose further that $q(v_i) \neq 0$ if $i \leq t$ and $q(v_i) = 0$ for all $i > t$. Show that $\{v_{t+1}, \ldots, v_n\}$ spans rad V.

(5) (i) Let V be a nonzero regular quadratic space. Show that if $\varphi : V \to V$ is a surjective function satisfying $B\big(\varphi(x), \varphi(y)\big) = B(x, y)$ for all $x, y \in V$, then φ is linear and hence an isometry of V onto itself.

(ii) Show that if V is a quadratic space whose quadratic form is nontrivial but not regular, then there is a function φ as described in part (i) that is *not* linear.

(6) Draw a picture of the hexagonal lattice A_2 in $\mathbb{R}^2$ described in the text. Include in your picture all the nonzero vectors of minimum length (equivalently, $q(v)$ is the minimum nonzero value represented by A_2), as well as whatever else is needed for the viewer to understand what elements are in A_2. (Be sure to include a proof that you have included all the minimal vectors in your picture.) Then sketch another lattice in $\mathbb{R}^2$ isometric to the hexagonal lattice.

(7) Show that the $\mathbb{Z}$-lattice A_2 has no orthogonal basis. Then do the same for the $\mathbb{Z}$-lattice $L \cong \begin{pmatrix} 0 & 1 \\ 1 & 0 \end{pmatrix}$.

(8) Suppose V is a binary quadratic $\mathbb{R}$-space, with $V \cong \begin{pmatrix} 2 & 1 \\ 1 & 2 \end{pmatrix}$ in the basis $\{v_1, v_2\}$. Obtain an orthogonal basis for V.

(9) Consider the $\mathbb{Z}$-lattice A_2 and its underlying quadratic $\mathbb{R}$-space $\mathbb{R}^2$. For each, determine whether the orthogonal group is finite or infinite.

2.3. Representation, Splitting, and Isotropy; Invariants $u(F)$ and $s(F)$

Recall that for an n-dimensional vector space V over F the **dual space** is the vector space $V^* = \mathrm{Hom}(V, F)$ of linear transformations $V \to F$. Given a basis $\{v_1, \dots, v_n\}$ for V, the **dual basis** of V^* is the set $\{v_1^*, \dots, v_n^*\}$, where the linear mapping $v_i^* : V \to F$ is defined by the equations $v_i^*(v_j) = \delta_{ij}$ for $1 \le j \le n$. The vector spaces V and V^* are isomorphic, since they have the same dimension, with the linear map given by $v_i \mapsto v_i^*$ as one example of an isomorphism. But this isomorphism depends on the choice of basis; for instance, if $\dim V = 2$ compare the isomorphisms from V to V^* determined in this way by a basis $\{v_1, v_2\}$ and by the basis $\{v_1, v_1 + v_2\}$.

Now let V be a *quadratic* space, and define $f_B : V \to V^*$ by $f_B(v) = B(v, \cdot)$. Notice that the definition of f_B is independent of any basis. In seeking to understand a mathematical object, it can be useful to view that object from different perspectives; and the following proposition links the Gram matrices of V to the mapping f_B that we have just defined.

Proposition 2.17. *Suppose $V \cong A = (a_{ij})$ in $\{v_1, \dots, v_n\}$. Then A is the matrix of the linear transformation $f_B : V \to V^*$ with respect to the bases $\{v_i\}$ and $\{v_i^*\}$. In particular, V is regular if and only if f_B is an isomorphism.*

Proof. Suppose $1 \le j \le n$. We must verify that $f_B(v_j) = \sum_{i=1}^n a_{ij} v_i^*$. But for $1 \le k \le n$ we have

$$(f_B(v_j))(v_k) = B(v_j, v_k) = a_{jk} = \left(\sum_{i=1}^n a_{ij} v_i^* \right)(v_k),$$

from which the result follows. □

Proposition 2.18. *If V is a regular quadratic space and $\sigma : V \to W$ is a representation, then σ is injective.*

Proof. If $\sigma(x) = 0$ then for all $v \in V$ we have $B(x, v) = B(\sigma x, \sigma v) = B(0, \sigma v) = 0$, hence $x \in \operatorname{rad} V = 0$. □

We note that the hypothesis of regularity is essential in the proposition. For example, given a nontrivial radical splitting $V = \operatorname{rad} V \perp V_1$, the endomorphism of V that collapses $\operatorname{rad} V$ to 0 and is the identity on V_1 is a noninjective representation of V into V.

Theorem 2.19. *If U is a regular subspace of V then $V = U \perp U^\perp$. If also $V = U \perp W$ then $W = U^\perp$.*

Proof. We know $U \cap U^\perp = 0$, because U is regular. Therefore, to verify the first statement it suffices to show that $V = U + U^\perp$. For this purpose, let $x \in V$. The mapping $B(x, \cdot)|_U$ is in U^*, hence (because U is regular) there exists $y \in U$ such that $B(y, \cdot)|_U = B(x, \cdot)|_U$. Therefore $x - y \in U^\perp$, and so $x = y + (x - y) \in U + U^\perp$ as desired. For the second statement we need only note that $W \subseteq U^\perp$, and the conclusion follows by comparing dimensions. □

Corollary 2.20. *Suppose $\alpha \neq 0$. Then there is a representation $\alpha \to V$ if and only if there is a splitting $V = \langle \alpha \rangle \perp W$ for some subspace W.*

In considering the isometry and representation problems involving a given regular quadratic F-space V, the first step is generally obtaining an orthogonal basis $\{v_1, \ldots, v_n\}$ and hence an expression $V \cong \langle \alpha_1, \ldots, \alpha_n \rangle$. Then we try to use properties of F to simplify further.

Examples 2.21. (i) If F is algebraically closed, replacing each v_i by $v_i/\sqrt{\alpha_i}$ gives $V \cong \langle 1, \ldots, 1 \rangle$.

(ii) If $F = \mathbb{R}$, then replacing v_i by $v_i/\sqrt{|\alpha_i|}$ yields $V \cong \langle 1, \ldots, 1, -1, \ldots, -1 \rangle$. We will see later that the numbers of positive and negative terms in this expression are invariants of V.

(iii) Suppose F is a finite field of odd order q. The group homomorphism $F^* \to F^*$ with action $\alpha \mapsto \alpha^2$ has kernel $\{\pm 1\}$. It follows that $(F^* : F^{*2}) = 2$. Therefore there is a splitting $V \cong \langle 1, \ldots, 1, \epsilon, \ldots, \epsilon \rangle$, where ϵ is a fixed nonsquare in F. We will see in Chapter 3 that the basis can be chosen so that ϵ appears at most once in this expression.

(iv) If $F = \mathbb{Q}$, then suitable scalar multiplication yields $V \cong \langle \alpha_1, \ldots, \alpha_n \rangle$ with each α_i a squarefree integer. With what is known about quadratic spaces over $\mathbb{Q}$ (we will discuss the details later), given two quadratic $\mathbb{Q}$-spaces, $V \cong \langle \alpha_1, \ldots, \alpha_n \rangle$ and $W \cong \langle \beta_1, \ldots, \beta_n \rangle$, it is a finite process to determine whether V and W are isometric. But, given that the answer is affirmative, we note wistfully that no effective algorithm is known for manipulating a given basis of V associated with Gram matrix $\langle \alpha_1, \ldots, \alpha_n \rangle$ to produce a new basis for V with respect to which $V \cong \langle \beta_1, \ldots, \beta_n \rangle$.

Now we will work with the notions of isotropy and anisotropy introduced at the start of this chapter.

Examples 2.22. (i) The trivial space $V = 0$ (over any field) is anisotropic.

(ii) $\langle \alpha, -\alpha \rangle$ is isotropic.

(iii) Over $\mathbb{R}$, the spaces $\langle 1, \ldots, 1 \rangle$ and $\langle -1, \ldots, -1 \rangle$ are anisotropic.

(iv) Over an algebraically closed field, every space of dimension ≥ 2 is isotropic.

(v) The $\mathbb{Q}$-space $\langle 1, -2 \rangle$ is anisotropic, while the $\mathbb{R}$-space $\langle 1, -2 \rangle$ is isotropic.

Theorem 2.23. *Let V be a regular binary (i.e., 2-dimensional) quadratic F-space. Then the following statements are equivalent:*

(i) *V is a hyperbolic plane.*

(ii) *V is isotropic.*

(iii) *$dV = -1$*

Proof. The implication (i)$\Longrightarrow$(ii) is trivial. To see that (ii)$\Longrightarrow$(iii) just extend an isotropic vector to a basis and consider the associated Gram matrix. Now assume (iii) holds. Then $V \cong \langle \alpha, \beta \rangle$, with $\alpha\beta = -\gamma^2$, say; hence $\beta \in -\alpha F^{*2}$. Therefore $V \cong \langle \alpha, -\alpha \rangle$ in some basis $\{v_1, v_2\}$. Then $V \cong \begin{pmatrix} 0 & 1 \\ 1 & 0 \end{pmatrix}$ in the basis $\left\{ v_1 + v_2, \dfrac{1}{2\alpha}(v_1 - v_2) \right\}$. $\square$

Now suppose H is a hyperbolic plane; say $H \cong \begin{pmatrix} 0 & 1 \\ 1 & 0 \end{pmatrix}$ with respect to the hyperbolic pair $\{x, y\}$. It is easily checked that H has exactly two isotropic lines, namely Fx and Fy. (By a **line** we mean a one-dimensional subspace.) Thus the set of isotropic vectors in H is $(Fx \cup Fy) - \{0\}$.

Definition 2.24. A quadratic F-space V is **universal** if it represents every element of F.

Proposition 2.25. (i) *A hyperbolic plane is universal.*

(ii) *If V is regular and isotropic then V is split by a hyperbolic plane and hence is universal.*

Proof. (i) Let $\{v_1, v_2\}$ be a hyperbolic pair in the given plane, and let $\alpha \in F$. Then $q(v_1 + \frac{\alpha}{2}v_2) = \alpha$.

(ii) Suppose x is an isotropic vector in V. Since V is regular, there exists $y \in V$ such that $B(x, y) \neq 0$. Then $H = Fx + Fy$ is a hyperbolic plane. Since H is regular it follows that H splits V. $\square$

Example 2.26. An orthogonal sum $V = H_1 \perp \cdots \perp H_t$ of hyperbolic planes is a **hyperbolic space**. Such a V is regular, since $dV = (-1)^t \neq 0$. Let $\{v_1, \ldots, v_t\}$ be a set of isotropic vectors in V, with $v_i \in H_i$. Then $\{v_1, \ldots, v_t\}$ spans a t-dimensional subspace W of V on which the quadratic form q is the zero mapping. Thus a regular space can have a totally isotropic subspace of large dimension. This raises the question of how large the dimension of a totally isotropic subspace of a regular space can be, and we will return to this later.

Proposition 2.27. *Let V be a regular quadratic F-space, and let $\alpha \in F^*$. Then*

$$\alpha \longrightarrow V \qquad \Longleftrightarrow \qquad V \perp \langle -\alpha \rangle \textit{ is isotropic.}$$

Proof. The necessity ("$\Longrightarrow$") is clear. Now suppose conversely that the space $V \perp \langle -\alpha \rangle$ is isotropic. Then there exist $x \in V$ and $\beta \in F^*$, not simultaneously zero, such that $q(x) - \beta^2\alpha = 0$. If $\beta \neq 0$ then $q\left(\dfrac{x}{\beta}\right) = \alpha$. And if $\beta = 0$ then V is isotropic and hence universal; so $\alpha \to V$. $\square$

The preceding short proposition shows that the issue of isotropy, which up to this point may have seemed to the reader to be of dubious interest, is actually central to the representation theory of quadratic spaces. In fact we will see later that we will have solved the question of when one given quadratic F-space represents another, and in particular the question of when two given spaces are isometric, once we can determine whether a space is isotropic. We have seen that determining whether a *binary* space is isotropic (and hence is a hyperbolic plane) reduces to a determinant calculation, as long as we know how to recognize squares in our coefficient field, but for spaces of higher dimension the problem may be much more difficult.

Suppose the regular quadratic F-space V is not universal; say $\alpha \not\longrightarrow V$. Then we know V is anisotropic, and in fact $V_1 = V \perp \langle -\alpha \rangle$ is anisotropic. If V_1 is also not universal then we can continue this process, getting an ascending chain

$$V \subset V_1 \subset V_2 \subset \cdots$$

of anisotropic F-spaces. If $(F^* : F^{*2}) < \infty$ this construction must terminate; and—especially considering the relevance of isotropy and anisotropy to the representation question—in general for a given field F it is reasonable to ask whether there is an upper bound to the dimension of anisotropic quadratic F-spaces. This leads to the following definition.

Definition 2.28. If there is an upper bound for the dimension of anisotropic quadratic F-spaces, the largest such dimension is the **u-invariant** of F, denoted $u(F)$. If no such upper bound exists, we write $u(F) = \infty$.

If $u(F) < \infty$ then every F-space of dimension $u(F)$ is universal. Clearly $u(F) = 1$ if F is algebraically closed. And $u(\mathbb{R}) = \infty$, since no $\mathbb{R}$-space of the form $\langle 1, 1, \ldots, 1\rangle$ is universal. The determination of the u-invariant of a field can be quite difficult. We will see in Section 2.8 that $u(F)$=2 if F is a finite field. And in Chapters 3 and 4 we will introduce the fields $\mathbb{Q}_p$ of **p-adic numbers** and show that $u(\mathbb{Q}_p) = 4$ for all p. (The $\mathbb{Q}_p$ are certain extensions of the field $\mathbb{Q}$ of rational numbers, and there is one such field corresponding to each prime p.) It was shown in the 1960s that every positive integer of the form 2^k is the u-invariant of some field. Then in 1991 Merkurjev showed that every even positive integer is a $u(F)$. The first example of field F with $u(F)$ odd (and greater than 1) was due to Izhboldin, who constructed a field with $u(F) = 9$; his result appeared in 2001. See Lam's book [**Lam**] for further details.

We have seen that representation by quadratic F-spaces is closely linked to isotropy, and in exploring the latter concept—and in particular in determining whether or not the invariant $u(F)$ is finite—a good place to start is to consider whether an m-dimensional space of the form $V \simeq \langle 1, \ldots, 1\rangle$ is isotropic if m is sufficiently large. From what we have seen, this happens only if there is a representation

$$-1 \to \underbrace{\langle 1, \ldots, 1\rangle}_{m-1}.$$

Equivalently, -1 is a sum of $m - 1$ squares in F. This leads us to the following definition.

Definition 2.29. If there is a positive integer n such that -1 is a sum of n squares in F, the smallest such n is called the **level** (or, in German, the **stufe**) of F, denoted $s(F)$. If no such n exists the field F is said to be **formally real**, and we write $s(F) = \infty$.

Clearly $s(F) = 1$ if -1 is a square in F, and we will see examples of fields satisfying $s(F) = 2$ and $s(F) = 4$. In particular, later in this chapter we will show that if F is a finite field of odd order, then $s(F)$ is 1 or 2, depending on the cardinality of F; and in Chapter 4 we will see that the field $\mathbb{Q}_2$ of 2-adic numbers satisfies $s(\mathbb{Q}_2) = 4$. It is therefore natural to ask for which positive integers k there is a field F with $s(F) = k$. The following proposition shows that $s(F) = 3$ is not a possibility. Interestingly, the proof employs an instance of an identity arguably due to Diophantus around the year 250:

$$(x^2 + y^2)(z^2 + t^2) = (xz \pm yt)^2 + (xt \mp yz)^2.$$

Proposition 2.30. *There is no field F with $s(F) = 3$.*

Proof. Suppose otherwise. Say $s(F) = 3$ and $-1 = a^2 + b^2 + c^2$, with $a, b, c \in F^*$. Then $0 = (1+a^2)+(b^2+c^2)$, and so upon multiplying by $1+a^2$ and applying Diophantus's identity we obtain

$$0 = (1 + a^2)^2 + (1 + a^2)(b^2 + c^2) = (1 + a^2)^2 + (b - ac)^2 + (c + ab)^2$$

and therefore

$$-1 = \left(\frac{b - ac}{1 + a^2}\right)^2 + \left(\frac{c + ab}{1 + a^2}\right)^2$$

contradicting the assumption that $s(F) = 3$. □

More elaborate arguments, but with the same flavor as the proof of the preceding proposition, can be used to show that $s(F)$ cannot be 5 or 7. In 1953 Kaplansky showed that $s(F)$ could be $1, 2, 4, 8$, or a multiple of 16; and then in 1965 Pfister showed that there is a field F with $s(F) = k$ if and only if k is a power of 2. Pfister's work in the 1960s formed the foundation of the algebraic theory of quadratic forms over fields.

2.4. The Orthogonal Group of a Space

Let V be a quadratic F-space. The **orthogonal group** of V, denoted $O(V)$, is the group (under composition) of isometries of V onto V. If $V \cong A$ in the basis $\{v_1, \ldots, v_n\}$, then $O(V)$ is isomorphic to the multiplicative group of matrices

$$\{T \in GL_n(F) \,|\, {}^tTAT = A\},$$

the group of **automorphs** of A; thus $\det \sigma = \pm 1$ for all $\sigma \in O(V)$ if V is regular. The subgroup $\{\sigma \in O(V) \,|\, \det \sigma = 1\}$ is the **rotation group** or **special orthogonal group** of V, denoted $O^+(V)$ or $SO(V)$. (The set of isometries of V of determinant -1 is sometimes denoted $O^-(V)$.) As with any group, we can ask for a set of particularly elementary generators for $O(V)$, we can look for normal subgroups, and so on.

Example 2.31. Suppose $H \cong \begin{pmatrix} 0 & 1 \\ 1 & 0 \end{pmatrix}$ in $\{x, y\}$. Then Fx and Fy are the only isotropic lines in H, and it follows that with respect to the basis $\{x, y\}$ a rotation has a matrix of the form $\begin{pmatrix} \alpha & 0 \\ 0 & \alpha^{-1} \end{pmatrix}$, with $\alpha \in F^*$, while an element of $O^-(V)$ has a matrix of the form $\begin{pmatrix} 0 & \alpha \\ \alpha^{-1} & 0 \end{pmatrix}$.

Definition 2.32. Let V be a quadratic space, and let $y \neq 0$ be an anisotropic vector in V. Define the **symmetry with respect to** y to be the mapping $\tau_y : V \to V$ by

$$x \longmapsto x - \frac{2B(x, y)}{q(y)} y \qquad \text{for all } x \in V.$$

The following properties are easily checked:

(i) $\tau_y \in O(V)$.

(ii) $\tau_{\alpha y} = \tau_y$ if $\alpha \neq 0$; in words: every nonzero vector in the line of v produces the same symmetry.

(iii) $\tau_y x = x \iff x \in (Fy)^\perp$, and $\tau_y y = -y$. Thus τ_y is the "reflection" in the hyperplane ($(n-1)$-dimensional subspace) $(Fy)^\perp$, and there is a one-to-one correspondence between symmetries and anisotropic lines. If $\{u_1, \ldots, u_{n-1}\}$ is a basis for $(Fy)^\perp$, then τ_y has matrix

$$\begin{pmatrix} -1 & & & \\ & 1 & & \\ & & \ddots & \\ & & & 1 \end{pmatrix}$$

with respect to the basis $\{y, u_1, \ldots, u_{n-1}\}$, and therefore $\det \tau_y = -1$.

(iv) $\tau_y^2 = 1_V$. That is, τ_y is an **involution** in the group $O(V)$.

(v) If $\sigma \in O(V)$ then $\sigma \tau_y \sigma^{-1} = \tau_{\sigma y}$.

Verification of the next proposition is left to the reader.

Proposition 2.33. *Let $x, y \in V$, with $q(x) = q(y)$ and $q(x - y) \neq 0$. Then $\tau_{x-y} x = y$.*

Example 2.34. Let H be a hyperbolic plane, with $\{x, y\}$ a hyperbolic pair in H, and let $\tau \in O(H)$ be a symmetry. Then $\tau = \tau_z$, where $z = x - \alpha y$ for some $\alpha \neq 0$, and τ has matrix $\begin{pmatrix} 0 & \alpha^{-1} \\ \alpha & 0 \end{pmatrix}$ with respect to $\{x, y\}$. Thus τ switches the two isotropic lines in H. Moreover, from the example preceding the definition of symmetry, the converse holds: If $\sigma \in O(H)$ switches the isotropic lines, then σ is a symmetry. If $\lambda \in O(H)$ and λ is *not* a symmetry, then λ has matrix $\begin{pmatrix} \alpha & 0 \\ 0 & \alpha^{-1} \end{pmatrix}$. In this case we can write λ as a product of two symmetries: $\lambda = \tau_{x-y}\tau_{x-\alpha y}$.

Proposition 2.35. *Let V be an n-dimensional quadratic space, and suppose $x, y \in V$, with $q(x) = q(y) \neq 0$. Then*

(i) *There exists $\varphi \in O(V)$ such that $\varphi x = y$ and φ is either a symmetry or a product of two symmetries.*

(ii) *If V is regular and $n > 1$, there is such a φ that is a product of two symmetries (and hence $\varphi \in O^+(V)$).*

Proof. (i) From the hypothesis, we have $q(x-y) + q(x+y) = 4q(x) \neq 0$, so either $q(x-y) \neq 0$ or $q(x+y) \neq 0$. If $q(x-y) \neq 0$ then $\varphi = \tau_{x-y}$ works, by the preceding proposition. If $q(x+y) \neq 0$ then $\tau_{x+y}x = -y$, and hence $\tau_y\tau_{x+y}x = y$.

(ii) If $q(x+y) \neq 0$ we are done, by part (i). Otherwise $q(x-y) \neq 0$, and in this case choose $w \in (Fy)^\perp$ such that $q(w) \neq 0$. (Such a w exists since V [and hence also $(Fy)^\perp$] is regular.) Then $\varphi = \tau_w\tau_{x-y}$ does the job. □

Proposition 2.36. *If V is a regular quadratic F-space, then $O(V)$ is generated by symmetries.*

Proof. We use induction on $n = \dim V$. The case $n = 1$ is trivial (why?), so assume that $n > 1$ and that the result is known in lower dimensions.

Let $\sigma \in O(V)$; we must show σ is a product of symmetries. Choose $x \in V$ such that $q(x) \neq 0$; thus $V = Fx \perp U$, say. Since $q(\sigma x) = q(x)$, from the preceding proposition there is an isometry ψ that is a product of symmetries and such that $\psi\sigma x = x$. Then also $\psi\sigma(U) = U$. From the induction hypothesis, $\psi\sigma|_U$ is a product of symmetries on U. Each of these symmetries extends (uniquely) to a symmetry on V, and then $\psi\sigma$ is the product of these extended symmetries. The conclusion follows. □

We remark that a stronger form of the preceding proposition holds: In expressing an isometry as a product of symmetries, at most $n = \dim V$ symmetries are needed. (This is the Cartan-Dieudonné Theorem.) For example, if V is $\mathbb{R}^3$ with the standard dot product as its symmetric bilinear form, and $1_V \neq \sigma \in O^+(V)$, then it follows that $\sigma = \tau_1\tau_2$, a product of two different symmetries. The intersection of the fixed planes of τ_1 and τ_2 is a line. In words: every nontrivial rotation of Euclidean 3-space has a unique axis of rotation.

The factorization of an isometry into symmetries is not unique. (For instance, one can insert $1_V = \tau_y^2$ anywhere in a given factorization.) So the question arises as to what if anything is invariant about an isometry's factorizations into symmetries. Of course if $\sigma = \tau_{y_1} \cdots \tau_{y_k}$ the parity of k is determined by the determinant of σ. But it is also true—though we will not prove it—that

$$\theta(\sigma) \stackrel{\text{def}}{=} q(y_1) \cdots q(y_k) \cdot F^{*2} \in F^*/F^{*2}$$

is independent of the factorization; in other words, the function

$$\theta : O(V) \to F^*/F^{*2}$$

is well defined. (Proof that θ is well defined can be carried out using the theory of **Clifford algebras**. See [**O'M1**], Chapter 5, for details.) This value $\theta(\sigma)$ is the **spinor norm** of σ. Once we accept that θ is well defined, it follows immediately that it is a group homomorphism. We define the **spinorial kernel** to be the group

$$O'(V) = \{\sigma \in O^+(V) \,|\, \theta(\sigma) = 1\} = \ker(\theta|_{O^+(V)}),$$

and we let $\Omega(V)$ denote the commutator subgroup of $O(V)$. Then we have the evident normal series

$$\Omega(V) \lhd O'(V) \lhd O^+(V) \lhd O(V),$$

and in fact each subgroup in this expression is normal in $O(V)$. Finally, let $Z(V)$ denote the center of $O(V)$. It is a fact (again, we omit the proof) that $Z(V) = \{\pm 1_V\}$ except in one instance: when V is a hyperbolic plane over the field $\mathbb{F}_3$ of three elements; in the exceptional case $Z(V) = O(V)$. The study of interesting subgroups of the general linear group $GL(V)$ (the group of automorphisms of V) belongs to the theory of the "classical groups". For instance, it is a theorem in that subject that, with our present notation, the group $\Omega(V)/\left(\Omega(V) \cap Z(V)\right)$ is a simple group if V is isotropic and $\dim V \geq 5$. (A good source for this material is Artin's *Geometric Algebra* [**A**].) In particular, if one restricts F to finite fields, this gives an infinite family of finite simple groups.

In Chapter 9 we will see that the spinorial kernel plays an important role in the struggle to classify lattices over the integers up to isometry.

2.5. Witt's Cancellation Theorem and Its Consequences

Recall from the start of this chapter that if R is a principal ideal domain that is not a field, then an R**-lattice** is a free R-module of finite rank.

Example 2.37. Consider a $\mathbb{Z}$-lattice L satisfying $L \cong \langle 1 \rangle \perp \begin{pmatrix} 0 & 1 \\ 1 & 0 \end{pmatrix}$ in the basis $\{v_1, v_2, v_3\}$. Then $L \cong \langle 1 \rangle \perp \begin{pmatrix} 1 & 1 \\ 1 & 0 \end{pmatrix}$ in $\{v_1 - v_3, v_1 + v_2, v_3\}$.

But $\mathbb{Z}$-lattices $\left\langle \begin{pmatrix} 1 & 1 \\ 1 & 0 \end{pmatrix} \right\rangle$ and $\left\langle \begin{pmatrix} 0 & 1 \\ 1 & 0 \end{pmatrix} \right\rangle$ are not isometric. (The former represents 1, while the latter represents only even integers.) Thus, given a congruence $M \perp J \cong M \perp K$ over $\mathbb{Z}$, one cannot just "cancel" the M and deduce that $J \cong K$.

The following important result shows that cancellation *is* allowed for quadratic spaces over fields.

Theorem 2.38 (Witt's Cancellation). *Let V be a quadratic F-space, and let U and W be regular subspaces, with $U \cong W$. Then $U^\perp \cong W^\perp$.*

Proof. We use induction on $m = \dim U = \dim W$.

If $m = 1$ then $U = Fx, W = Fy$, with $q(x) = q(y) \neq 0$. Then there is an isometry $\varphi \in O(V)$ such that $\varphi x = y$, and hence $\varphi U^\perp = W^\perp$.

Now suppose $m > 1$, and suppose further that the lower dimensional cases have been checked. Say $U \cong W \cong \langle \alpha_1, \ldots, \alpha_m \rangle$. By the case $m = 1$ there is an isometry

$$\underbrace{\langle \alpha_2, \ldots, \alpha_m \rangle}_{Z} \perp U^\perp \quad \cong \quad \underbrace{\langle \alpha_2, \ldots, \alpha_m \rangle}_{Z'} \perp W^\perp.$$

Now let $\lambda : Z \perp U^\perp \to Z' \perp W^\perp$ be such an isometry. Then

$$\lambda(Z) \perp \lambda(U^\perp) = Z' \perp W^\perp$$

with $\lambda(Z) \cong Z'$. Applying the induction hypothesis now yields

$$U^\perp \cong \lambda(U^\perp) \cong W^\perp$$

and we are done. □

Remark 2.39. In the preceding proof, having obtained the congruence $Z \perp U^\perp \cong Z' \perp W^\perp$, with $Z \cong Z'$, the reader may have felt a temptation to invoke the induction hypothesis and thereby cancel Z and Z' to get the conclusion. The fly in this ointment lies in the fact that the induction hypothesis deals with cancellation of two isometric subspaces of the *same* space, whereas there is no reason to believe that the spaces $Z \perp U^\perp$ and $Z' \perp W^\perp$ are the same. Once the isometry λ has been applied, however, this objection is overcome, so the cancellation (based on the induction hypothesis) becomes acceptable.

Theorem 2.40 (Sylvester's Law of Inertia). *Let V be a regular n-dimensional quadratic space over $\mathbb{R}$. Then V has an orthogonal splitting of the form*

$$V \cong \langle \underbrace{1, \ldots, 1}_{r}, \underbrace{-1, \ldots, -1}_{s} \rangle$$

Moreover, the dimensions r and s are invariants of V; that is, if also

$$V \cong \langle \underbrace{1, \dots, 1}_{r'}, \underbrace{-1, \dots, -1}_{s'} \rangle,$$

then $r = r'$ and $s = s'$.

Proof 1. Argue by induction. The case $n = 1$ is trivial, so assume that $n > 1$ and that the lower-dimensional cases have been covered. If $r = n$ then also $r' = n$, since $q(x) > 0$ for all $x \neq 0$. [This is the **positive definite** case. Similarly, V and q would be said to be **negative definite** if $q(x) < 0$ for all $x \neq 0$.] Otherwise each of the given splittings can be written in the form $\langle \dots \rangle \perp \langle -1 \rangle$. Apply Witt's Cancellation Theorem and the induction hypothesis. □

It's worth noting that Sylvester's law can be proved without invoking Witt's Theorem:

Proof 2. Write $V = P_1 \perp N_1 = P_2 \perp N_2$, with q positive definite on P_1 and P_2 and negative definite on N_1 and N_2. We must prove that $\dim P_1 = \dim P_2$; by symmetry it suffices to prove $\dim P_1 \leq \dim P_2$. Let $\varphi : V \to V$ be the projection of V onto the direct summand P_2. Then $\ker \varphi = N_2$, and hence

$$\ker(\varphi|_{P_1}) = P_1 \cap N_2 = 0.$$

Thus $\varphi|_{P_1} : P_1 \to P_2$ is an injection, and hence we are finished. □

Definition 2.41. The invariants r and s appearing in Sylvester's theorem are denoted $\text{ind}^+ V$ and $\text{ind}^- V$, the **positive index** and **negative index** of V, respectively. The **signature** of V, denoted $\text{sig}\, V$, is the difference $\text{ind}^+ V - \text{ind}^- V$. Thus V is positive definite if $\text{ind}^+ V = n$ and negative definite if $\text{ind}^- V = n$. If both $\text{ind}^+ V$ and $\text{ind}^- V$ are nonzero, then V is said to be **indefinite**.

Thus two regular n-dimensional $\mathbb{R}$-spaces are isometric if and only if they have the same positive and negative indices or, equivalently, the same signature.

Now let V be a regular F-space. If V is isotropic it has a hyperbolic plane component H_1; if $H_1^\perp$ is isotropic it too splits off a hyperbolic plane, and so on. Thus V has a splitting

$$V = H_1 \perp \cdots \perp H_k \perp V_a$$

in which the H_i are hyperbolic planes, $k \geq 0$, and V_a is anisotropic. Such a splitting for V is called a **Witt decomposition** for V. The number k, which, by Witt's Theorem, is independent of the particular Witt decomposition, is the **Witt index** of V, denoted $\text{ind}\, V$. The subspace V_a is called

"the" anisotropic part of V. In general V_a is not uniquely determined, but by Witt's Theorem its isometry class is uniquely determined.

Note that for a regular $\mathbb{R}$-space V, we have $\operatorname{ind} V = \min\{\operatorname{ind}^+ V, \operatorname{ind}^- V\}$.

Proposition 2.42. *Let V be a regular n-dimensional space, and let U be a subspace. Then* $\dim U + \dim U^\perp = n$.

Proof. (First note that the result is clear if U is regular, since then $V = U \perp U^\perp$.)

The mapping $f_B : V \to V^*$ given by $x \mapsto B(x, \cdot)$ is an isomorphism. Let U^0 be the annihilator of U. That is,

$$U^0 = \{\lambda \in V^* \mid \lambda(u) = 0 \text{ for all } u \in U\}.$$

From linear algebra we know $\dim U^0 = n - \dim U$. But $f_B U^\perp = U^0$, and hence the conclusion. □

Question. What is an upper bound for the dimension of a totally isotropic subspace of a regular n-dimensional quadratic space? An equivalent in matrix language: How big can a square block of zeroes be if it is located on the diagonal of a nonsingular symmetric matrix? An n-dimensional hyperbolic space contains a totally isotropic subspace of dimension $\frac{n}{2}$. The following result shows that this is biggest possible.

Proposition 2.43. *Let V be regular, and let U be a totally isotropic subspace with basis $\{u_1, \ldots, u_r\}$. Then V has an orthogonal component of the form*

$$H_1 \perp \cdots \perp H_r$$

in which the H_i are hyperbolic planes and $u_i \in H_i$. Therefore

$$\dim U \leq \frac{1}{2} \dim V.$$

Proof. We argue by induction on $r = \dim U$. The case $r = 1$ is trivial (note that in this case $\dim V \geq 2$). Now assume $r > 1$ with the appropriate induction hypothesis. Define $U_{r-1} = Fu_1 + \cdots + Fu_{r-1}$. So $U_{r-1} \subset U$ and hence $U_{r-1}^\perp \supset U^\perp$. Pick $y \in U_{r-1}^\perp - U^\perp$, and let $H_r = Fu_r + Fy$. Then H_r is a hyperbolic plane, and so $V = H_r \perp H_r^\perp$. Notice that $H_r \subseteq U_{r-1}^\perp$, and so $U_{r-1} \subseteq H_r^\perp$. Now U_{r-1} is an $(r-1)$-dimensional totally isotropic subspace of the regular space $H_r^\perp$, so by the induction hypothesis $H_r^\perp$ has an orthogonal component of the form

$$H_1 \perp \cdots \perp H_{r-1},$$

with $u_i \in H_i$ for $1 \leq i \leq r-1$. The result follows. □

Theorem 2.44 (Witt's Isometry Extension). *Suppose V and V' are isometric regular F-spaces, U is a subspace of V, and $\sigma : U \to V'$ is an isometry into V'. Then σ extends to an isometry $V \to V'$.*

Proof. First check that the result is easy to achieve if U is regular. [In fact, if U is not regular the proof consists largely of working to imbed U in a regular subspace of V whose special construction makes an extension of σ to it easily achieved, and then the result will follow from the initial statement in the proof.]

Write $U = U_0 \perp \text{rad } U$, and let $\{u_1, \dots, u_r\}$ be a basis for rad U. Since U_0 is a regular subspace of V, so is $U_0^\perp = \text{rad } U \perp \cdots$. As in the preceding proposition, there is a splitting

$$U_0^\perp = \underbrace{H_1 \perp \cdots \perp H_r}_{H} \perp S$$

in which the H_i are hyperbolic planes and $u_i \in H_i$. Thus $V = U_0 \perp H \perp S$. Now write

$$\sigma U = U', \quad \sigma U_0 = U_0', \quad \text{and} \quad \sigma u_i = u_i' \text{ for } 1 \leq i \leq r.$$

Then rad $U' = \sigma(\text{rad } U) = Fu_1' + \cdots + Fu_r'$, and $U' = U_0' \perp \text{rad } U'$. Thus $V' = U_0' \perp H' \perp S'$, where $H' = H_1' \perp \cdots \perp H_r'$ is hyperbolic and $u_i' \in H_i'$. It is now an easy exercise to extend σ to an isometry

$$\widehat{\sigma} : U_0 \perp H \to U_0' \perp H',$$

and, since $U_0 \perp H$ is regular, $\widehat{\sigma}$ extends to an isometry of V onto V'. □

Definition 2.45. Let (V, B) be a quadratic F-space, and let $\alpha \in F^*$. Define $V^\alpha = (V, B^\alpha)$, where $B^\alpha : V \times V \to F$ is given by $B^\alpha(x, y) = \alpha B(x, y)$ for all $x, y \in V$; similarly, define $q^\alpha(x) = \alpha q(x)$ for all $x \in V$. In moving from V to V^α we say that we have **scaled** V by α.

The proof of the following proposition is left to the reader.

Proposition 2.46. (i) *If H is a hyperbolic space, then $H^\alpha \cong H$ for all $\alpha \neq 0$.*

(ii) *If $V \cong A \in M_n(F)$, then $V^\alpha \cong \alpha A$. Hence $dV^\alpha = dV$ if n is even, and $dV = \alpha dV$ if n is odd.*

(iii) *V is isotropic if and only if V^α is isotropic.*

(iv) *If V is regular then $V \perp V^{-1}$ is hyperbolic.*

(v) *If V and W are regular and of the same dimension, then $V \cong W$ if and only if $V \perp W^{-1}$ is hyperbolic.*

(vi) *If V and W are regular spaces, then there is a representation $V \to W$ if and only if* $\text{ind}(\mathrm{W} \perp \mathrm{V}^{-1}) \geq \dim \mathrm{V}$.

Thus, developing criteria to determine when a quadratic F-space is isotropic is the key to the classification and representation problems for quadratic spaces over F.

2.6. Witt's Chain Equivalence Theorem

The dimension and discriminant are important invariants of a regular quadratic space, and in fact we will see later in this chapter that they are enough to classify spaces over finite fields. But over a great many fields of number-theoretic interest these two invariants need to be supplemented with something more; that is, there can exist non-isometric spaces having the same dimension and discriminant. This is evident over $\mathbb{R}$, by Sylvester's Law of Inertia (2.40), and it is also the case over the fields $\mathbb{Q}_p$ of p-adic numbers, as we will see in Chapter 4. The following important theorem shows that to check whether an element or structure associated with a quadratic space via a given orthogonal basis is an invariant of the space, it suffices to check that the given element or structure remains unchanged whenever at most two elements of the orthogonal basis are changed. We will apply this theorem in Chapter 4 to define a new invariant, the **Hasse symbol**, for $\mathbb{Q}_p$-spaces; and this, in conjunction with the dimension and discriminant, will be the final ingredient needed to classify $\mathbb{Q}_p$-spaces. This in turn is a major step in the classification of quadratic spaces over the field $\mathbb{Q}$ of rational numbers, to be completed in Chapter 5.

Theorem 2.47 (Witt's Chain Equivalence). *Let V be a regular quadratic space over a field F, and suppose $\mathbb{B}$ and $\mathbb{B}'$ are orthogonal bases for V. Then there is a "chain" of orthogonal bases $\mathbb{B} = \mathbb{B}_0, \mathbb{B}_1, \dots, \mathbb{B}_k = \mathbb{B}'$ such that each $\mathbb{B}_i$ is obtained from $\mathbb{B}_{i-1}$ by changing at most two basis elements.*

Proof. We use induction on $n = \dim V$. The case $n \leq 2$ is trivial, so assume that $n \geq 3$ and that the result has been proved for spaces of dimension less than n.

Suppose $\mathbb{B}_0 = \{u_1, \dots, u_n\}$ and $\mathbb{B}' = \{v_1, \dots, v_n\}$. Say

$$v_1 = \alpha_1 u_1 + \cdots + \alpha_r u_r,$$

with $\alpha_i \neq 0$ for each i. (Resubscript the u's if necessary.) Now we argue by induction on r.

Case (i): $r = 1$. So $v_1 = \alpha_1 u_1$. In this case put $\mathbb{B}_1 = \{v_1, u_2, \dots, u_n\}$. Then

$$Fu_2 \perp \cdots \perp Fu_n = Fv_2 \perp \cdots \perp Fv_n$$

and the induction hypothesis (on n) finishes the job.

Case (ii): $r = 2$. Here $v_1 = \alpha_1 u_1 + \alpha_2 u_2$. Then $Fu_1 \perp Fu_2 = Fv_1 \perp Fz$, say. Now take $\mathbb{B}_1 = \{v_1, z, u_3, \ldots, u_n\}$. Then $Fz \perp Fu_3 \perp \cdots \perp Fu_n = Fv_2 \perp \cdots \perp Fv_n$, and again apply the induction on n.

Case (iii): Finally, suppose $v_1 = \alpha_1 u_1 + \cdots + \alpha_r u_r$ with $r \geq 3$, and assume we have already covered the cases in which v_1 is a linear combination of fewer than r of the u_j. Put $q(u_i) = \lambda_i$ for $1 \leq i \leq r$. (So $\lambda_i \neq 0$, since $\mathbb{B}$ is an orthogonal basis of a regular space.) If

$$\underbrace{\alpha_1^2\lambda_1 + \alpha_2^2\lambda_2}_{A} = \underbrace{\alpha_1^2\lambda_1 + \alpha_3^2\lambda_3}_{B} = \underbrace{\alpha_2^2\lambda_2 + \alpha_3^2\lambda_3}_{C} = 0$$

then $0 = A - B + C = 2\alpha_2^2\lambda_2$, a contradiction. Therefore we can assume that $\alpha_1^2\lambda_1 + \alpha_2^2\lambda_2 \neq 0$; so if we set $w_1 = \alpha_1 u_1 + \alpha_2 u_2$ then $q(w_1) \neq 0$. Write $Fu_1 \perp Fu_2 = Fw_1 \perp Fz$, and then put $\mathbb{B}_1 = \{w_1, z, u_3, \ldots, u_n\}$. We can now write $v_1 = w_1 + \alpha_3 u_3 + \cdots + \alpha_r u_r$; and since this sum has length $r - 1$, by the induction hypothesis (on r) we can get from $\mathbb{B}_1$ to $\mathbb{B}'$ by a basis chain of the required type. □

2.7. Tensor Products of Quadratic Spaces; the Witt ring of a field

Recall that every regular quadratic F-space has a **Witt decomposition**: $V = H \perp V_a$, with H hyperbolic (possibly 0) and V_a anisotropic (possibly 0). We will say two spaces V and W are **similar** if their anisotropic parts V_a and W_a are isometric; we also say that V and W are in the same **similarity class**, denoted $[V]$. Because $[V] = [V_a]$, the collection of similarity classes can be identified with the collection of isometry classes of anisotropic F-spaces, and we write $W(F)$ for this collection. We are going to make $W(F)$ into a ring, but before defining operations we had better be certain that we are talking about a *set*. Recall from foundations of mathematics that a collection (we hesitate to use the word "class" because that word has other meanings for us here) is known to be a set if it is in one-to-one correspondence with a set. Now, congruence is an equivalence relation on the set of all symmetric F-matrices, so clearly the congruence classes of symmetric F-matrices form a set. But the isometry classes of quadratic F-spaces are in one-to-one correspondence with the set of congruence classes of symmetric F-matrices, so the collection of all isometry classes of quadratic F-spaces is a set, of which $W(F)$ is a subset. Now we can define an addition operation on $W(F)$:

$$[V] + [W] \quad \stackrel{\text{def}}{=} \quad [V \perp W].$$

If we think informally of the elements of $W(F)$ as anisotropic spaces, agreeing not to distinguish between isometric spaces, then the addition law can be stated like this: the sum of two anisotropic spaces is the anisotropic part of their orthogonal sum. It is left to the reader to check that we have given a well-defined operation; that is, if $[V] = [V']$ and $[W] = [W']$ then $[V \perp W] = [V' \perp W']$. We claim that with this operation $W(F)$ is an abelian group. It is easy to see that our "+" is associative and commutative. The zero element: $0 = [0] = [H]$ for any hyperbolic space H. Finally, the orthogonal sum of a space V and the scaled space V^{-1} is hyperbolic; hence

$$[V^{-1}] = -[V]$$

and so $W(F)$ is an abelian group, the **Witt group** of F.

Remark 2.48. In the algebraic theory of quadratic forms the focus is more on the *similarity* class $[V] \in W(F)$ of a quadratic F-space V than on its *isometry* class. In that setting the definition of discriminant is sometimes modified so that hyperbolic spaces have discriminant 1. (In other words: spaces trivial in $W(F)$ have trivial discriminant.) For instance in [**Sz**] the discriminant of an n-dimensional space V is taken to be $(-1)^{\frac{n(n-1)}{2}} dV$, where dV is the discriminant as we have defined it. In Watson [**W1**], where the focus is on quadratic forms over $\mathbb{Z}$, the discriminant has a further 2-power factor to guarantee that the discriminant of an integral quadratic form is an integer. For example, if $f(x, y) = ax^2 + bxy + cy^2$, with $a, b, c \in \mathbb{Z}$, then the associated lattice has Gram matrix

$$\begin{pmatrix} a & \frac{b}{2} \\ \frac{b}{2} & c \end{pmatrix}$$

for which our discriminant is $ac - \frac{b^2}{2}$, while Watson's is $b^2 - 4ac$. Incidentally, Watson's discriminant in the binary case conforms with standard practice in the literature of binary quadratic forms going back to Gauss [**Ga**]. (Also see [**BV**].)

Next we introduce a multiplication operation that will make $W(F)$ into a ring. For this we require the **tensor product** of two quadratic spaces. There are two approaches to this subject, and we will start with the more concrete of the two—the matrix version—for readers who have not seen the tensor product in its abstract formulation.

Tensor product (I). Let R be a ring, and let $X = (x_{ij}) \in M_m(R)$ and $Y = (y_{ij}) \in M_n(R)$. Then the **tensor product** or **Kronecker product** $X \otimes Y$ is the $mn \times mn$ matrix

$$X \otimes Y = \begin{pmatrix} x_{11}Y & x_{12}Y & \dots & x_{1m}Y \\ \vdots & & & \\ x_{m1}Y & \dots & \dots & x_{mm}Y \end{pmatrix}.$$

Now suppose X and Y are symmetric, and suppose further that $Y \cong Y'$; say $Y' = {}^tTYT$ with $T \in GL_n(R)$. Then $X \otimes Y \cong X \otimes Y'$, because

$X \otimes Y' = {}^tS(X \otimes Y)S$, with

$$S = \begin{pmatrix} T & & \\ & \ddots & \\ & & T \end{pmatrix} \in GL_{mn}(R).$$

A similar argument with X instead of Y leads more generally to this:

$$X \cong X' \text{ and } Y \cong Y' \quad \Longrightarrow \quad X \otimes Y \cong X' \otimes Y'.$$

Now suppose V and W are quadratic F-spaces, with $V \cong A \in M_m(F)$ and $W \cong C \in M_n(F)$ in respective bases $\{v_1, \ldots, v_m\}$ and $\{w_1, \ldots, w_n\}$. Then for the **tensor product** $V \otimes W$ take any mn-dimensional vector space over F and any basis for it, and denote the elements of that basis by

$$\{v_i \otimes w_j\}_{1 \le i \le m, 1 \le j \le n}.$$

Finally, define a symmetric bilinear form B on $V \otimes W$ by putting

$$B(v_i \otimes w_j, v_k \otimes w_l) = a_{ik}c_{jl}.$$

In other words, if B_V and B_W are the symmetric bilinear forms on V and W respectively, then

$$B(v_i \otimes w_j, v_k \otimes w_l) = B_V(v_i, v_k) \cdot B_W(w_j, w_l).$$

It follows that $V \otimes W \cong A \otimes C$ in the basis

$$\{v_1 \otimes w_1, v_1 \otimes w_2, \ldots, v_1 \otimes w_n, v_2 \otimes w_1, \ldots, v_2 \otimes w_n, \ldots, v_m \otimes w_1, \ldots, v_m \otimes w_n\}$$

with the indicated ordering. From the preceding remarks on matrices, using different Gram matrices for V and W would yield a new Gram matrix for $V \otimes W$ congruent to $A \otimes C$. Therefore the isometry class of $V \otimes W$ has been well defined by the procedure we have given.

Tensor product (II). (For those familiar with tensor products of vector spaces.) Given quadratic F-spaces V, W with respective symmetric bilinear forms B_V, B_W, the map

$$\varphi : V \times W \times V \times W \to F$$

given by $(v, w, v', w') \mapsto B_V(v, v') \cdot B_W(w, w')$ is multilinear. Hence there is a unique linear map $\hat{\varphi}$ making the following diagram commutative:

$$\begin{array}{ccc} V \otimes W \otimes V \otimes W & & \\ \uparrow{\scriptstyle \otimes} & \searrow^{\hat{\varphi}} & \\ V \times W \times V \times W & \xrightarrow[\varphi]{} & F \end{array}$$

But there is an isomorphism $V \otimes W \otimes V \otimes W \approx (V \otimes W) \otimes (V \otimes W)$ and hence a unique bilinear map $B : (V \otimes W) \times (V \otimes W) \to F$ satisfying

$$B(v \otimes w, v' \otimes w') = \varphi(v, w, v', w') = B_V(v, v') \cdot B_W(w, w')$$

for all $v, v' \in V$ and $w, w' \in W$. The vector space $V \otimes W$ with this bilinear map B is the **tensor product** of the quadratic spaces V and W. Once bases are chosen for V and W, this algebraic construction of $V \otimes W$ yields a space with the same Gram matrix as the previous matrix-based approach to $V \otimes W$.

Example 2.49. If $V \cong \langle \alpha_1, \ldots, \alpha_m \rangle$ and $W \cong \langle \beta_1, \ldots, \beta_n \rangle$, then

$$V \otimes W \cong \perp_{i,j} \langle \alpha_i \beta_j \rangle.$$

In particular, if H is a hyperbolic space then so is $V \otimes H$.

Whichever of the two constructions of $V \otimes W$ has been used, without ambiguity—in view of the preceding example—we can now define a product in $W(F)$ by

$$[V] \cdot [W] = [V \otimes W].$$

It is easy to check that with addition and multiplication as we have defined them, $W(F)$ is a commutative ring with unit element $[\langle 1 \rangle]$: the **Witt ring** of F.

Examples. (i) Over the field $\mathbb{C}$ of complex numbers, since every element is a square it follows that every regular space of even dimension is hyperbolic, and $W(\mathbb{C}) = \{0, [\langle 1 \rangle]\}$. In fact the mapping $W(\mathbb{C}) \to \mathbb{F}_2$ given by $[V] \longmapsto \dim V \pmod 2$ is a ring isomorphism.

(ii) Over $\mathbb{R}$ the nonzero elements of the Witt ring are of the form

$$[\langle 1, \ldots, 1 \rangle] \quad \text{and} \quad [\langle -1, \ldots, -1 \rangle],$$

and the mapping $W(\mathbb{R}) \to \mathbb{Z}$ given by $[V] \longmapsto \operatorname{sig} V$ is an isomorphism.

Remark 2.50. Having introduced the Witt ring $W(F)$ of a field F, the natural and healthy algebraic impulse is to explore its ring structure. What are its units, its torsion elements, its nilpotent elements, and its zero divisors? What can one say about its ideals, particularly its prime ideals and maximal ideals? How do the answers to these and other ring-theoretic questions relate to the properties of F, and—especially—what must one know about quadratic forms over F in order to understand the structure of $W(F)$? These are substantial and fascinating issues that have received extensive investigation, but they belong more to the *algebraic* theory of quadratic forms—which started with Pfister's work in the 1960s—than to the *arithmetic* theory—which has its roots in classical number theory and is the primary subject of this book. Moreover, the algebraic theory has received thorough and up-to-date treatment elsewhere. (See Lam's recent book [**Lam**] for the most current treatment; also see the books by Szymiczek [**Sz**] and Milnor–Husemoller [**MH**].) So, except for a brief glimpse in three of this section's exercises and sporadic mention in later sections, we will leave the analysis of Witt rings for further study.

Exercises

(1) Let V be a regular quadratic F-space, and let $\alpha, \beta \in F^*$. Prove that

$$-\beta \longrightarrow V \perp \langle\alpha\rangle \qquad \text{if and only if} \qquad -\alpha \longrightarrow V \perp \langle\beta\rangle.$$

(2) (i) Show that a 4-dimensional quadratic space V satisfying $dV = 1$ is hyperbolic if it is isotropic.
(ii) Deduce that if the element $a \in F^*$ is a sum of two squares, then the space $\langle 1, 1, -a, -a\rangle$ is hyperbolic.

(3) (i) Find $x, y, z \in \mathbb{Q}$ such that $5x^2 - 3y^2 - 2z^2 = 10$. [This can be done by trial and error. But try to do it in a more deliberate way.]
(ii) Are there infinitely many solutions to the equation in part (i)?

(4) Let K be a field, and let $\alpha_1, \ldots, \alpha_n \in K^*$. Show that if the quadratic K-space $\langle\alpha_1, \ldots, \alpha_n\rangle$ is anisotropic, then the quadratic $K(x)$-space $\langle\alpha_1, \ldots, \alpha_n\rangle$ is anisotropic over the rational function field $K(x)$. Conclude that $u(K) \leq u(K(x))$.

(5) Let V be a binary (i.e., 2-dimensional) anisotropic quadratic F-space, say $V \cong \langle\alpha_1, \alpha_2\rangle$; and suppose E/F is a finite extension. (i) If $[E : F]$ is odd, show that the quadratic E-space $\langle\alpha_1, \alpha_2\rangle$ is anisotropic. (ii) Can the conclusion of part (i) be made (or refuted) with confidence if $[E : F]$ is even?

(6) Show that if $u(F) < \infty$, then also $s(F) < \infty$, and in fact $s(F) \leq u(F)$.

(7) Show that $s\big(\mathbb{Q}(\sqrt{-2})\big) = 2$.

(8) Give an example of a field extension E/F satisfying $u(E) < u(F)$, then give an example of an extension L/K satisfying $u(L) > u(K)$.

(9) Suppose $V \cong \begin{pmatrix} 0 & 1 \\ 1 & 0 \end{pmatrix} \perp \langle 1\rangle$ in the basis $\{v_1, v_2, v_3\}$. Put $H = Fv_1 + Fv_2$. Show that V has another splitting of the form $H' \perp \langle 1\rangle$ in which H' is a hyperbolic plane $\neq H$.

(10) Let V be a regular quadratic $\mathbb{R}$-space; say $V \cong A \in M_n(\mathbb{R})$. Let D_t denote the determinant of the upper left $t \times t$ submatrix of A. Assume that $D_1, \ldots, D_n$ are all nonzero. Show that the signs of $D_1, \ldots, D_n$ determine the isometry class of V.

(11) Let V be a regular nonzero F-space, and let

$$G(V) = \{\, \alpha \in F^* \mid V^\alpha \cong V \,\}.$$

(i) Show that $G(V)$ is a subgroup of F^*.
(ii) Show that if there is a representation $1 \longrightarrow V$ then every element of $G(V)$ is represented by V.
(iii) Show that if $\dim V = 2$ and $1 \longrightarrow V$ then every nonzero element represented by V is in the group $G(V)$.

(12) (i) Show that if $\alpha \longrightarrow \langle 1, a_1\rangle$ then $\langle 1, a_1\rangle \otimes \langle 1, a_2\rangle \cong \langle 1, a_1\rangle \otimes \langle 1, \alpha a_2\rangle$.
(ii) Show that if $\beta \longrightarrow \langle a_1, a_2\rangle$ then $\langle 1, a_1\rangle \otimes \langle 1, a_2\rangle \cong \langle 1, \beta\rangle \otimes \langle 1, a_1 a_2\rangle$.

(13) Show that $W(F)$ is finite if and only if $u(F) < \infty$ and $(F^* : F^{*2}) < \infty$.

(14) Show that the similarity classes of even-dimensional quadratic F-spaces constitute a maximal ideal in the Witt ring $W(F)$.

(15) Suppose F is a field in which the sum of any two squares is a square. Prove that there is a group isomorphism $W(F) \approx \mathbb{Z}_2$ if and only if -1 is a square in F.

(16) If A is a set, a **word** or **string** of **length** $n \geq 1$ over the "alphabet" A is a finite sequence $(a_1, \dots, a_n) \in A^n$. Use Witt's Chain Equivalence Theorem to formulate the classification problem for regular n-dimensional quadratic spaces over a field F as a problem on transformations of words of length n over the alphabet F^*.

2.8. Quadratic Spaces over Finite Fields

Unless otherwise indicated, **throughout this section F will denote a finite field $\mathbb{F}_q$ of odd order q.** To avoid notational confusion, we will use Q (instead of q) as our generic notation for a quadratic form. We have already seen that $(F^* : F^{*2}) = 2$; hence if δ is any nonsquare in F, then any other nonsquare is of the form $\delta\varepsilon^2$ for some $\varepsilon \in F^*$. Combining this with the Gram-Schmidt algorithm has shown us that every regular quadratic F-space has the form $\langle 1, \dots, 1, \delta, \dots, \delta\rangle$. The following proposition will show that at most one δ is needed in this expression.

Proposition 2.51. *If $\lambda \in F^*$, then $\langle\lambda, \lambda\rangle \cong \langle 1, 1\rangle$.*

Proof. The result is clear if $\lambda \in F^{*2}$ or if $-1 \in F^{*2}$. (In the latter case, both spaces under inspection are hyperbolic planes.) Therefore without loss of generality we can assume $\lambda \notin F^{*2}$ and $-1 \notin F^{*2}$. The sets

$$F^{*2} \qquad \text{and} \qquad 1 + F^{*2} = \{1 + \alpha^2 \mid \alpha \in F^*\}$$

have the same cardinality. But $1 \in F^{*2}$ while $1 \notin 1 + F^{*2}$. Therefore there is a nonsquare $\beta \in (1 + F^{*2})$; so $\beta \to \langle 1, 1\rangle$ and hence (since $\lambda \in \beta F^{*2}$) also $\lambda \to \langle 1, 1\rangle$. It follows that $\langle 1, 1\rangle \cong \langle\lambda, \cdot\rangle$, hence by a discriminant argument we must actually have $\langle 1, 1\rangle \cong \langle\lambda, \lambda\rangle$. □

Theorem 2.52. *Let V and W be regular quadratic F-spaces. Then*

(i) $V \cong W \iff \dim V = \dim W$ *and* $dV = dW$.

(ii) *If* $\dim V < \dim W$ *then there is a representation* $V \to W$.

Proof. (i) The necessity is clear. For the sufficiency, note that from the preceding proposition it follows that every space has an orthogonalization of the form $\langle 1, \dots, 1, dV \rangle$.
(ii) It follows from part (i) that $W \cong \langle 1, \dots, 1, dV, dV dW \rangle$. □

Recall from earlier in this chapter that if K is a field then $u(K)$ is the largest dimension of an anisotropic quadratic K-space. (And $u(K) = \infty$ if those dimensions are unbounded.)

Corollary 2.53. $u(F) = 2$.

Proof. We first note that $u(F) \geq 2$, since for any nonsquare δ we know that the plane $\langle 1, -\delta \rangle$ is anisotropic. On the other hand, if V is regular and $\dim V \geq 3$ then $V \cong \langle 1, \dots, 1, -1, -dV \rangle$ and hence is isotropic. □

Corollary 2.54. *The Witt ring $W(F)$ of a finite field F has four elements:*

$$[0],\ [\langle 1 \rangle],\ [\langle \delta \rangle],\ [\langle 1, -\delta \rangle].$$

(Here δ is a nonsquare in F.)

From our work up to here, it is now an easy matter to determine whether a given element is represented by a given F-space, provided that we can recognize which elements of our field are squares. It will be especially useful to know whether -1 is a square, and that is the purpose of the next lemma.

Lemma 2.55. *If $q \equiv 1 \pmod 4$ then $-1 \in F^{*2}$. If $q \equiv 3 \pmod 4$ then $-1 \notin F^{*2}$.*

Proof. The group F^* is cyclic of order $q - 1$, and -1 is its only element of order 2. (The equation $x^2 = 1$ has only ± 1 as roots.) Recall that a finite abelian group G of order $p_1^{\alpha_1} \cdots p_r^{\alpha_r}$, with the p_i distinct primes, can be expressed as a direct sum: $G = \bigoplus_i G_i$, where G_i has order $p_i^{\alpha_i}$. (The subgroup G_i is the *p_i-primary component* of the decomposition.) Since F^* is cyclic the primary components are also cyclic; let ε be a generator for the 2-primary component. Now $q \equiv 1 \pmod 4$ if and only if ε has order 2^k for some $k \geq 2$. But in this case we have

$$(\varepsilon^{2^{k-1}})^2 = \varepsilon^{2^k} = 1 \qquad \text{and hence} \qquad \varepsilon^{2^{k-1}} = -1 = (\varepsilon^{2^{k-2}})^2.$$

On the other hand, if $q \equiv 3 \pmod 4$ then $4 \nmid \#F^*$, and hence ± 1 are the only elements of 2-power order. So -1 cannot be a square. □

It follows from the preceding lemma that the unique binary anisotropic F-space is $\langle 1,1\rangle$ if and only if $q \equiv 3 \pmod 4$.

Definition 2.56. Let V be a regular quadratic F-space, and let $\alpha \in F$. Define the associated **representation number**

$$r(\alpha, V) = \#\{x \in V \mid Q(x) = \alpha\}.$$

The goal of this section is to find formulas for representation numbers.

Example 2.57. Let H be a hyperbolic plane, with $\{v_1, v_2\}$ a hyperbolic pair.

(i) $r(0,H) = 2q-1$. This is immediate upon noticing that each isotropic line Fv_i has q elements and $Fv_1 \cap Fv_2 = \{0\}$.

(ii) Let $\alpha \neq 0$. Then $r(\alpha, H) = q-1$. This follows from the fact that $Q(\lambda v_1 + \mu v_2) = \alpha$ if and only if $\lambda \neq 0$ and $\mu = \dfrac{\alpha}{2\lambda}$.

Lemma 2.58. *If V is a regular quadratic F-space of even dimension $2k$, and $\alpha \in F^*$, then*

(i) $V \cong V^\alpha$;

(ii) $r(\alpha, V) = \dfrac{1}{q-1}\big(q^{2k} - r(0,V)\big)$.

Proof. (i) This is immediate, since $dV = dV^\alpha$.
(ii) If $V \cong A \in M_{2k}(F)$ in the basis $\{v_1, \dots, v_{2k}\}$ then $V \cong \alpha A$ in some basis $\{w_1, \dots, w_{2k}\}$, by part (i); and the mapping in $GL(V)$ with action $v_i \mapsto w_i$ for $1 \le i \le 2k$ gives a bijection between the sets

$$\{\, v \mid Q(v) = 1\,\} \quad \text{and} \quad \{\, v \mid Q(v) = \alpha\,\}.$$

Thus the value $r(\alpha, V)$ is independent of α. It follows that

$$r(\alpha, V) = \frac{1}{q-1} \sum_{\beta \in F^*} r(\beta, V).$$

But the sum in the preceding expression is just the total number of nonzero anisotropic vectors in V, from which the result follows. □

Theorem 2.59. *Let V be a regular nonzero quadratic space of even dimension $2k$.*

(i) *If V is hyperbolic then* $r(\alpha, V) = \begin{cases} q^{k-1}(q^k + q - 1) & \text{if } \alpha = 0, \\ q^{k-1}(q^k - 1) & \text{if } \alpha \neq 0. \end{cases}$

(ii) *If V is not hyperbolic then* $r(\alpha, V) = \begin{cases} q^{k-1}(q^k - q + 1) & \text{if } \alpha = 0, \\ q^{k-1}(q^k + 1) & \text{if } \alpha \neq 0. \end{cases}$

Proof. In both parts it is enough to prove the first statement, since the second will then follow immediately from that and the lemma.

(i) Let H_k denote a hyperbolic space of dimension $2k$, and we will argue by induction on k. The case $k = 1$ was done in the example preceding the lemma. Now suppose the formula holds for some k, and we will check it for $k + 1$.

Write $H_{k+1} = H_k \perp H$, with H a hyperbolic plane. Let $v = v_1 + v_2$, with $v_1 \in H_k$ and $v_2 \in H$. There are $r(0, H_k)(2q - 1)$ possibilities for v in which $Q(v_1) = Q(v_2) = 0$. On the other hand, for each $\alpha \neq 0$ there are $(q - 1)r(\alpha, H_k)$ isotropic v for which $Q(v_1) = \alpha$, since $r(-\alpha, H) = q - 1$. Adding these results for all $\alpha \in F^*$ and applying the lemma gives

$$\begin{aligned} r(0, H_{k+1}) &= r(0, H_k)(2q - 1) + (q - 1)^2 r(1, H_k) \\ &= r(0, H_k)(2q - 1) + (q - 1)\big(q^{2k} - r(0, H_k)\big) \\ &= q r(0, H_k) + q^{2k+1} - q^{2k} \end{aligned}$$

representations of 0 by H_{k+1}. [Recall from the lemma that the value $r(\alpha, H_k)$ is the same for all $\alpha \in F^*$; so we can write this common value as $r(1, H_k)$. We have done this in the displayed material immediately preceding this remark.] The conclusion now follows immediately from the induction hypothesis.

(ii) If V is not hyperbolic, then $V = H_{k-1} \perp W$, with H_{k-1} hyperbolic and W binary and anisotropic. We are counting the isotropic vectors $v + w$ with $v \in H_{k-1}$ and $w \in W$. There are $r(0, H_{k-1})$ such vectors in which $w = 0$. Now let w be an arbitrary nonzero vector in W; say $Q(w) = -\alpha \in F^*$. Then there are $r(1, H_{k-1})$ representations of α by H_{k-1}; that is, there are $r(1, H_{k-1})$ choices of $v \in H_{k-1}$ such that $Q(v + w) = 0$. But there are $q^2 - 1$ nonzero vectors $w \in W$, and therefore in all $(q^2 - 1)r(1, H_{k-1})$ representations of 0 by vectors $v + w$ of this second kind. Thus the total number of representations of 0 by $V = H_{k-1} \perp W$ is

$$r(0, H_{k-1}) + (q^2 - 1)r(1, H_{k-1}).$$

Now apply part (i) to conclude the proof. □

Theorem 2.60. *Let V be a regular quadratic space of odd dimension $2k+1$. Say $V = \langle \varepsilon \rangle \perp H_k$, with $\varepsilon \in F^*$ and H_k hyperbolic of dimension $2k$. Then*

(i) $r(0, V) = q^{2k}$.

(ii) *Suppose $\alpha \in F^*$. Then $r(\alpha, V) = q^{2k} + q^k$ if $\alpha \in \varepsilon F^{*2}$, and $r(\alpha, V) = q^{2k} - q^k$ if $\alpha \notin \varepsilon F^{*2}$.*

Proof. (i) Consider the representations of 0 by vectors $v = v_1 + v_2$ with $v_1 \in \langle \varepsilon \rangle$ and $v_2 \in H_k$. If $v_1 = 0$ there are $r(0, H_k)$ possibilities for v_2. On the other hand, for each of the $q-1$ nonzero possibilities for v_1 there are $r(1, H_k)$

possibilities for v_2. Thus $r(0,V) = r(0,H_k) + (q-1)r(1,H_k)$. Plugging in the values from the preceding theorem now gives the conclusion.

(ii) We continue to use the notation of part (i). First suppose $\alpha \in \varepsilon F^{*2}$. Then there are two choices for $v_1 \in \langle\varepsilon\rangle$ representing α, and for each of these there are $r(0,H_k)$ possibilities for v_2 to yield $Q(v_1+v_2) = \alpha$. Also, there are $q-2$ vectors in $\langle\varepsilon\rangle$ *not* representing α, and when one of these is used as v_1 there are $r(1,H_k)$ ways to choose $v_2 \in H_k$ so that $Q(v) = \alpha$. In all, this makes

$$r(\alpha, \langle\varepsilon\rangle \perp H_k) = 2r(0,H_k) + (q-2)r(1,H_k)$$

possible choices for v, and the conclusion now follows from the preceding theorem.

Now suppose $\alpha \notin \varepsilon F^{*2}$. For each $v_1 \in \langle\varepsilon\rangle$ we have $Q(v_1) \neq \alpha$, and there are then $r\big(\alpha - Q(v_1), H_k\big) = r(1,H_k)$ ways to choose $v_2 \in H_k$ to make $Q(v_1+v_2) = \alpha$. Thus $r(\alpha, V) = qr(1,H_k)$, and again the preceding theorem gives the result. $\square$

2.9. Hermitian Spaces

While our focus in this book is on quadratic forms and their associated symmetric bilinear forms on spaces and modules, this is just part of the broader theory of inner products. One can pursue the theory over wider classes of coefficient rings, on infinite-dimensional spaces, and so on; but in many cases the fundamental problems in those areas can be related and reduced to corresponding problems on quadratic spaces of the sort we have been considering. In this section we look at one instance of this phenomenon: the theory of Hermitian forms over fields. Our guide here will be a pioneering 1940 paper by Nathan Jacobson [**J**].

Let E be a field of characteristic not 2, and suppose $* : E \to E$ (given by $\alpha \longmapsto \alpha^*$) is a field automorphism and also a nontrivial **involution**; that is, $*$ is not the identity, but its square is: $\alpha^{**} = \alpha$ for all $\alpha \in E$. Then from Galois theory we know that E is a quadratic extension of the fixed field

$$F = \{\, \alpha \in E \mid \alpha^* = \alpha \,\}$$

of $*$. Hence there are elements $i \in E - F$ and $\lambda \in F$ with these properties: $E = F(i)$, with $i^2 = -\lambda \in F$; every element $\alpha \in E$ has a unique expression $\alpha = a + bi$ with $a, b \in F$, and then $(a+bi)^* = a - bi$. In particular, $i^* = -i$. [Of course the world's most popular example of this setup is when $E = \mathbb{C}$, the mapping $*$ is complex conjugation, $F = \mathbb{R}$, and $i^2 = -1$; but in fact our theory will apply to any quadratic extension E/F of fields of characteristic not 2.] The map $*$ extends to involutions on the matrix rings $M_n(E)$ via $X = (x_{ij}) \longmapsto X^* = (x_{ij}^*)$.

Associated with $*$ are two maps from E to F; the **norm** $N_{E/F}$ and the **trace** $T_{E/F}$, abbreviated N and T, respectively:

$$N(\alpha) = \alpha\alpha^* \quad \text{and} \quad T(\alpha) = \alpha + \alpha^*.$$

Notice in particular that N restricts to a group homomorphism $E^* \to F^*$; moreover $N\alpha = \alpha^2$ if $\alpha \in F$, and $Ni = \lambda$. Throughout this section, unless otherwise indicated, the symbols $E, F, *, i, \lambda, N,$ and T will be as described above, and all our spaces should be understood to be finite-dimensional.

Definition 2.61. Let V be a vector space over E. A mapping $s : V \times V \to E$ is a **sesquilinear form** with respect to $*$ if for all $x, y, z \in V$ the following properties hold:

(i) $s(\cdot, z) : V \to E$ is linear; that is, $s(\alpha x + y, z) = \alpha s(x, z) + s(y, z)$ for all $\alpha \in E$;

(ii) $s(x, y) = (s(y, x))^*$. [Note: we will usually write this as $s^*(y, x)$.]

It follows from (i) and (ii) that s is only **semi**linear (also called *conjugate-linear*) in the second variable; by this we mean that we have the additive property $s(x, y + z) = s(x, y) + s(x, z)$; but $s(x, \alpha y) = \alpha^* s(x, y)$ for all $\alpha \in E$. Also observe that were we not to insist that $*$ be nontrivial—though we do—then sesquilinearity of s would include bilinearity as a special case.

Definition 2.62. A map $H : V \to E$ is a **Hermitian form** on V if there is a sesquilinear form s on V such that $H(v) = s(v, v)$ for all $v \in V$, and then the pair (V, H) $\big($or $(V, s)\big)$ is a **Hermitian E-space**, usually written more simply as V if the form is understood.

Proposition 2.63. *Let V be a Hermitian E-space, with Hermitian form H and associated sesquilinear form s. Then*

(i) *$H(v) \in F$ for all $v \in V$.*

(ii) *$H(\alpha v) = N(\alpha)H(v)$ for all $\alpha \in E$ and $v \in V$.*

(iii) *$H(v + w) = H(v) + H(w) + T\left(s(v, w)\right)$; and, more generally,*

$$H(\sum_i \alpha_i v_i) = \sum_i N(\alpha_i)H(v_i) + T\left(\sum_{i<j} \alpha_i \alpha_j^* s(v_i, v_j)\right)$$

for all $\alpha_i \in E$ and $v_i \in V$.

(iv) *s is the only sesquilinear form on V that induces H.*

Proof. Only part (iv) needs work; namely, one can perform the tedious arithmetic required to check that

$$s(v, w) = \frac{1}{2i}\left(H(v - iw) - \lambda H(w) - H(v) + i\left(H(v + w) - H(v) - H(w)\right)\right),$$

from which the result follows. □

An **isometry** of Hermitian E-spaces V and W is an isomorphism (as vector spaces over E) that preserves H (or, equivalently, preserves s), completely analogous to the situation for quadratic spaces. We seek a complete set of invariants for isometry of Hermitian spaces. The isometries of a Hermitian space V onto itself constitute a group, denoted $U(V)$, under composition. This is the **unitary group** of V. [Remark: were we to allow the involution $*$ to be the identity map, then we would have $E = F$ and $U(V) = O(V)$.]

If the Hermitian E-space V has basis $\mathbb{B} = \{v_1, \ldots, v_n\}$, then, as for quadratic spaces, there is an associated **Gram matrix**: $V \cong A = (s(v_i, v_j))$. Here the matrix A is generally not symmetric, but instead ${}^tA = A^*$. Now if $\mathbb{B}'$ is another basis, and the transition matrix $\mathbb{B} \to \mathbb{B}'$ is T, then $V \cong {}^tTAT^*$ in the basis $\mathbb{B}'$, and $|{}^tTAT^*| = N(|T|) \cdot |A|$. A Gram-Schmidt process yields an orthogonal basis for V over E; and since $H(v) \in F$ for all v in that basis, it follows that $|A| \in F$. Therefore we define the **discriminant** dV to be the coset of $|A|$ in $(F^*/NE^*) \cup \{0\}$. So in the theory of Hermitian spaces norms play the role that squares played in the quadratic theory.

Just as in the theory of quadratic spaces, and therefore unworthy of detailed repetition here, one can go on to define the **radical** of a Hermitian space and the associated concept of **regularity** of spaces (or **nondegeneracy** of s and H) and note that V is regular if and only if $dV \neq 0$.

If V is an n-dimensional Hermitian E-space with basis $\{v_1, \ldots, v_n\}$, then simultaneously V is a $2n$-dimensional vector space over F with basis $\{v_1, \ldots, v_n, iv_1, \ldots, iv_n\}$, and we now define a map $B : V \times V \to F$ by

$$B(v, w) = \frac{1}{2} T\left(s(v, w)\right).$$

Thus if $s(v, w) = a + bi$, with $a, b \in F$, then $B(v, w) = a$. (For example, if E/F is the extension $\mathbb{C}/\mathbb{R}$ and $*$ is complex conjugation, then B is just the "real part" of s.) This B is easily checked to be a symmetric bilinear form on V viewed as a vector space over F, **the symmetric bilinear form induced by** s; and then $q(x) = B(x, x)$ is the **induced quadratic form** on V. The following proposition—whose checking we leave to the reader—indicates some properties of the induced form B with respect to the E-space structure of V.

Proposition 2.64. *For all $\alpha \in E$ and $v, w \in V$,*

(i) $B(\alpha v, \alpha w) = N(\alpha) \cdot B(v, w)$;

(ii) $B(\alpha^* v, w) = B(v, \alpha w)$;

(iii) *If $\alpha^* = -\alpha$ (so $\alpha = ai$ for some $a \in F$), then $B(v, \alpha v) = 0$.*

Now suppose conversely that we are given an n-dimensional vector space over E that, when viewed as a $2n$-dimensional space over F, has the structure of a quadratic F-space, with symmetric bilinear form B and associated quadratic form q. We ask: is there a sesquilinear form on V that induces B? Clearly a necessary condition on B is that it satisfy the conditions of the preceding proposition. And if it does, then it is a tedious, possibly irritating, but ultimately trivial exercise to check that the mapping $s : V \times V \to E$ defined by

$$s(v, w) = B(v, w) - \frac{i}{\lambda} B(iv, w)$$

is a sesquilinear form that induces B. Moreover the radical of V as a Hermitian space is equal to its radical as a quadratic space; hence V is regular (resp., trivial) as a Hermitian E-space if and only if it is regular (resp., trivial) as a quadratic F-space with respect to the induced symmetric bilinear form. Finally, the reader should check that for all $v \in V$ we have $H(v) = q(v)$; in particular, the mappings $H : V \to F$ and $q : V \to F$ represent the same elements of F. From this it follows that the sesquilinear form s inducing B is unique.

It is natural to ask how isometry of Hermitian E-spaces relates to isometry of quadratic F-spaces, and the answer is in the following theorem.

Theorem 2.65 (Jacobson). *Two Hermitian E-spaces are isometric if and only if they are isometric with respect to their induced quadratic F-space structures.*

Proof. Denote by $V_{1E} = (V_1, s_1, H_1)$ and $V_{2E} = (V_2, s_2, H_2)$ the Hermitian spaces and by $V_{1F} = (V_1, B_1, q_1)$ and $V_{2F} = (V_2, B_2, q_2)$ their induced quadratic space structures. Clearly an isometry $V_{1E} \cong V_{2E}$ is automatically an isometry $V_{1F} \cong V_{2F}$. Proof in the other direction is less obvious, however. (In particular, an F-linear map need not be E-linear.) We can assume that all the forms under discussion are nontrivial, since the B_i are nontrivial if and only if the s_i are.

We will argue by induction on $n = \dim V_{iE}$. If $n = 1$ the result is clear, so we can suppose that $n > 1$ and that the result has been checked for Hermitian E-spaces of dimension $< n$.

Choose x_1, x_2 in V_1, V_2, respectively, such that $q_1(x_1) = q_2(x_2) = \beta \neq 0$. Then as Hermitian spaces there are splittings $V_j = Ex_j \perp U_j$, for $j = 1, 2$. Here $Ex_j = Fx_j + F(ix_j)$, and we claim that U_j is actually the orthogonal complement of Ex_j with respect to B_j. Inclusion "$\subseteq$" is clear, since if $s_j(w, x_j) = 0$ then $B_j(w, x_j) = 0$. Conversely, if $B_j(x_j, w) = B_j(ix_j, w) = 0$ then

$$s_j(x_j, w) = B_j(x_j, w) - \frac{i}{\lambda} B_j(ix_j, w) = 0$$

as claimed.

Thus as quadratic F-spaces we have

$$V_j \cong (Fx_j + F(ix_j)) \perp U_j, \quad \text{with} \quad Fx_j + F(ix_j) \cong \begin{pmatrix} \beta & 0 \\ 0 & \lambda\beta \end{pmatrix}.$$

By Witt's Theorem there is a quadratic space isometry $U_1 \cong U_2$, and hence by the induction hypothesis there is also a *Hermitian* space isometry $U_1 \cong U_2$, and hence Hermitian isometry

$$V_1 \;\cong\; Ex_1 \perp U_1 \;\cong\; Ex_2 \perp U_2 \;\cong\; V_2. \qquad \square$$

Example 2.66. Let $*$ be complex conjugation on $\mathbb{C}$; so the fixed field is $\mathbb{R}$. Let V be a regular n-dimesional Hermitian $\mathbb{C}$-space. Say $V \cong \langle a_1, \ldots, a_n \rangle$, with $a_i \in \mathbb{R}$, with respect to basis $\{v_1, \ldots, v_n\}$. Then as a quadratic $\mathbb{R}$-space we have

$$V \cong \langle a_1, \ldots, a_n \rangle \perp \langle a_1, \ldots, a_n \rangle$$

with respect to $\mathbb{R}$-basis $\{v_1, \ldots, v_n, iv_1, \ldots, iv_n\}$. Therefore, if

$$W \cong \langle b_1, \ldots, b_n \rangle$$

is another regular n-dimensional Hermitian $\mathbb{C}$-space, then $V \cong W$ if and only if the number of positive a_i is equal to the number of positive b_i.

Now we return to the general Hermitian setting as at the start of this section.

Theorem 2.67. *Let V and W be regular n-dimensional Hermitian E-spaces, and suppose $u(F) \leq 4$. Then*

$$V \cong W \quad \Longleftrightarrow \quad dV = dW.$$

Proof. Only the sufficiency requires proof, and the case $n = 1$ is trivial. So suppose $n \geq 2$ and also suppose that the result has been checked for spaces of lower dimension. As a $2n$-dimensional quadratic F-space, V is universal; therefore, as a Hermitian E-space, V has a component of the form $\langle 1 \rangle$. Iteration of this observation yields a splitting

$$V \cong \langle 1, \ldots, 1 \rangle \perp \langle dV \rangle,$$

and the same holds for W, giving the result. $\square$

Examples. We saw in Section 2.8 that if F is a finite field then $u(F) = 2$, and it is shown in [**Lam**] (Chapter 2, Prop. 3.8) that $u(k(x)) = 2$ if k is an algebraically closed field. Later on we will see that $u(\mathbb{Q}_p) = 4$ (here $\mathbb{Q}_p$ denotes the field of *p-adic numbers*), and we will also see that $u(\mathbb{F}_q(x)) = 4$.

Corollary 2.68. *Let E be a finite field with nontrivial involution $*$, and let $n \in \mathbb{N}$. Then there is only one isometry class of regular Hermitian E-spaces of dimension n.*

Proof. As usual, let F denote the fixed field of $*$; so $E = F(i)$ with $i^2 = -\lambda \in F$. If V is an n-ary Hermitian E-space there is a splitting $V \cong \langle a_1, \ldots, a_n \rangle$ with all $a_i \in F$. So as a quadratic F-space we have

$$V \cong \langle a_1, \ldots, a_n \rangle \perp \langle \lambda a_1, \ldots, \lambda a_n \rangle$$

and hence $dV = \lambda^n$. The same argument being true for any other regular n-ary Hermitian E-space W forces a quadratic space isometry $V \cong W$ (since dimension and discriminant constitute a full set of invariants for quadratic spaces over finite fields), and hence a Hermitian space isometry by Jacobson's theorem. □

Jacobson's theorem reduces the classification of n-dimensional Hermitian E-spaces to the classification of certain $2n$-dimensional quadratic F-spaces, namely those of the form $\langle 1, \lambda \rangle \otimes W$, where W is an n-dimensional F-space and $\lambda \in NE^*$. So these F-spaces are not "generic" F-spaces; in particular, not only must they have even dimension but they have very special discriminants. The moral of this story is that one can generally expect a question for Hermitian spaces to be easier than the analogous question for quadratic spaces. The preceding corollary is a perfect instance of this phenomenon. Namely, it shows that over a finite field the dimension is the only fundamental isometry invariant of a Hermitian space, though we know that for *quadratic* spaces over a finite field both the dimension and discriminant are essential (and independent) invariants for isometry.

Exercises

(1) If F is a finite field of odd order q, verify the ring isomorphisms

$$W(F) \approx \begin{cases} \mathbb{Z}_4 \text{ if } q \equiv 3 \pmod 4, \\ \mathbb{Z}_2 \times \mathbb{Z}_2 \text{ if } q \equiv 1 \pmod 4). \end{cases}$$

(2) Construct a field $\mathbb{F}_9$ of nine elements. [Outline: Take as the underlying set the collection of ordered pairs of elements from $\mathbb{Z} \pmod 3$. Define addition coordinatewise, and define multiplication with the help of an irreducible quadratic polynomial over $\mathbb{Z} \pmod 3$.] Then, working with your constructed $\mathbb{F}_9$, do the following:
(i) Obtain generator(s) for the primary component(s) of $\mathbb{F}_9^*$.
(ii) Determine the square roots of -1.
(iii) Display the addition and multiplication tables for the Witt ring $W(\mathbb{F}_9)$.

(3) (i) Over the finite field $\mathbb{F}_q$, with q odd, determine the number of points on the **unit n-sphere**. That is, determine the number of solutions of the equation $x_1^2 + \cdots + x_n^2 = 1$.
(ii) Answer the same question if q is *even*. (Hint: You don't need the present chapter to solve this one.)

(4) Again with q odd, determine the number of elements in the orthogonal group $O(V)$ if $V \cong \langle 1,1,1\rangle$ over $\mathbb{F}_q$.

(5) Let $t \in \mathbb{N}$, and suppose V is a quadratic F-space satisfying

$$V \cong \langle 1, \alpha_1\rangle \otimes \cdots \otimes \langle 1, \alpha_t\rangle,$$

for some $\alpha_1, \ldots, \alpha_t \in F^*$. (The quadratic form on such a space and—by abuse of terminology—the space itself are both called **Pfister forms**.)
(i) Show that if $t \geq 2$ and F is a finite field then V is hyperbolic.
(ii) If $F = \mathbb{R}$, show that V is isotropic if and only if it is hyperbolic.

(6) Let $E = \mathbb{Q}(\sqrt{-2})$, with $*$ the nontrivial involution on E, and let $F = \mathbb{Q}$. Suppose V is a regular Hermitian E-space satisfying $V \cong \langle 1, \ldots, 1\rangle$ with respect to $\{v_1, \ldots, v_n\}$. Let $V_{\mathbb{Q}}$ denote V carrying its induced structure as a quadratic $\mathbb{Q}$-space. Determine a Gram matrix for $V_{\mathbb{Q}}$ and find the discriminant $dV_{\mathbb{Q}}$. Is $V_{\mathbb{Q}}$ isotropic or anisotropic?

(7) Repeat the preceding problem, but now with $E = \mathbb{Q}(\sqrt{2})$.

(8) Let E/F be a quadratic field extension, let V be a regular Hermitian E-space of dimension $n \geq 1$, and let V_F denote V viewed as a $2n$-dimensional quadratic F-space. Show that the unitary group $U(V)$ is a proper subgroup of the orthogonal group $O(V_F)$.

(9) Let E/F be a quadratic extension of finite fields. Use our material on Hermitian spaces to determine the group index $(F^* : NE^*)$.

Chapter 3

Valuations, Local Fields, and p-adic Numbers

3.1. Introduction to Valuations

Definition 3.1. Let F be a field. A mapping $|\ \ |: F \to \mathbb{R}$ is a **valuation** on F if the following properties hold for all $\alpha, \beta \in F$:

(i) $|\alpha| \geq 0$; $\quad |\alpha| = 0 \Longleftrightarrow \alpha = 0$.

(ii) $|\alpha\beta| = |\alpha||\beta|$; thus $|\ \ |$ induces a homomorphism $F^* \to \mathbb{R}^*$, and the image of this homomorphism is called the **value group** of the valuation, sometimes denoted by $|F^*|$.

(iii) $|\alpha + \beta| \leq |\alpha| + |\beta|$. (This is the **triangle inequality**.)

A field on which a valuation has been specified is a **valuated field**. A valuation that satisfies the **strong triangle inequality** $|\alpha + \beta| \leq \max\{|\alpha|, |\beta|\}$ is said to be a **nonarchimedean** valuation, while a valuation that is not nonarchimedean is (*surprise!*) **archimedean**.

Examples 3.2. (i) The ordinary absolute value on $\mathbb{R}, \mathbb{Q}$, or $\mathbb{C}$ is clearly an archimedean valuation; to distinguish it from other valuations that will be defined shortly, the absolute value of α is often written $|\alpha|_\infty$.

(ii) On any field F define

$$|\alpha| = \begin{cases} 1 & \text{if } \alpha \neq 0, \\ 0 & \text{otherwise.} \end{cases}$$

This is the **trivial valuation** on F. As the reader may suspect, this mapping is too "undiscriminating" to be really useful. But it is clearly a valuation, and it is nonarchimedean.

(iii) Let p be a prime number. Define the **p-adic valuation** $|\ \ |_p$ on the field $\mathbb{Q}$ of rational numbers as follows: Set $|0|_p = 0$; and if $\alpha \in \mathbb{Q}^*$, write α in the form $p^\nu \frac{a}{b}$, with $a, b \in \mathbb{Z}$ relatively prime to p, and then define $|\alpha|_p = \frac{1}{p^\nu}$. The value $\nu \in \mathbb{Z}$ occurring in this expression is the **p-adic order** of α, usually written $\operatorname{ord}_p \alpha$ or $\nu_p(\alpha)$. (We put $\operatorname{ord}_p 0 = \infty$). It is easily checked that $|\ \ |_p$ is a nonarchimedean valuation. Note that for an integer α, the value $|\alpha|_p$ is small (equivalently, $\operatorname{ord}_p \alpha$ is large) if and only if α is highly divisible by p.

(iv) Let k be a field, and let p be any prime (i.e., monic and irreducible) polynomial over k. Define a **p-adic valuation** by fixing a real number λ satisfying $0 < \lambda < 1$ and then proceeding along the lines of the previous example: Set $|0|_p = 0$; and if $\alpha \in F^*$, write α in the form $p^\nu \frac{a}{b}$, with $a, b \in k[x]$ relatively prime to p, and then define $|\alpha|_p = \lambda^\nu$. The same reasoning as in example (iii) shows this to be a nonarchimedean valuation, and again one has a well-defined **p-adic order** function. Note that the map $|\ \ |_p$ as defined here is "a" p-adic valuation, not "the" p-adic valuation because of the flexibility in the choice of λ. If k is a *finite* field, say $k = \mathbb{F}_q$, then with the choice $\lambda = \frac{1}{q}$ the valuation $|\ \ |_p$ is the **normalized** p-adic valuation. (The word "normalized" is usually omitted.) We will see later why it is useful to agree on a particular choice for λ.

(v) There is a way to define a nonarchimedean valuation on $k(x)$ that is not defined via a prime polynomial in $k[x]$, and this valuation will be denoted $|\ \ |_\infty$. Start as usual by defining $|0|_\infty = 0$, as the definition requires. Then fix a real number $c > 1$, and for nonzero $f, g \in k[x]$ define

$$\left|\frac{f}{g}\right|_\infty = c^{\deg f - \deg g}.$$

Notice that this new valuation is just a disguised version of the degree function, in the sense that for nonzero $f, g \in k[x]$ we have $|f|_\infty > |g|_\infty$ if and only if $\deg f > \deg g$. In particular, $|\ \ |_\infty$ is trivial on the coefficient field k. Paralleling what was done in (iv), when $k = \mathbb{F}_q$ we choose $c = q$ to get the *normalized* valuation $|\ \ |_\infty$.

The proof of the following proposition, giving elementary properties of valuations, is an exercise for the reader.

Proposition 3.3. *Let* $|\ \ |$ *be a valuation on* F. *Then*

(i) $|\pm 1| = 1$; *more generally,* $|\varepsilon| = 1$ *if* ε *is any root of unity in* F. *In particular, if* F *is finite then the valuation is trivial.*

(ii) $|-\alpha| = |\alpha|$, *and if* $\alpha \neq 0$ *then* $|\alpha^{-1}| = |\alpha|^{-1}$.

(iii) $\big||\alpha| - |\beta|\big| \leq |\alpha - \beta|$. *[Note: On the left side of this inequality, the outer valuation is the ordinary absolute value on* $\mathbb{R}$, *while the inner one is the given valuation on* F.*]*

(iv) $|\alpha_1 + \cdots + \alpha_n| \leq |\alpha_1| + \cdots + |\alpha_n|$; *and if the valuation is nonarchimedean then* $|\alpha_1 + \cdots + \alpha_n| \leq \max_i |\alpha_i|$.

It is now reasonable to ask for some kind of characterization of the valuations on a given field. Of course a trivial valuation can be defined on any field, but what is there beyond that? Is there an easy test to determine whether a given valuation is archimedean or nonarchimedean? (After all, if the underlying field is large one may not be able to directly test whether the strong triangle inequality holds in all cases.) And, finally, how is all this relevant to the theory of quadratic spaces? A partial answer to this last question: it may be possible to use a valuation on F to construct an extension field E over which quadratic spaces are more tractable than over F itself, and then to use the results over E to draw inferences about spaces over F.

Definition 3.4. The **natural numbers** of a field F are the elements

$$1, 1+1, 1+1+1, \ldots,$$

that is, the elements of the form $n \cdot 1_F$ with $n \in \mathbb{N}$. In particular, if F has characteristic p, then the natural numbers of F are $1, \ldots, p-1$: the nonzero elements of the prime field of F.

Proposition 3.5. *A valuation on* F *is nonarchimedean if and only if it is bounded on the natural numbers of* F.

Proof. If $|\ \ |$ is nonarchimedan then $|t| \leq 1$ for every natural number $t \in F$. (Argue by induction.) Conversely, suppose $|t| \leq B$ for every natural number $t \in F$. Then for every $n \in \mathbb{N}$ and $\alpha, \beta \in F$ we have (by the binomial theorem)

$$|\alpha + \beta|^n \leq (n+1)B \left(\max\{|\alpha|, |\beta|\}\right)^n.$$

Now take nth roots and let $n \to \infty$. The result follows from the fact that

$$\lim_{n\to\infty} \left((n+1)B\right)^{\frac{1}{n}} = 1. \qquad \square$$

Corollary 3.6. (i) *If* $\operatorname{char} F > 0$ *then every valuation on* F *is nonarchimedean.*

(ii) *Let E/F be a field extension. Then a valuation on E is nonarchimedean if and only if its restriction to F is nonarchimedean.*

While every valuation shares properties with the ordinary absolute value, each valuation has its own distinct personality, and one must be on guard against over-applying intuitions based on our childhood experiences with $|\ |_\infty$. The following result makes this danger more explicit. It tells us that in a nonarchimedean universe, when a hippopotamus and a rabbit simultaneously step on the same scale their combined weight is equal to the weight of the hippo.

Proposition 3.7 (Principle of Domination). *Let $|\ |$ be a nonarchimedean valuation on F, and suppose $|\alpha| > |\beta|$. Then $|\alpha+\beta| = |\alpha|$. More generally, given $\alpha_1, \ldots, \alpha_n \in F$, with $|\alpha_1| > |\alpha_i|$ for all $i > 1$, then*

$$|\alpha_1 + \cdots + \alpha_n| = |\alpha_1|.$$

Proof. It suffices to prove the first statement, since the second then follows immediately by induction. Clearly $|\alpha+\beta| \le |\alpha|$, by the nonarchimedean property. On the other hand, since $\alpha = (\alpha+\beta) + (-\beta)$ we have $|\beta| < |\alpha| \le \max\{|\alpha+\beta|, |\beta|\}$ and hence $|\alpha| \le |\alpha+\beta|$, proving the claim. □

Actually, it should be no surprise that a valuation is not an additive function. After all, for any valuation $|\ |$ we have $|1+(-1)| = 0$; and while $|47|_5 = |3|_5 = 1$, we have $|47+3|_5 = \frac{1}{25} < 1$. The principle of domination tells us that from $|47|_5 = 1$ and $\left|\frac{10}{13}\right|_5 < 1$ we can immediately conclude that $\left|47\frac{10}{13}\right|_5 = 1$ without doing any further arithmetic.

3.2. Equivalence of Valuations; Prime Spots on a Field

If $|\ |$ is a valuation on F, we can define the associated *distance* between α and β to be the number $d(\alpha,\beta) = |\alpha-\beta|$. Then this function d is a *metric* on F, and this gives our field a topological structure. More explicitly, for $\varepsilon > 0$ define the **open disk of radius** ε **and center** α to be the set

$$D_\varepsilon(\alpha) = \{\beta \in F \mid d(\alpha,\beta) < \varepsilon\} = \{\beta \in F \mid |\alpha-\beta| < \varepsilon\}$$

and then define a set A to be open in the topology if A contains an open disk around each of its points.

Given the valuation-induced metric on F we can immediately discuss the behavior of sequences. In particular, for $\{a_n\}$ a sequence in F we have the notion of **convergence** with respect to the valuation $|\ |$:

$$\lim_{n\to\infty} a_n = L \iff \forall \varepsilon > 0, \quad \exists N \in \mathbb{N} \text{ such that } |a_n - L| < \varepsilon \quad \forall n \ge N.$$

And we can speak of **Cauchy sequences**: $\{a_n\}$ is Cauchy (again with respect to $|\ \ |$) if

$$\forall \varepsilon > 0, \quad \exists N \in \mathbb{N} \text{ such that } |a_n - a_m| < \varepsilon \quad \forall n, m \geq N.$$

Example 3.8. We will now consider some further phenomena involving nonarchimedean valuations, illustrating some of their differences from the ordinary absolute value.

(i) Let $|\ \ |$ be the trivial valuation on F. Then $\lim_{n\to\infty} a_n = L$ if and only if there is an N such that $a_n = L$ for all $n > N$; that is, the sequence is "eventually constant."

(ii) With respect to the p-adic valuation on $\mathbb{Q}$, we have $\lim_{n\to\infty} p^n = 0$; and the sequence

$$1, \quad 1+p, \quad 1+p+p^2, \quad \ldots, \quad 1+p+\cdots+p^n, \quad \ldots$$

is Cauchy.

Proposition 3.9. *Let $|\ \ |$ and $\|\ \ \|$ be valuations on F. Then the following conditions are equivalent:*

(i) *The metrics associated with $|\ \ |$ and $\|\ \ \|$ induce the same topology.*

(ii) $|\alpha| < 1 \Longleftrightarrow \|\alpha\| < 1$.

(iii) *There is a positive real number λ such that $|\alpha|^\lambda = \|\alpha\|$ for all $\alpha \in F$.*

Proof. First note that if $|\ \ |$ is the trivial valuation and $\|\ \ \|$ satisfies any of the given conditions, then $\|\ \ \|$ must also be trivial, so there is nothing to be proved. Therefore without loss of generality we can assume that the given valuations are nontrivial.

(i)$\Longrightarrow$(ii): Assuming (i), suppose $|\alpha| < 1$; then $\alpha^n \to 0$ with respect to $|\ \ |$. Consider the disk

$$D = \{\beta \in F \mid \|\beta\| < 1\}.$$

This disk contains 0 and is open in the $\|\ \ \|$-topology, hence is $|\ \ |$-open as well, so D contains some $|\ \ |$-disk $D_\varepsilon(0)$, with $0 < \varepsilon < 1$. But $D_\varepsilon(0)$ contains α^n if n is large enough, so $\alpha^n \in D$, and hence $|\alpha_2| < 1$. By symmetry this proves the result.

(ii)$\Longrightarrow$(iii): With (ii) in hand, pick $\alpha \in F^*$ with $|\alpha| < 1$; then also $\|\alpha\| < 1$, and so $\|\alpha\| = |\alpha|^\rho$, with

$$\rho = \frac{\log \|\alpha\|}{\log |\alpha|} > 0.$$

We claim that $\|\beta\| = |\beta|^\rho$ for all $\beta \in F^*$. Were this not the case, then (taking inverses if necessary) there would be a β with $\|\beta\| < |\beta|^\rho$. Then choose a rational number $\frac{m}{n}$, with $n > 0$, such that

$$\|\beta\| < \|\alpha\|^{\frac{m}{n}} = |\alpha|^{\rho\frac{m}{n}} < |\beta|^{\rho}.$$

The left inequality gives $\left\|\frac{\beta^n}{\alpha^m}\right\| < 1$, while the right inequality gives $\left|\frac{\beta^n}{\alpha^m}\right| > 1$, a contradiction of (ii).

(iii)$\Longrightarrow$(i): The open disks form a base for a metric topology, and from (iii) every open $|\ \ |$-disk is also an open $\|\ \ \|$-disk, so the result follows. □

Definition 3.10. Valuations $|\ \ |$ and $\|\ \ \|$ on F are **equivalent** if the conditions of the preceding proposition hold. Equivalence is an equivalence relation on the set of valuations on F, and an equivalence class of valuations is called a **prime spot**, **spot**, **prime**, or (by some authors) a **place** on F.

Corollary 3.11. *If $|\ \ |$ and $\|\ \ \|$ are equivalent valuations, then every Cauchy sequence with respect to $|\ \ |$ is Cauchy with respect to $\|\ \ \|$.*

Remark 3.12. On $\mathbb{Q}$, the spot containing $|\ \ |_p$ is the **p-adic spot**. The p-adic spots (as p varies through the primes) are often called the **finite spots** on $\mathbb{Q}$, while the spot containing the ordinary absolute value "$|\ \ |_\infty$" is the **infinite spot**. The symbol "∞" is often called the **infinite prime**. Similar terminology holds for the spots on $k(x)$ that we have defined.

In view of the preceding proposition, the following result is now clear.

Proposition 3.13. (i) *The only valuation equivalent to the trivial valuation is the trivial valuation.*

(ii) *If a valuation is nonarchimedean, so is any equivalent valuation.*

Accordingly, without ambiguity a spot can be said to be **trivial, nonarchimedean**, or **archimedean**.

We remark that if $|\ \ |$ is a valuation and $\lambda > 1$, then $|\ \ |^\lambda$ need not be a valuation. For instance if $\lambda > 1$ and $|\ \ |$ is the ordinary absolute value, then $|1+1|^\lambda = 2^\lambda > 2 = |1|^\lambda + |1|^\lambda$, so $|\ \ |^\lambda$ fails the triangle inequality. However if $\lambda \le 1$ we do get a valuation, as the following proposition shows.

Proposition 3.14. *If $|\ \ |$ is a valuation on F and $0 < \lambda \le 1$, then $|\ \ |^\lambda$ is a valuation.*

Proof. Only the triangle inequality is nontrivial. We have

$$|\alpha + \beta|^\lambda \le (|\alpha| + |\beta|)^\lambda,$$

so it is enough to prove that $(|\alpha| + |\beta|)^\lambda \le |\alpha|^\lambda + |\beta|^\lambda$. But since $0 < \lambda \le 1$ we have

$$1 = \frac{|\alpha|}{|\alpha| + |\beta|} + \frac{|\beta|}{|\alpha| + |\beta|} \le \left(\frac{|\alpha|}{|\alpha| + |\beta|}\right)^\lambda + \left(\frac{|\beta|}{|\alpha| + |\beta|}\right)^\lambda.$$

Now multiplying through by $(|\alpha| + |\beta|)^\lambda$ does the job. $\square$

Most of the quadratic forms we will consider in this book will have coefficients in the field $\mathbb{Q}$ of rational numbers, or a rational function field $K(x)$—especially with finite coefficient field K, or fields constructed from these via valuation theory. It is therefore important that we know what valuations exist on these fields, and the following theorem (along with the subsequent remark) tells us that we have already been introduced to all of these valuations.

Theorem 3.15. (i) *A nontrivial nonarchimedan valuation on $\mathbb{Q}$ is equivalent to $|\ \ |_p$ for some prime number p.*

(ii) *A nontrivial valuation on a rational function field $K(x)$ that is trivial on K is equivalent either to $|\ \ |_p$ for some prime polynomial p or to $|\ \ |_\infty$.*

Proof. (i) Let $|\ \ |$ be nontrivial and nonarchimedean on $\mathbb{Q}$. So $|m| \leq 1$ for all $m \in \mathbb{N}$, and in fact $|m| < 1$ for some $m \in \mathbb{N}$. Let p be the smallest positive integer such that $|p| < 1$; clearly p is prime.

Now suppose $m \in \mathbb{N}$ with $|m| < 1$. Use the division algorithm to write $m = pq + r$ with $0 \leq r < p$. If we had $r \neq 0$ then we would have $|r| = 1$, by the minimality of p. But then $|m| = 1$, by the principle of domination, a contradiction. Hence $r = 0$. Thus for $m \in \mathbb{N}$ we have

$$|m| < 1 \Longleftrightarrow p \mid m.$$

Now suppose $0 \neq \alpha \in \mathbb{Q}$, and write $\alpha = p^\nu \frac{a}{b}$, with $(a,p) = (b,p) = 1$. Then $|\alpha| = |p|^\nu$ by what we have already shown. Now choose $\lambda > 0$ such that

$$|p| = \left(\frac{1}{p}\right)^\lambda.$$

(In fact, take $\lambda = \dfrac{\log|p|}{\log\frac{1}{p}}$.) Then $|\ \ | = |\ \ |_p^\lambda$.

(ii) A nontrivial valuation on $K(x)$ that is trivial on K is nonarchimedean, since it is bounded on the natural numbers of the field. If $|x| > 1$ and $f(x) \in K[x]$, then $|f(x)| = |x|^{\deg f}$ by the principle of domination; therefore if $r(x) = \dfrac{f(x)}{g(x)}$ then $|r(x)| = |x|^{\deg f - \deg g}$, so $|\ \ |$ is equivalent to $|\ \ |_\infty$. (Choose $\lambda \in \mathbb{R}$ such that $|x| = |x|_\infty^\lambda$, and then $|\ \ | = |\ \ |^\lambda$.)

On the other hand, if $|x| \leq 1$ then $|f(x)| \leq 1$ for all $f(x) \in K(x)$ by the principle of domination, and $M = \{\, f(x) \in K[x] \mid |f| < 1 \,\}$ is a prime ideal in $K[x]$; so $M = \big(p(x)\big)$ for some prime polynomial $p(x)$. Any nonzero rational function $r \in K(x)$ can be written in the form $r = p^\nu \frac{a}{b}$ with

$a, b \in K[x]$ relatively prime to p. Then $|r| = |p|^{\nu}$, and therefore $|\ \ |$ is equivalent to $|\ \ |_p$. $\square$

Corollary 3.16. *The only nontrivial spots on* $\mathbb{F}_q(x)$ *are the p-adic spots and* ∞.

Remark 3.17. It can be shown that the only nontrivial spot on $\mathbb{Q}$ other than the p-adic spot is ∞, the spot containing the ordinary absolute value $|\ \ |_\infty$. We omit the proof, but the interested reader can find one in [**Ba**] and [**O'M1**].

Exercises

In Exercises 3, 8, 9, and 10, assume x is transcendental over the relevant field.

(1) Let R be an integral domain with quotient field F. Suppose the mapping $|\ \ | : R \to \mathbb{R}$ satisfies properties (i), (ii), (iii) in the valuation definition when applied to elements of R. Prove that $|\ \ |$ extends to a unique valuation on F.

(2) (i) A ring R is said to be a **local ring** if and only if its nonunits form an ideal. Prove that R is a local ring if and only if it has a unique maximal ideal.
(ii) A subring R of a field F is a **valuation ring** if

$$\alpha \in F \Longrightarrow \alpha \in R \text{ or } \alpha^{-1} \in R.$$

Prove that if R is a valuation ring then R is a local ring. [Hint: If α and β are nonunits of R and $\frac{\alpha}{\beta} \in R$, consider the element $1 + \frac{\alpha}{\beta}$.]
(iii) If $|\ \ |$ is a valuation on F, show that $R = \{\alpha \in F \mid |\alpha| \leq 1\}$ is a ring (and in fact a valuation ring) if and only if $|\ \ |$ is nonarchimedean.

(3) Let K be a field, and let $|\ \ |$ be a nonarchimedean valuation on K. Show that the mapping $|\ \ |$ extends to a valuation on $K(x)$ by first extending it to $K[x]$ as follows: If $f(x) = \sum_i a_i x^i$, with $a_i \in K$, then define $|f(x)| = \max |a_i|$; and then apply Exercise 1. [Suggestion: To deduce that $|f(x)g(x)| = |f(x)| \cdot |g(x)|$, pay special attention to the coefficient a_{i_0} of $f(x)$ having the smallest subscript i_0 such that $|a_{i_0}| = \max |a_i|$, along with a similarly chosen coefficient b_{j_0} of $g(x)$.]

(4) Let $|\ \ |$ be a nonarchimedean valuation on F.
(i) By a "triangle" in F we mean three distinct points $\alpha, \beta, \gamma \in F$. Using the distance function defined by the valuation, prove that every triangle in F is "isosceles."

(ii) Let $\beta \in D_\varepsilon(\alpha)$. Show that $D_\varepsilon(\beta) = D_\varepsilon(\alpha)$. That is, every point of an open disk is the center!

(5) If $\lim_{n\to\infty} a_n = L$ with respect to some valuation $|\quad|$, show that

$$\lim_{n\to\infty} |a_n| = |L|.$$

(6) Given a valuated field F, let $U = \{\alpha \in F \mid |\alpha| = 1\}$. Define homomorphisms $F^* \to F^*/U \to |F^*|$. If this short sequence of homomorphisms is exact (the image of one homomorphism is the kernel of the next), what does it tell you about the given valuation? Assuming $\operatorname{char} F \neq 2$, is it possible for the slightly longer sequence $1 \to F^* \to F^*/U \to |F^*| \to 1$ to be exact? If not, erase a minimal number of arrows so that an exact sequence remains.

(7) Let E/F be an algebraic extension of fields, and suppose there is a valuation on E whose restriction to F is the trivial valuation. Show that the valuation is trivial. Also: Is the word "algebraic" needed here?

(8) We first recall the **stereographic projection** onto $\mathbb{C}$. Imagine a sphere E (for "Earth") sitting on the complex plane $\mathbb{C}$ with the South Pole coincident with 0. Then there is a bijection between $E - \{\text{North Pole}\}$ and $\mathbb{C}$. Namely, map each point $P \in E - \{\text{North Pole}\}$ to the point on $\mathbb{C}$ hit by the line through P and the North Pole. Use the stereographic projection to justify the terms "finite spot" and "infinite spot" on $\mathbb{C}(x)$. [This terminology is traditionally extended to other rational function fields as well, as has been done in the text.]

(9) If $F = k(x)$, one can also write $F = k(y)$ for appropriate $y \neq x$. With this in mind, show that there is ambiguity in speaking of "the valuation $|\quad|_\infty$ on F" (one person's $|\quad|_\infty$ may be another person's $|\quad|_p$) without further comment, and suggest a way to avoid the ambiguity.

(10) (i) With $F = k(x)$, show that the valuation $|\quad|_\infty$ (as defined in the text) is nonarchimedean.
(ii) Let $f(x) = \sum_{i=0}^n a_i x^i \in \mathbb{F}_q[x]$, with $a_i \in \mathbb{F}_q$ for all i. Using $c = q$ in the definition of $|\quad|_\infty$, determine $|f(x)|_\infty$.

3.3. Completions, $\mathbb{Q}_p$, Residue Class Fields

Definition 3.18. A field F is **complete** with respect to a valuation $|\quad|$ if every Cauchy (with respect to $|\quad|$) sequence in F converges in F. Given a sequence $\{a_n\}$, if the associated sequence of "partial sums"

$$s_1 = a_1, \quad s_2 = a_1 + a_2, \quad \dots, \quad s_n = a_1 + \cdots + a_n$$

converges to L, we write $\sum_1^\infty a_i = L$ and use the language of infinite series as in elementary calculus.

The following facts related to valuations on F are elementary; we leave most of the proofs as exercises for the reader.

(i) If the series $\sum a_n$ converges then $a_n \to 0$. (The converse need not be true, as we know from our dealings with $| \ \ |_\infty$ on $\mathbb{R}$.)

(ii) If $\sum |a_n|$ converges then $\sum a_n$ converges.

(iii) If $\lim_{n\to\infty} a_n = L$ with respect to valuation $| \ \ |_1$, then the same equation holds with respect to any valuation $| \ \ |_2$ equivalent to $| \ \ |_1$. Therefore F is complete with respect $| \ \ |_1$ if and only if it is complete with respect to $| \ \ |_2$; so without ambiguity one can speak of F's being complete or not *at* or *with respect to* a given spot $\mathfrak{p}$ on F.

(iv) ("Freshman Dream") Suppose F is complete at the nonarchimedean spot $\mathfrak{p}$. Then $\sum a_n$ converges if and only if $a_n \to 0$.

Proof of (iv). In view of (i), only the sufficiency requires proof here. Pick a valuation $| \ \ | \in \mathfrak{p}$, let $\varepsilon > 0$, and choose N such that $|a_n| < \varepsilon$ for all $n \geq N$. Then for $n > m \geq N$ the associated sequence $\{s_n\}$ of partial sums satisfies

$$|s_n - s_m| = |a_{m+1} + \cdots + a_n| \leq \max_{m+1 \leq i \leq n} |a_i| < \varepsilon$$

by the principle of domination, and hence is Cauchy. So we are done, since F is complete at $\mathfrak{p}$. □

Suppose E/F is a field extension, $\mathfrak{P}$ is a spot on E, and $| \ \ | \in \mathfrak{P}$. Then the restriction $| \ \ |\Big|_F$ is a valuation on F. If also $| \ \ |' \in \mathfrak{P}$, then $| \ \ |\Big|_F$ and $| \ \ |'\Big|_F$ are equivalent valuations on F; call their spot $\mathfrak{p}$. We say $\mathfrak{P}$ **divides** $\mathfrak{p}$, written $\mathfrak{P} \mid \mathfrak{p}$.

Definition 3.19. Let $\mathfrak{p}$ be a spot on F. A **completion** of F at $\mathfrak{p}$ is a field $E \supseteq F$ with a spot $\mathfrak{P} \mid \mathfrak{p}$ such that E is complete at $\mathfrak{P}$ and F is dense in E.

The proof of the following theorem is an imitation of the construction of the real numbers from the rational numbers, so we will be content to sketch the proof rather than repeat all the details. However we give two such sketches.

Theorem 3.20. *Let $\mathfrak{p}$ be a spot on F. Then F has a completion at $\mathfrak{p}$.*

Proof sketch 1. Let $| \ \ | \in \mathfrak{p}$, and let $d(\alpha, \beta) = |\alpha - \beta|$ be the associated metric. Then F has a completion as a metric space; call it E. We need to give E an appropriate structure as a valuated field. Given $\alpha, \beta \in E$, there are sequences $\{a_n\}, \{b_n\}$ in F such that $a_n \to \alpha$ and $b_n \to \beta$. Then $\{a_n\}$ and $\{b_n\}$ are Cauchy, and it follows that the sequences $\{a_n + b_n\}$ and $\{a_n b_n\}$ are Cauchy; hence these new sequences converge in E. Call their limits $\alpha+\beta$ and

$\alpha\beta$, respectively. With these operations, E is a field. Now define a valuation $|\ \ |'$ on E as follows: with α and $\{a_n\}$ as above, define $|\alpha|' = \lim_{n\to\infty} |a_n|$; this limit exists because $|a_n|$ is a Cauchy sequence of real numbers, from first principles of valuations (part (iii) of the first proposition of this chapter). Then E with the spot $\mathfrak{p}$ containing $|\ \ |'$ is the desired completion. □

Proof sketch 2. Make the set C of Cauchy sequences of elements of F into a commutative ring with unit by defining

$$\{a_n\} + \{b_n\} = \{a_n + b_n\} \quad \text{and} \quad \{a_n\}\{b_n\} = \{a_n b_n\}.$$

Let M be the set of null (that is, converging to 0) sequences. One checks that M is a maximal ideal in C; therefore C/M is a field, and this is our E. The mapping $F \to E$ given by $a \mapsto \{a\} + M$ is an injective homomorphism (here $\{a\}$ denotes a constant sequence). Identify F with its image in E under this map. Finally, if $x \in E$, say $x = \{a_n\} + M$, then define $|x| = \lim_{n\to\infty} |a_n|$. One checks that this makes E into the desired completion of F. □

As is evident from our proof sketches, there is more than one way to construct a completion of a field F with respect to a given valuation. But it is a straightforward exercise to verify that there is *essentially* only one way, in the sense that if E and E' are two such completions, with extended valuations $|\ \ |$ and $|\ \ |'$, then there is an isomorphism $\lambda : E \to E'$ that restricts to the identity map on F and such that $|\lambda(\alpha)|' = |\alpha|$ for all $\alpha \in E$. A completion for F at $\mathfrak{p}$ is generally denoted by $F_{\mathfrak{p}}$, and it is customary to also use $\mathfrak{p}$ for the unique spot on $F_{\mathfrak{p}}$ extending the spot $\mathfrak{p}$ on F. Similarly we use the same symbol for a valuation (in $\mathfrak{p}$) on F and its unique extension to a valuation on $F_{\mathfrak{p}}$.

The following proposition is an exercise for the reader.

Proposition 3.21. *If $\mathfrak{p}$ is a nonarchimedean spot on F and $|\ \ | \in \mathfrak{p}$, then the value groups $|F^*|$ and $|F_{\mathfrak{p}}^*|$ are equal.*

Note that the "nonarchimedean" condition is essential for the preceding proposition. For example, $\mathbb{R}$ is the completion of $\mathbb{Q}$ with respect to $|\ \ |_\infty$, and $|\pi|_\infty = \pi \notin |\mathbb{Q}^*|$.

Definition 3.22. Let p be a prime number. The field $\mathbb{Q}_p$ of ***p*-adic numbers** is the completion of the field of rational numbers with respect to the p-adic valuation.

Example 3.23. (Cassels [**C3**]) Consider the behavior of the following sequence of integers with respect to the 5-adic valuation:

$$a_1 = 4, \quad a_2 = 34, \quad a_3 = 334, \quad a_4 = 3334, \quad \ldots, \quad a_n = \underbrace{3333\ldots3}_{n-1 \text{ digits}}4, \quad \ldots$$

Then for $m > n$ we have

$$|a_m - a_n|_5 = |3\dots3\underbrace{0\dots0}_{n \text{ zeroes}}|_5 = \frac{1}{5^n}.$$

Therefore $\{a_n\}$ is a 5-adically Cauchy sequence in $\mathbb{Q}$, so it certainly converges in $\mathbb{Q}_5$. But does it converge in $\mathbb{Q}$? If so, to what? [Guess before proceeding!] Notice that

$$a_n = \underbrace{33\dots3}_{n \text{ digits}}+1 = \frac{10^n+2}{3}.$$

But $10^n \to 0$ (5-adically) as $n \to \infty$, and therefore $\lim_{n\to\infty} a_n = \dfrac{2}{3} \in \mathbb{Q}$.

Definition 3.24. Let $\mathfrak{p}$ be a nonarchimedean spot on F, and let $|\ \ | \in \mathfrak{p}$. Then put

$$\begin{aligned} \mathfrak{o} = \mathfrak{o}(\mathfrak{p}) &= \{\alpha \in F \mid |\alpha| \le 1\}, \\ \mathfrak{u} = \mathfrak{u}(\mathfrak{p}) &= \{\alpha \in F \mid |\alpha| = 1\}, \\ \mathfrak{m} = \mathfrak{m}(\mathfrak{p}) &= \{\alpha \in F \mid |\alpha| < 1\}. \end{aligned}$$

Here, because $\mathfrak{p}$ is nonarchimedean, $\mathfrak{o}$ is a ring, called the **valuation ring** or **ring of integers** of F at $\mathfrak{p}$; $\mathfrak{u}$ is the group of units of $\mathfrak{o}$ or "of F at $\mathfrak{p}$"; finally, $\mathfrak{m}$ is an ideal, and it consists of all the nonunits of $\mathfrak{o}$. Therefore $\mathfrak{m}$ is the unique maximal ideal of $\mathfrak{o}$, making $\mathfrak{o}$ a **local ring**: a ring with a unique maximal ideal. The field $\mathfrak{o}/\mathfrak{m}$, or any isomorphic copy, is called the **residue class field of** F **at** $\mathfrak{p}$, denoted by $F(\mathfrak{p})$ or $\overline{F(\mathfrak{p})}$ or (if $\mathfrak{p}$ is understood) $\overline{F}$. A subset of $\mathfrak{o}$ consisting of one element from each coset in $\mathfrak{o}/\mathfrak{m}$, with 0 chosen as the representative of $\mathfrak{m}$, is said to be a **representative set** for the residue class field of F at $\mathfrak{p}$. Notice that for all $\alpha \in F$, either $\alpha \in \mathfrak{o}$ or $\alpha^{-1} \in \mathfrak{o}$; and F is the quotient field of $\mathfrak{o}$. Finally, check that if $\mathfrak{p}_1$ and $\mathfrak{p}_2$ are nonarchimedean spots on F, then

$$\mathfrak{p}_1 = \mathfrak{p}_2 \iff \mathfrak{o}(\mathfrak{p}_1) = \mathfrak{o}(\mathfrak{p}_2) \iff \mathfrak{m}(\mathfrak{p}_1) = \mathfrak{m}(\mathfrak{p}_2).$$

Proposition 3.25. *Let E/F be a field extension, with nonarchimedean spots $\mathfrak{P} \mid \mathfrak{p}$, and let $\overline{E}, \overline{F}$ be the associated residue class fields. Then $\overline{F}$ is (isomorphic to) a subfield of $\overline{E}$.*

Proof. Define a homomorphism $\overline{F} \to \overline{E}$ by $\alpha + \mathfrak{m}(\mathfrak{p}) \mapsto \alpha + \mathfrak{m}(\mathfrak{P})$, where $\alpha \in \mathfrak{o}(\mathfrak{p})$. This is injective because $\mathfrak{m}(\mathfrak{p}) = F \cap \mathfrak{m}(\mathfrak{P})$. □

Corollary 3.26. *$\overline{F} = \overline{F_{\mathfrak{p}}}$ when $\mathfrak{p}$ is nonarchimedean.*

Proof. Inclusion "$\subseteq$" is clear from the proposition. For "$\supseteq$," to avoid confusion write $\mathfrak{P}$ for the spot on $F_{\mathfrak{p}}$ extending $\mathfrak{p}$. Let $\alpha \in \mathfrak{o}(\mathfrak{P})$. We must find $\beta \in \mathfrak{o}(\mathfrak{p})$ such that $\beta + \mathfrak{m}(\mathfrak{P}) = \alpha + \mathfrak{m}(\mathfrak{P})$. If $|\alpha| < 1$ choose $\beta = 0$. If $|\alpha| = 1$, since F is dense in $F_{\mathfrak{p}}$ there exists $\beta \in F$ such that $|\beta - \alpha| < 1$.

Then $|\beta| = |\alpha| = 1$ by the principle of domination, and $\beta - \alpha \in \mathfrak{m}(\mathfrak{P})$, so $\beta + \mathfrak{m}(\mathfrak{P}) = \alpha + \mathfrak{m}(\mathfrak{P})$ as desired. □

3.4. Discrete Valuations

Definition 3.27. A nonarchimedean valuation is **discrete** if its value group is an infinite cyclic subgroup of $\mathbb{R}^*$. The valuation ring determined by a discrete valuation is said to be a **discrete valuation ring**.

Clearly if one valuation in a spot $\mathfrak{p}$ is discrete so are they all; hence without ambiguity we can speak of a discrete *spot*. For example, if $F = \mathbb{Q}$ or $k(x)$ and p is prime (a prime number or a prime polynomial, resp.) then the p-adic spot on F is discrete.

Example 3.28. (This example is for readers who have seen some algebraic number theory.) Let K be an algebraic number field, with R its ring of algebraic integers; and let S be the set of nonzero prime ideals of R. Recall that a **fractional R-ideal** is a nonzero R-module in K of the form λA, with $\lambda \in K^*$ and A a nonzero ideal of R. (In particular, every nonzero R-ideal is a fractional ideal.) The *product* of fractional ideals A, B is defined as for ordinary ideals and is again a fractional ideal: AB is the set of all finite sums $\sum \alpha\beta$ with $\alpha \in A, \beta \in B$. Every fractional R-ideal has a unique expression as a product

$$\prod_{\mathfrak{p}\in S} \mathfrak{p}^{\nu_\mathfrak{p}},$$

with $\nu_\mathfrak{p} \in \mathbb{Z}$ and $\nu_\mathfrak{p} = 0$ for almost all $\mathfrak{p} \in S$. Now associate a valuation $|\ \ |_\mathfrak{p}$ with each $\mathfrak{p} \in S$ as follows: define $|0|_\mathfrak{p} = 0$, and for $\lambda \in K^*$ express the principal fractional ideal $(\lambda) = \lambda R$ in the form $\prod_{\mathfrak{p}\in S} \mathfrak{p}^{\nu_\mathfrak{p}}$ and then define

$$|\lambda|_\mathfrak{p} = \left(\frac{1}{\#(R/\mathfrak{p})}\right)^{\nu_\mathfrak{p}}.$$

This is a discrete valuation on the number field K. One usually lets the symbol $\mathfrak{p}$ also represent the spot containing $|\ \ |_\mathfrak{p}$, and then $\operatorname{ord}_\mathfrak{p} \lambda = \nu_\mathfrak{p}$. Clearly $R \subset \mathfrak{o}(\mathfrak{p})$, and in fact $R = \bigcap_{\mathfrak{p}\in S} \mathfrak{o}(\mathfrak{p})$.

The proof of the following proposition is an exercise for the reader.

Proposition 3.29. *Let $\mathfrak{p}$ be a discrete spot on a field F, and let $|\ \ | \in \mathfrak{p}$. Then the following four conditions are equivalent.*

(i) *π is an irreducible element of the valuation ring $\mathfrak{o}(\mathfrak{p})$.*

(ii) $|\pi| = \max\{\, |\alpha| \mid \alpha \in \mathfrak{m}(\mathfrak{p}) \,\}$.

(iii) *$|\pi| < 1$ and $|\pi|$ generates the value group $|F^*|$.*

(iv) $(\pi) = \mathfrak{m}(\mathfrak{p})$.

Definition 3.30. An element π satisfying the conditions of the preceding proposition is said to be a **prime element** (or **uniformizing element**) of F at $\mathfrak{p}$.

To see that a prime element exists at the discrete spot $\mathfrak{p}$, take any element α whose value $|\alpha|$ generates $|F^*|$. If $|\alpha| < 1$ then set $\pi = \alpha$, otherwise take $\pi = \alpha^{-1}$.

Example 3.31. Let p be a prime number. Then p is a prime element for $\mathbb{Q}$ at the p-adic spot, and so is any rational number α satisfying $\operatorname{ord}_p \alpha = 1$.

If π is a prime element of F at $\mathfrak{p}$, and $\pi' \in F$, then π' is prime if and only if $\pi' = \varepsilon\pi$ for some $\varepsilon \in \mathfrak{u}(\mathfrak{p})$. In fact every $\alpha \in F^*$ has a unique expression in the form $\alpha = \varepsilon\pi^\nu$, where $\varepsilon \in \mathfrak{u}$ and $\nu \in \mathbb{Z}$. The exponent ν is the **$\mathfrak{p}$-adic order** of α, written $\operatorname{ord}_\mathfrak{p} \alpha$ or $\nu_\mathfrak{p}(\alpha)$. We also put $\operatorname{ord}_\mathfrak{p} 0 = \infty$, with the understanding that the symbol ∞ satisfies $\infty > n$ for all $n \in \mathbb{Z}$.

The following properties of the $\mathfrak{p}$-adic order function are easy to check:

(i) $\operatorname{ord}_\mathfrak{p} \alpha = \log_{|\pi|} |\alpha|$ if $\alpha \neq 0$.

(ii) $|\alpha| < |\beta|$ if and only if $\operatorname{ord}_\mathfrak{p} \alpha > \operatorname{ord}_\mathfrak{p} \beta$.

(iii) $\operatorname{ord}_\mathfrak{p} \alpha > 0$ if and only if $\alpha \in \mathfrak{m}$.

(iv) $\operatorname{ord}_\mathfrak{p} \alpha \geq 0$ if and only if $\alpha \in \mathfrak{o}$.

(v) $\operatorname{ord}_\mathfrak{p}(\alpha\beta) = \operatorname{ord}_\mathfrak{p} \alpha + \operatorname{ord}_\mathfrak{p} \beta$.

(vi) $\mathfrak{o}(\mathfrak{p})$ is a principal ideal domain with ideals

$$\mathfrak{o} \supset (\pi) \supset (\pi^2) \supset (\pi^3) \supset \cdots \supset (0).$$

3.5. The Canonical Power Series Representation

We have defined the field $\mathbb{Q}_p$ of p-adic numbers to be the completion of $\mathbb{Q}$ with respect to the p-adic valuation. But let's be frank: up to now p-adic numbers have seemed to be rather abstract entities. To work with $\mathbb{Q}_p$ effectively, for instance in connection with the theory of quadratic forms, we need a concrete way to represent p-adic numbers that lends itself to computations, and that is our next goal.

Notation. If A is an ideal in some ring, the expression $a \equiv b \pmod{A}$ means $a - b \in A$. If $A = (c)$, a principal ideal, we usually write $a \equiv b \pmod{c}$ instead.

Theorem 3.32. *Let $\mathfrak{p}$ be a discrete spot on F, with C a representative set for the residue class field at $\mathfrak{p}$, and π a fixed prime element of F at $\mathfrak{p}$. Then*

(i) *Every element $\alpha \in F^*$ has a unique expression $\alpha = \sum_{i=\nu}^{\infty} c_i\pi^i$, with $c_i \in C$ and $\nu = \operatorname{ord}_\mathfrak{p} \alpha$.*

(ii) *If $\mathfrak{p}$ is complete then every such power series converges in F.*

Proof. Only the first part requires proof, since the second part is an instance of the "freshman dream" considered earlier.

Write $\alpha = \varepsilon\pi^{\nu}$, with $\varepsilon \in \mathfrak{u}$ and $\nu \in \mathbb{Z}$. Choose $c_\nu \in C$ such that $\varepsilon \equiv c_\nu \pmod{\mathfrak{m}}$; say $\varepsilon = c_\nu + \lambda$, with $\lambda \in \mathfrak{m}$. Then $\alpha = c_\nu\pi^\nu + \alpha_1$, with $\operatorname{ord}_\mathfrak{p} \alpha_1 \geq \nu + 1$. Now repeat this process on α_1, and continue. After $m+1$ iterations we get

$$\alpha = c_\nu\pi^\nu + c_{\nu+1}\pi^{\nu+1} + \cdots + c_{\nu+m}\pi^{\nu+m} + \alpha_{m+1},$$

with $\operatorname{ord}_\mathfrak{p} \alpha_{m+1} > \nu + m$. Thus there is a sequence $\{c_\nu, c_{\nu+1}, \dots\} \subseteq C$ such that

$$\lim_{m\to\infty} \sum_{i=\nu}^{m} c_i\pi^i = \alpha,$$

and this proves the "existence" claim.

Now suppose $\alpha = \sum_{i=n}^{\infty} c_i\pi^i = \sum_{i=t}^{\infty} d_i\pi^i$, with $c_i, d_i \in C$. That is,

$$\alpha = \lim_{m\to\infty} \sum_{i=n}^{m} c_i\pi^i = \lim_{m\to\infty} \sum_{i=t}^{m} d_i\pi^i.$$

Fix a valuation $|\ \ | \in \mathfrak{p}$. From the principle of domination we have $n = t = \operatorname{ord}_p \alpha$; also,

$$\sum_{i=n}^{\infty} (c_i - d_i)\pi^i = \lim_{m\to\infty} \sum_{i=n}^{m} (c_i - d_i)\pi^i = 0.$$

Suppose there were an index k for which $c_k \neq d_k$, and we seek a contradiction. Choose a minimal such k. Then $|c_k - d_k| = 1$, since c_k and d_k represent different cosets in the residue class field $\mathfrak{o}(\mathfrak{p})/\mathfrak{m}(\mathfrak{p})$. But then, again by the principle of domination, we would have

$$\left|\sum_{i=n}^{\infty} (c_i - d_i)\pi^i\right| = |\pi^k| \neq 0,$$

the desired contradiction. $\square$

Proposition 3.33. (i) *Let p be a prime number, and let $\mathfrak{p}$ be the p-adic spot on $\mathbb{Q}$. Then*

$$\{0, 1, \dots, p-1\}$$

is a representative set for the residue class field of $\mathbb{Q}$ (and therefore also of $\mathbb{Q}_p$) at $\mathfrak{p}$.

(ii) *Let $F = k(x)$, let $p(x)$ be a prime polynomial in $k[x]$, and let $\mathfrak{p}$ be the p-adic spot on F. Then the polynomials of degree less than $\deg p(x)$ are a representative set for the residue class field of F (and therefore of $F_\mathfrak{p}$) at $\mathfrak{p}$. The coefficient field k is a representative set for F (or F_∞) at the spot ∞.*

Proof. (i) Suppose $0 \neq \alpha \in \mathfrak{o}$. So $\alpha = p^{\nu}\dfrac{a}{b}$ for some integers a, b relatively prime to p and $\nu \geq 0$. We seek an integer $x \in \{0, 1, \ldots, p-1\}$ such that $x - \alpha \in \mathfrak{m}$; that is, $|x-\alpha|_p < 1$. If $\nu \geq 1$ we can choose $x = 0$, so without loss of generality we can assume $\nu = 0$. But for $x \in \{0, 1, \ldots, p-1\}$ we have

$$|x-\alpha|_p < 1 \quad \Longleftrightarrow \quad |bx-a|_p < 1 \quad \Longleftrightarrow \quad bx \equiv a \pmod{p},$$

and this last congruence is solvable in $\mathbb{Z}$ for x because $(b,p) = 1$. Reducing x mod p as needed puts it into the desired set.

(ii) For the finite spots, imitate the proof of part (i). As for ∞, notice that $F = k(y)$ with $y = \frac{1}{x}$, and then use the fact that while the given spot ∞ on F is "infinite" with respect to the subring $k[x]$, it becomes a "finite" spot—in fact the y-adic spot—on F with respect to the subring $k[y]$. □

Corollary 3.34. *Let $\alpha \in \mathbb{Q}_p^*$. Then α has a unique representation in the form*

$$\alpha = \sum_{i=\nu}^{\infty} c_i p^i,$$

with $\nu = \operatorname{ord}_p \alpha$ and $c_i \in \{0, 1, \ldots, p-1\}$.

Notation 3.35. If $\mathfrak{p}$ is the p-adic spot on $\mathbb{Q}_{\mathfrak{p}}$, the associated valuation ring $\mathfrak{o}(\mathfrak{p})$ is usually denoted $\mathbb{Z}_p$ and called the ring of ***p*-adic integers**. (Of course one must be careful not to confuse this ring with the ring $\mathbb{Z}$ mod $p = \{0, 1, \ldots, p-1\}$, which is also denoted $\mathbb{Z}_p$ in many books.)

We note in passing the contrast of the canonical p-adic representation with the decimal representation of real numbers: The fact that $2.999\ldots = 3.000\ldots$ shows that the equality

$$\sum_{i=n}^{\infty} a_i \left(\frac{1}{10}\right)^i = \sum_{i=n}^{\infty} b_i \left(\frac{1}{10}\right)^i$$

with $a_i, b_i \in \{0, 1, \ldots, 9\}$ does not imply that $a_i = b_i$ for all i.

Consider p-adic numbers $A = \sum_{i=\nu}^{\infty} a_i p^i$ and $B = \sum_{i=\nu}^{\infty} b_i p^i$, with $a_i, b_i \in \mathbb{Z}$ (so A and B might not be in canonical form), and put $A_n = \sum_{i=\nu}^{n} a_i p^i$ and $B_n = \sum_{i=\nu}^{n} b_i p^i$. Then $A_n \to A$ and $B_n \to B$ as $n \to \infty$. Therefore $A + B = \lim_{n\to\infty} A_n + B_n$, and similarly for subtraction and multiplication. Thus

$$A \pm B = \sum_i (a_i \pm b_i) p^i \qquad \text{and} \qquad AB = \sum_k c_k p^k,$$

with $c_k = \sum_{i+j=k} a_i b_j$.

The p-adic expansion of a natural number α is essentially the representation of α in the base p, which can be quickly obtained via the division algorithm as follows. Find integers q_0, a_0 with $0 \leq a_0 \leq p-1$ such that

$\alpha = pq_0 + a_0$; then a_0 is the rightmost digit of α in the base p. If $q_0 \neq 0$ replace α with q_0 and repeat the process; that is, obtain $q_0 = pq_1 + a_1$, with $0 \leq a_1 \leq p-1$. Then a_1 is the next digit of α. Iterate until achieving $\alpha = \sum_{i=0}^{t} a_i p^i$, with $0 \leq a_i \leq p-1$ for all i. But computing the p-adic expansion of other rational numbers is a greater challenge, and here are some examples.

Examples 3.36.

(1) Let us determine the canonical 3-adic expansion of $\alpha = -\frac{13}{8}$. We model our solution after the proof of the theorem. The observation that $|\frac{13}{8}|_3 = 1$ tells us that the solution will have the form $\sum_{i=0}^{\infty} c_i 3^i$ with $c_i \in \{0,1,2\}$ and $c_0 \neq 0$; because if $c_0 = 0$ then every partial sum $A_n = \sum_{i=0}^{n} c_i 3^i$ would satisfy $|A_n|_3 < \frac{1}{3}$, making the convergence $A_n \to \alpha$ an impossibility. We want $|-\frac{13}{8} - c_0|_3 < 1$; since $|8|_3 = 1$ this is equivalent to $|-13 - 8c_0|_3 < 1$ and hence to the congruence $8c_0 \equiv -13 \pmod 3$. Therefore $c_0 = 1$, from which we obtain $-\frac{21}{8} = \sum_{i=1}^{\infty} c_i 3^i$, or $-\frac{7}{8} = \sum_{i=1}^{\infty} c_i 3^{i-1}$. Proceeding as for c_0 we obtain $c_1 = 1$ and then $c_2 = 2$ (in other words, these are the first two digits in the 3-adic expansion of $-\frac{7}{8}$), and therefore

$$-\frac{13}{8} = 1 + 1\cdot 3 + 2\cdot 3^2 + \sum_{i=3}^{\infty} c_i 3^i,$$

which simplifies to $-\frac{7}{8} = \sum_{i=3}^{\infty} c_i 3^{i-3}$. But now we see that our computational program has entered a little loop, endlessly repeating the first two digits of $-\frac{7}{8}$. In summary

$$-\frac{13}{8} = 1 + \sum_{i=1}^{\infty} c_i 3^i, \text{ with } c_i = \begin{cases} 1 \text{ if } i \equiv 1 \pmod 2, \\ 2 \text{ if } i \equiv 0 \pmod 2. \end{cases}$$

(2) Let p be a prime. What is the canonical representation of -1 in $\mathbb{Q}_p$? Since $\operatorname{ord}_p(-1) = 0$, the answer will have the form

$$-1 = a_0 + a_1 p + a_2 p^2 + \cdots$$

with $a_i \in \{0, 1, \ldots, p-1\}$ and $a_0 \neq 0$. To get $|-1 - a_0|_p < 1$ we must have $a_0 = (p-1)$. Therefore

$$-1 = (p-1) + a_1 p + a_2 p^2 + \cdots.$$

But in fact trivially $-1 = (p-1) + p(-1)$, so upon iterating what we have already done we obtain

$$-1 = \sum_{n=0}^{\infty} (p-1)p^n.$$

More formally, suppose it has been shown that $a_i = p-1$ whenever $0 \le i \le n$. Thus

$$\begin{aligned}-1 &= (p-1) + (p-1)p + \cdots + (p-1)p^n + a_{n+1}p^{n+1} + \cdots \\ &= (p^{n+1} - 1) + a_{n+1}p^{n+1} + \cdots\end{aligned}$$

and so $-p^{n+1} = p^{n+1}(a_{n+1} + a_{n+2}p + \cdots)$. It follows that

$$-1 = a_{n+1} + a_{n+2}p + a_{n+3}p^2 + \cdots,$$

and hence $a_{n+1} = p-1$ by the induction hypothesis.

Now that we have derived the answer, we will restate the result and give a quick proof. The claim: $-1 = \sum_{n=0}^{\infty}(p-1)p^n$. The proof: Consider the sequence of partial sums $s_0, s_1, s_2, \ldots$, where $s_k = \sum_{n=0}^{k}(p-1)p^n$. We have

$$s_k = \sum_{n=0}^{k}(p^{n+1} - p^n) = -1 + p^{k+1} \to -1 \text{ as } k \to \infty.$$

As an alternative check of the claim that $-1 = \sum_{n=0}^{\infty}(p-1)p^n$ (and to risk being accused of beating this topic into the ground), we imitate the high school procedure for finding the rational number represented by a repeating decimal. [Recall the process from those happy days: if the digits eventually repeat with period k, multiply the decimal by 10^k and subtract.] We know the series $\sum_{n=0}^{\infty}(p-1)p^n$, converges in $\mathbb{Q}_p$, say to α. (Apply the Cauchy criterion to the partial sums.) With $s_k = \sum_{n=0}^{k}(p-1)p^n$ as before, we have $ps_k = \sum_{n=1}^{k+1}(p-1)p^n \to p\alpha$ as $k \to \infty$; that is, $p\alpha = \sum_{n=1}^{\infty}(p-1)p^n$. Subtraction now gives $\alpha - p\alpha = p-1$, and hence $\alpha = -1$ as claimed.

(3) The preceding example displays an instance of negating an element (in this case the element is 1) of $\mathbb{Q}_p$ whose canonical expansion is known. More generally, suppose the nonzero element $\alpha \in \mathbb{Q}_p$ has canonical expansion $\alpha = \sum_{i=\nu}^{\infty} c_i p^i$, with $\nu = \operatorname{ord}_p(\alpha)$. We leave it to the reader to prove (by induction on the coefficient subscripts) that

$$-\alpha = \sum_{i=\nu}^{\infty} d_i p^i, \text{ with } \begin{cases} d_\nu = p - c_\nu, \\ d_i = p - c_i - 1 \text{ if } i > \nu. \end{cases}$$

Proof of this formula provides further confirmation for the expansion of -1 exhibited in the preceding example; and—upon going back to the example before that—the reader can use the formula to give the 3-adic expansion

$$\frac{13}{8} = 2 + \sum_{i=1}^{\infty} d_i 3^i, \text{ with } d_i = \begin{cases} 1 \text{ if } i \equiv 1 \pmod 2, \\ 0 \text{ if } i \equiv 0 \pmod 2). \end{cases}$$

Now we check this result. Put $\alpha = \sum_{i=1}^{\infty} d_i 3^i$ as above. Then $\alpha - 3^2\alpha = 3$; that is, $\alpha = -\dfrac{3}{8}$, as needed.

3.6. Hensel's Lemma, the Local Square Theorem, and Local Fields

Given an interesting ring, it is only a matter of time before we are called upon to determine whether a given polynomial with coefficients in that ring has any zeros. For this purpose the following criterion for existence of a root is very important. It involves the **formal derivative** of a polynomial: If $f(x) = \sum_{i=1}^{n} a_i x^i$ then $f'(x) = \sum_{i=1}^{n} i a_i x^{i-1}$. (Another "freshman dream": no limits!)

Theorem 3.37 (Hensel's Lemma). *Let $\mathfrak{p}$ be a complete discrete spot on F, let $\mathfrak{o}$ be the associated valuation ring, and let $|\ \ | \in \mathfrak{p}$. For a polynomial $f(x) \in \mathfrak{o}[x]$, suppose there is an element $a \in \mathfrak{o}$ such that*

$$|f(a)| < |f'(a)|^2.$$

Then there is an element $b \in F$ such that $f(b) = 0$.

Proof. We may assume that $f(a) \neq 0$, since otherwise we are done. If $\alpha, \beta \in F$ then

$$f(\alpha + \beta) = f(\alpha) + \beta f'(\alpha) + \beta^2 g(\alpha, \beta)$$

for some $g(x, y) \in \mathfrak{o}[x, y]$. Let $\beta = \dfrac{-f(a)}{f'(a)}$; from the factorization

$$\beta = -f'(a) \cdot \frac{f(a)}{f'(a)^2}$$

it follows that $\beta \in \mathfrak{o}$. With $\alpha = a$, the first two terms of the preceding expression for $f(\alpha + \beta)$ add to 0, and we have

$$|f(a + \beta)| \leq |\beta^2| = \left|\frac{f(a)}{f'(a)}\right|^2 = |f(a)| \frac{|f(a)|}{|f'(a)|^2} < |f(a)|.$$

Also,

$$f'(a + \beta) = f'(a) + \beta \cdot (\text{polynomial in } a, \beta \text{ with coefficients in } \mathfrak{o}).$$

Thus

$$|f'(a + \beta) - f'(a)| \leq |\beta| = \frac{|f(a)|}{|f'(a)|} < \frac{|f'(a)|^2}{|f'(a)|} = |f'(a)|,$$

and hence $|f'(a+\beta)| = |f'(a)|$ by the principle of domination. Because it is easy to get lost in the inequalities, here is a digest of the argument to this point: Starting with $a \in \mathfrak{o}$ such that $0 < |f(a)| < |f'(a)|^2$, we have shown the existence of $\beta \in \mathfrak{o}$ such that $|f(a+\beta)| < |f(a)|$ and $|f'(a+\beta)| = |f'(a)|$.

Now iterate this process. That is, construct an infinite sequence $a = a_0, a_1, a_2, \ldots$ in $\mathfrak{o}$, where $a_{i+1} = a_i + \beta_i$, with

$$\beta_i = \frac{-f(a_i)}{f'(a_i)}.$$

Here $|f'(a_i)| = |f'(a)|$ for all i, and

$$|f(a)| = |f(a_0)| > |f(a_1)| > |f(a_2)| > \cdots ;$$

therefore $\beta_i \to 0$ as $i \to \infty$. (Remember: the value group is an infinite cyclic subgroup of $\mathbb{R}^*$.) The sequence $\{a_i\}_{i\geq 0}$ is Cauchy because $|a_{i+1} - a_i| = \beta_i \to 0$, hence converges to b, say. But $f(a_i) \to 0$ as $i \to \infty$. It follows that $f(b) = 0$. Let's look more closely at that final "it follows that." If we write $f(x) = \sum_{i=0}^{n} c_i x^i$, and we put $b = a_i + \varepsilon_i$ (with $\varepsilon_i \to 0$), then the binomial expansion gives $f(b) = f(a_i) + \varepsilon_i K$, with $|K| \leq 1$ (because $c_i \in \mathfrak{o}$ and the valuation is nonarchimedean). Hence $f(a_i) \to f(b)$ as claimed, and so $f(b) = 0$. □

Example 3.38. The polynomial $f(x) = x^2 + 1$ is solvable in $\mathbb{Q}_5$. To see this, first note that $|f(2)|_5 = |5|_5 < 1$. Also $f'(x) = 2x$, and so $|f'(2)|_5 = |4|_5 = 1$. Now apply Hensel's Lemma.

In the field of real numbers, the nonzero squares are the positive numbers; so if a is a nonzero square and b is sufficiently close to a, then b is also a square. The following result, the Local Square Theorem—our main application of Hensel's lemma—will show that the same kind of statement holds for a field with a complete discrete spot, in particular for $\mathbb{Q}_p$. Mathematicians who work with quadratic forms are necessarily obsessed with squares, and the Local Square Theorem is one of our most fundamental tools in grappling with this obsession.

Theorem 3.39 (Local Square Theorem). *Assume $\mathfrak{p}$ is a complete discrete spot on the field F. Then every element of the form $1 + 4\pi\alpha$, with π prime and $|\alpha| \leq 1$, is a square in F. In particular, if $|2| = 1$ (as in $\mathbb{Q}_p$ when $p \neq 2$) then $1 + \lambda \in F^{*2}$ whenever $|\lambda| < 1$.*

Proof. Write $\lambda = 4\pi\alpha$, with $\alpha \in \mathfrak{o}$ and π prime in $\mathfrak{o}$, and consider the polynomial $f(x) = \pi x^2 + x - \alpha \in \mathfrak{o}[x]$. By Hensel's Lemma (with $a = \alpha$), f has a root $\beta \in \mathfrak{o}$, and by the quadratic formula β has the form $\dfrac{-1 \pm \sqrt{1+\lambda}}{2\pi}$; so $\sqrt{1+\lambda} \in F$ and $1 + \lambda = (1 + 2\pi\beta)^2$. □

Corollary 3.40. *In the setting of the Local Square Theorem, let $\varepsilon, \delta \in \mathfrak{u}(\mathfrak{p})$, with $\varepsilon \equiv \delta \pmod{4\pi}$. Then $\varepsilon \in \delta\mathfrak{u}^2$.*

Proof. The hypothesis gives $\varepsilon = \delta + 4\pi\alpha$ for some $\alpha \in \mathfrak{o}$. Now divide by δ and apply the Local Square Theorem. □

Example 3.41. Again let us see the solvability of $f(x) = x^2 + 1$ in $\mathbb{Q}_5$: since $-1 = 4 + 5(-1)$, the fact that $-1 \in \mathbb{Q}_5^{*2}$ is now immediate from the corollary.

The following corollary, again in the setting of the Local Square Theorem, formally confirms that elements close to nonzero squares are squares. (So in this respect the fields under consideration are similar to $\mathbb{R}$.)

Corollary 3.42. *The group F^{*2} is an open subgroup of F^*.*

Proof. Let $\alpha \in F^{*2}$; say $\alpha = \varepsilon^2\pi^{2\nu}$, with $\varepsilon \in \mathfrak{u}$. Suppose $|\beta - \alpha| < |4\pi^{2\nu+1}|$. Then by the principle of domination we have $\beta = \mu\pi^{2\nu}$, with $|\varepsilon^2 - \mu| \leq |4\pi|$; hence μ is square by the preceding corollary. □

Definition 3.43. If $\mathfrak{p}$ is a complete discrete spot on the field F, and the associated residue class field $\overline{F(\mathfrak{p})}$ is finite, then the pair $(F, \mathfrak{p})$ is said to be a **local field**. (If $\mathfrak{p}$ is understood, then for brevity we just say F is a local field.) The symbol $N\mathfrak{p}$ denotes the cardinality of $\overline{F(\mathfrak{p})}$.

Thus if p is a prime number then every $\mathbb{Q}_p$ is a local field; and $N\mathfrak{p} = p$.

Exercise 3.44. Show that if $\mathbb{F}_q$ is a finite field and p is either a prime polynomial in $\mathbb{F}_q[x]$ or the infinite prime ∞, then the completion $\mathbb{F}_q(x)_p$ is a local field. (Suggestion: Imitate the proof that $\mathbb{Q}_p$ is a local field.)

Notation 3.45. When confusion is unlikely, from now on if $\mathfrak{p}$ is a nonarchimedean spot on F, we will also use $\mathfrak{p}$ for the maximal ideal $\mathfrak{m}(\mathfrak{p}) = \pi\mathfrak{o}$ of $\mathfrak{o}(\mathfrak{p})$. And if F is a local field and $r \in \mathbb{N}$ we write

$$\mathfrak{u}_r = 1 + \mathfrak{p}^r = \{\, 1 + \pi^r\alpha \mid \alpha \in \mathfrak{o} \,\},$$

a subgroup of $\mathbb{Q}_p^*$. Here $\mathfrak{p}^r$ denotes the $\mathfrak{o}$-ideal generated by the products of r elements of $\mathfrak{p}$, which in this case is equal to the set of elements of the form $\pi^r\alpha$, since $\mathfrak{p} = (\pi)$. This yields a strictly descending chain of subgroups of $\mathbb{Q}_p^*$, namely

$$\mathfrak{u} > \mathfrak{u}_1 > \mathfrak{u}_2 > \cdots .$$

Our coming work on quadratic forms will require some information on squares, and the following proposition will be helpful. We use this notation: for an abelian multiplicative group G, the symbol G^2 denotes its subgroup of squares.

For much of the balance of this section we follow the treatment in O'Meara [**O'M1**].

Proposition 3.46. *Let $r > 0$. Then*

(i) $(\mathfrak{u} : \mathfrak{u}_r) = (N\mathfrak{p} - 1)(N\mathfrak{p})^{r-1}$;

(ii) $\mathfrak{u}_r^2 = 1 + 2\mathfrak{p}^r$ *if* $r > \operatorname{ord}_{\mathfrak{p}} 2$.

Proof. The residue class mapping takes $\mathfrak{u}$ onto the multiplicative group $\overline{F}^*$ of nonzero elements of the residue class field and has kernel $1+\mathfrak{p}$. Hence

$$(\mathfrak{u} : 1+\mathfrak{p}) = N\mathfrak{p} - 1.$$

Now fix a prime element π, assume $r \geq 1$, and define $\varphi : \mathfrak{u}_r \to \overline{F}$ by

$$1 + \pi^r \alpha \mapsto \overline{\alpha}$$

for $\alpha \in \mathfrak{o}$. It is straightforward to check that φ is a surjective group homomorphism onto the *additive* group $\overline{F}$, with $\ker \varphi = \mathfrak{u}_{r+1}$. Thus $(\mathfrak{u}_r : \mathfrak{u}_{r+1}) = N\mathfrak{p}$, and the conclusion to part (i) follows upon multiplying successive group indices.

Now for the proof of (ii). The hypothesis gives $\mathfrak{p}^r \subseteq 2\mathfrak{p}$, from which the inclusion "$\subseteq$" follows immediately. For the opposite inclusion, suppose $x \in 1 + 2\mathfrak{p}^r$, say $x = 1 + 2\pi^r\alpha$, with $\alpha \in \mathfrak{o}$. Then

$$x = (1+\pi^r\alpha)^2 - \pi^{2r}\alpha^2 = (1+\pi^r\alpha)^2(1+\pi^{2r}\beta)$$

for some $\beta \in \mathfrak{o}$. Moreover $(1+\pi^{2r}\beta) = (1+\gamma)^2$ for some $\gamma \in 2\mathfrak{p}$ by the Local Square Theorem. (Here again we have used the assumption that $r > \operatorname{ord}_{\mathfrak{p}} 2$.) Then

$$|2\gamma| = |2||\gamma| = |\gamma + 2||\gamma| = |\gamma^2 + 2\gamma| = |\pi^{2r}\beta| \leq |2\pi^{r+1}|$$

and hence $\gamma \in \mathfrak{p}^{r+1} \subset \mathfrak{p}^r$. It follows that $x \in (1+\mathfrak{p}^r)^2$. □

Lemma 3.47. *Let A be an abelian group, and let φ be a homomorphism with* $\operatorname{dom}\varphi = A$ *and* $\ker\varphi = K$. *Finally, suppose B is a subgroup of A of finite index. Then*

$$(A : B) = (\varphi A : \varphi B) \cdot (K : K \cap B).$$

Proof. We have $(A : B) = (A : BK)(BK : B) = (A : BK)(K : K \cap B)$. Therefore, since $\varphi(BK) = \varphi B$, to prove the lemma it will suffice to show that $(A : BK) = (\varphi A : \varphi(BK))$.

The map $\psi : A \to \varphi A/\varphi(BK)$ given by $a \mapsto \overline{\varphi(a)}$ is clearly a surjective homomorphism. If $a \in \ker\psi$ then $\varphi(a) = \varphi(bk) = \varphi(b)$ for some $b \in B, k \in K$. Therefore $ab^{-1} \in K$ and hence $a \in BK$; thus $\ker\psi \subseteq BK$ and so in fact $\ker\psi = BK$, which completes the proof. □

Theorem 3.48. *Let $(F, \mathfrak{p})$ be a local field. Then*

$$(F^* : F^{*2}) = 2(\mathfrak{u} : \mathfrak{u}^2) = 4(N\mathfrak{p})^{\operatorname{ord}_{\mathfrak{p}} 2}.$$

Proof. We begin by proving the second equality, and for this we will freely use the notation and results of the lemma and the proposition immediately preceding it. Let $A = \mathfrak{u}$ and $B = \mathfrak{u}_r$, with $r > \operatorname{ord}_{\mathfrak{p}} 2$, and define $\varphi : \mathfrak{u} \to \mathfrak{u}$ by

$\alpha \mapsto \alpha^2$. So $K = \{\pm 1\}$, and since $2 \notin \mathfrak{p}^r$ it follows that $-1 = 1 + (-2) \notin B$; thus $K \cap B = \{1\}$. We now have

$$(\mathfrak{u} : \mathfrak{u}_r) = (\mathfrak{u}^2 : \mathfrak{u}_r^2) \cdot 2 = 2(\mathfrak{u}^2 : 1 + 2\mathfrak{p}^r) = 2(\mathfrak{u}^2 : \mathfrak{u}_{r+\mathrm{ord}_\mathfrak{p} 2}).$$

If we now multiply the expressions on the left and right of the preceding line by $(\mathfrak{u} : \mathfrak{u}^2)$ and then cancel $(N\mathfrak{p} - 1)(N\mathfrak{p})^{r-1}$ from both, we are left with $(\mathfrak{u} : \mathfrak{u}^2) = 2(N\mathfrak{p})^{\mathrm{ord}_\mathfrak{p} 2}$, proving the second equation in the theorem's statement. The first equation then follows from the observation that if $\{\varepsilon_1, \ldots, \varepsilon_t\}$ is a representative set for $\mathfrak{u}/\mathfrak{u}^2$ then $\{\varepsilon_1, \ldots, \varepsilon_t, \pi\varepsilon_1, \ldots, \pi\varepsilon_t\}$ is a representative set for F^*/F^{*2}. □

A local field $(F, \mathfrak{p})$ is said to be **nondyadic** if $2 \in \mathfrak{u}(\mathfrak{p})$ (two equivalent conditions: $|2| = 1$, $\mathrm{ord}_\mathfrak{p} 2 = 0$); and $(F, \mathfrak{p})$ is **dyadic** otherwise (two equivalent conditions: $|2| < 1$, $\mathrm{ord}_\mathfrak{p} 2 > 0$). Thus if F is a nondyadic local field then $(F^* : F^{*2}) = 4$; so if Δ is any nonsquare unit then $\{1, \Delta, \pi, \pi\Delta\}$ is a complete set of representatives of F^*/F^{*2}. On the other hand, if F is dyadic then the finite group F^*/F^{*2} can be arbitrarily large. (See the example following the next paragraph.)

Corollary 3.49. *If p is an odd prime number, then $\mathbb{Q}_p$ is a nondyadic local field; and $\mathbb{Q}_2$ is a dyadic local field. If q is an odd prime-power, then all the completions $\mathbb{F}_q(x)_\mathfrak{p}$ of a rational function field $\mathbb{F}_q(x)$ are nondyadic local fields. Therefore*

$$(\mathbb{Q}_p^* : \mathbb{Q}_p^{*2}) = \begin{cases} 4 & \text{if } p \neq 2, \\ 8 & \text{if } p = 2, \end{cases} \quad \text{and} \quad (\mathbb{F}_q(x)_\mathfrak{p}^* : \mathbb{F}_q(x)_\mathfrak{p}^{*2}) = 2.$$

The local fields of main interest for us will be the fields of p-adic numbers and the fields $\mathbb{F}_q(x)_p$. Are there other local fields? To answer this question, we first need to introduce **global fields**. These are of two kinds: the **algebraic number fields**, which are the finite extensions of the field $\mathbb{Q}$ of rational numbers; and the **algebraic function fields in one variable over finite constant fields**. And by this latter mouthful we mean fields that are finite extensions of rational function fields $k(x)$, with k a finite field. It was shown by Emil Artin and George Whaples in 1945 that the local fields are precisely the completion of global fields with respect to their discrete spots.

Example 3.50. Without proof, here is an example of a dyadic local field with a large group of square classes. Let F be an algebraic number field containing $\sqrt[998]{2}$, and suppose $\mathfrak{p}$ is a spot on F extending the 2-adic spot on $\mathbb{Q}$. Then $\mathrm{ord}_\mathfrak{p} 2 \geq 998$, and hence the local field $F_\mathfrak{p}$ satisfies

$$(F_\mathfrak{p}^* : F_\mathfrak{p}^{*2}) = 4(N\mathfrak{p})^{\mathrm{ord}_\mathfrak{p} 2} \geq 2^{1000}.$$

3.7. The Legendre Symbol; Recognizing Squares in $\mathbb{Q}_p$

In seeking to understand quadratic forms over a global field F and its subrings, an important approach over the years has been to first consider quadratic forms over the completions of F with respect to F's discrete spots, and then to try to tie together whatever is learned in the local settings to obtain conclusions over F. Our favorite global field here will be $\mathbb{Q}$; this being the case, our local fields of primary interest will be the fields $\mathbb{Q}_p$ of p-adic numbers. We conclude this chapter with some very useful observations on square classes in $\mathbb{Q}_p$, with the goal of showing how the square class of an element $a \in \mathbb{Q}_p$ is determined by a's canonical representation. In particular, when $p \neq 2$ the square class in $\mathbb{Z} \pmod p$ of the leading coefficient of a is of special importance. With this in mind, we recall a bit of classical number theory.

Definition 3.51. Let p be an odd prime number. For $a \in \mathbb{Z}$ with $(a,p) = 1$ the **Legendre symbol** is

$$\left(\frac{a}{p}\right) = \begin{cases} 1 & \text{if } a \equiv x^2 \pmod p \text{ is solvable in } \mathbb{Z}, \\ -1 & \text{otherwise.} \end{cases}$$

Of course the ring of integers mod p is just the finite field $\mathbb{F}_p$; so, upon identifying an integer a with its residue class in $\mathbb{F}_p$, the Legendre symbol can be given the following equivalent definition: For $a \in \mathbb{F}_p^*$, define

$$\left(\frac{a}{p}\right) = \begin{cases} 1 & \text{if } a \in \mathbb{F}_p^{*2}, \\ -1 & \text{otherwise.} \end{cases}$$

In practice, how do we compute the Legendre symbol? (Equivalently, how can we efficiently determine whether a given element of $\mathbb{F}_p^*$ is a square?) Here are the basic ideas. Since $\#\mathbb{F}_p^* = p-1$, we have $a^{p-1} = 1$ for all $a \in \mathbb{F}_p^*$, and hence $a^{\frac{p-1}{2}} = \pm 1$; in fact if $a = b^2$ then $a^{\frac{p-1}{2}} = 1$. We know from Chapter 2 that $(\mathbb{F}_p^* : \mathbb{F}_p^{*2}) = 2$ and hence $\#\mathbb{F}_p^{*2} = \frac{p-1}{2}$, so the squares in $\mathbb{F}_p^*$ are precisely the roots of the polynomial $x^{\frac{p-1}{2}} - 1$ in $\mathbb{F}_p$. Therefore $a^{\frac{p-1}{2}} = -1$ if a is not a square. Thus

$$\left(\frac{a}{p}\right) = a^{\frac{p-1}{2}} \qquad \text{for all } a \in \mathbb{F}_p^*.$$

Equivalently, $\left(\frac{a}{p}\right) \equiv a^{\frac{p-1}{2}} \pmod{p}$ for all $a \in \mathbb{Z}$ satisfying $(a, p) = 1$. In particular,

$$\left(\frac{-1}{p}\right) = \begin{cases} 1 & \text{if } p \equiv 1 \pmod{4}, \\ -1 & \text{if } p \equiv 3 \pmod{4}. \end{cases}$$

The following properties of the Legendre symbol are immediate from the definition and the preceding remarks:

(i) $\left(\frac{1}{p}\right) = 1.$

(ii) If $a \equiv b \pmod{p}$ then $\left(\frac{a}{p}\right) = \left(\frac{b}{p}\right).$

(iii) $\left(\frac{ab}{p}\right) = \left(\frac{a}{p}\right)\left(\frac{b}{p}\right).$

(iv) $\left(\frac{ac^2}{p}\right) = \left(\frac{a}{p}\right).$

More difficult to prove is the following result. (A proof can be found in most textbooks on so-called "elementary" number theory. For a more sophisticated treatment, see *A Course in Arithmetic* by J.-P. Serre [**Se**].)

(v) $\left(\frac{2}{p}\right) = (-1)^{\frac{p^2-1}{8}}.$

And *much* more difficult to prove is **Gauss's Law of Quadratic Reciprocity**, which amounts to an agreement among different odd primes p and q to take in each other's laundry. More precisely, the law says that (but for one class of exceptional cases) p is a square mod q if and only if q is a square mod p. Spelled out in terms of Legendre symbols: If p and q are distinct odd primes, then $\left(\frac{p}{q}\right) = \left(\frac{q}{p}\right)$ unless $p \equiv q \equiv 3 \pmod{4}$, in which case $\left(\frac{p}{q}\right) = -\left(\frac{q}{p}\right)$. In a more compact form, Quadratic Reciprocity can be stated as follows:

(vi) If p and q are distinct odd primes, then

$$\left(\frac{p}{q}\right)\left(\frac{q}{p}\right) = (-1)^{\frac{p-1}{2}\frac{q-1}{2}}.$$

By judicious application of rules (i) through (vi) the value of a given Legendre symbol can usually be obtained quite quickly, even if the given numbers are large.

Now we return to our consideration of square classes in $\mathbb{Q}_p^*$. First suppose p is odd, and let $\varepsilon \in \mathfrak{u}$, say $\varepsilon = \sum_{i \geq 0} a_i p^i$, with $0 \leq a_i \leq p - 1$ for all i and

$a_0 \neq 0$. Then from the Local Square Theorem $\varepsilon \in a_0\mathfrak{u}^2$. But

$$\begin{aligned} a_0 \in \mathfrak{u}^2 &\iff a_0 = (b_0 + b_1p + \cdots)^2 = b_0^2 + p(\dots) \\ &\iff a_0 \equiv b_0^2 \pmod{p} \\ &\iff \left(\frac{a_0}{p}\right) = 1, \end{aligned}$$

where $\left(\frac{\cdot}{p}\right)$ is the Legendre symbol. Thus the square class of $\varepsilon \in \mathfrak{u}$ is determined by the quadratic character of the leading coefficient in ε's canonical representation.

Now suppose $p = 2$. Then every element of $\mathfrak{u}$ has a canonical representation of the form

$$\varepsilon = 1 + a_1 \cdot 2 + a_2 \cdot 4 + 8\lambda$$

with $\lambda \in \mathbb{Z}_2$ and $a_i \in \{0, 1\}$ for $i = 1, 2$. Therefore, upon considering the four possible combinations of a_1, a_2, we see from the Local Square Theorem that $\varepsilon \in \beta\mathfrak{u}^2$ with $\beta \in \{1, 3, 5, 7\}$. But $(\mathfrak{u} : \mathfrak{u}^2) = 4$ when $p = 2$, hence $\{1, 3, 5, 7\}$ is a representative set for $\mathfrak{u}/\mathfrak{u}^2$; and $\{1, 3, 5, 7, 2, 6, 10, 14\}$ is a representative set for $\mathbb{Q}_2^*/\mathbb{Q}_2^{*2}$. We can now sum up our observations on recognizing squares in $\mathbb{Q}_p^*$ as follows.

Theorem 3.52. *Let $\alpha \in \mathbb{Q}_p^*$; say $\alpha = p^\nu \cdot \sum_{i=0}^\infty a_ip^i$, with $0 \le a_i \le p-1$ and $a_0 \neq 0$. Then*

(i) *If $p \neq 2$, then $\alpha \in \mathbb{Q}_p^{*2} \iff \nu \equiv 0 \pmod 2$ and $\left(\frac{a_0}{p}\right) = +1$.*

(ii) *If $p = 2$, then $\alpha \in \mathbb{Q}_2^{*2} \iff \nu \equiv 0 \pmod 2$ and $a_1 = a_2 = 0$.*

Corollary 3.53. *If p is a prime number, then*

$$-1 \in \mathbb{Q}_p^{*2} \quad \textit{if and only if} \quad p \equiv 1 \pmod 4.$$

Exercises

(1) (i) Briefly justify these basic rules for doing arithmetic in $\mathbb{Q}_p$:

$$\sum_{n=0}^\infty a_np^n \pm \sum_{n=0}^\infty b_np^n = \sum_{n=0}^\infty (a_n \pm b_n)p^n$$

and

$$\Big(\sum_{n=0}^\infty a_np^n\Big)\Big(\sum_{n=0}^\infty b_np^n\Big) = \sum_{n=0}^\infty c_np^n \quad \text{with } c_n = \sum_{i=0}^n a_ib_{n-i}.$$

(ii) Show that if $\alpha = \sum_{n=\nu}^\infty a_np^n \in \mathbb{Q}_p$ then $p^m\alpha = \sum_{n=\nu}^\infty a_np^{n+m}$.

(iii) $\left(\sum_{n=\mu}^{\infty} a_n p^n\right)\left(\sum_{n=\nu}^{\infty} b_n p^n\right) = \ ?$
[Remark: The claim is **not** being made in (i), (ii), and (iii) that the given or resulting p-adic expansions are *canonical.*]

(2) (i) Prove that the canonical p-adic expansion of any negative integer is infinite in length, in the sense that infinitely many of its coefficients are nonzero. (Give a one- or two-line argument using almost nothing in this chapter.)
(ii) Which rational numbers have finite canonical p-adic expansion?
(iii) Show that if M is a positive integer then there is a sequence $s_1, s_2, \ldots$ of positive integers with the property that $\lim_{n\to\infty} s_n = \infty$ in the usual archimedean sense but yet p-adically the sequence converges to M.

(3) Let F be a field, let $\mathfrak{p}$ be a complete discrete spot on F, and suppose $\alpha \in F^*$. Assume that for every real number $\varepsilon > 0$ there exist $\beta = \beta_\varepsilon$ and $\gamma = \gamma_\varepsilon \in F$ such that $\alpha = \beta^2 + \gamma$, with $|\gamma| < \varepsilon$. Show that $\alpha \in F^{*2}$.

(4) (i) Show that if $\alpha = \sum_{n=\nu}^{\infty} a_n p^n$ and $\beta = \sum_{n=\mu}^{\infty} b_n p^n$ are p-adic numbers and there is an integer N such that $a_n = b_n$ for all $n \geq N$, then $\alpha - \beta \in \mathbb{Q}$.
(ii) Call a p-adic expansion $\alpha = \sum_{n=\nu}^{\infty} a_n p^n$ **eventually periodic** if there are integers N and k, with $k > 0$, such that $a_n = a_{n+k}$ for all $n \geq N$. Prove that if α is eventually periodic then $\alpha \in \mathbb{Q}$.

(5) (i) Find the canonical 7-adic expansion for the 7-adic number

$$\alpha = 3 + 8 \cdot 7 + \sum_{n=2}^{\infty} 6 \cdot 7^n.$$

(ii) Do the same for the 7-adic number $\beta = \sum_{n=0}^{\infty} 8 \cdot 7^n$.

(6) *Without writing anything down*, determine which of the numbers

$$70, 71, 72, 73$$

are squares in $\mathbb{Q}_5$. Then, for the benefit of future generations, write a brief proof of your claim.

(7) Show that if p is an odd prime, then exactly half the integers in the set $\{1, 2, \ldots, p-1\}$ are squares in $\mathbb{Q}_p$. (Use the methods of this chapter, rather than an argument that appears in whatever undergraduate number theory book may be in your possession.)

(8) (i) Use Hensel's Lemma (rather than the theorem at the end of the section) to show that -1 is a square in $\mathbb{Q}_5$.
(ii) In the canonical representation $\sqrt{-1} = \sum_{n=0}^{\infty} a_n 5^n \in \mathbb{Q}_5$, determine coefficients a_0 through a_4.

(iii) In the answer to part (ii), why were there two possible answers for a_0, while once a value for a_0 was fixed there was only one possibility for the subsequent a_i?

(9) The goal of this problem is to prove that if p and q are distinct primes, then $\mathbb{Q}_p$ and $\mathbb{Q}_q$ are not isomorphic.
(i) First prove the result if $p = 2$ and q is odd.
(ii) Next prove the result if p and q are both odd and either $\left(\dfrac{p}{q}\right) = 1$ or $\left(\dfrac{q}{p}\right) = 1$.
(iii) Finally, modify your argument in part (ii) to deal with the case in which $p < q$ and $\left(\dfrac{q}{p}\right) = \left(\dfrac{p}{q}\right) = -1$. [Suggestion: Begin by choosing a positive integer $n < p$ such that $\left(\dfrac{n}{p}\right) = -1$.]

(10) It follows from our criterion involving Legendre symbols that 2 is a square in $\mathbb{Q}_7$. Use the Local Square Theorem to show that $2 \in \mathbb{Q}_7^{*2}$ without using anything about Legendre symbols.

(11) Let $a \in \mathbb{Z}$. Obtain a necessary and sufficient condition on a for the equation $x^2 + x + a = 0$ to be solvable in the field $\mathbb{Q}_2$ of 2-adic numbers.

(12) If S is a nonempty set of nonarchimedean spots on a field F, define $\mathfrak{o}(S) = \bigcap_{\mathfrak{p} \in S} \mathfrak{o}(\mathfrak{p})$. This is clearly a subring of F.
(i) Express $\mathbb{Z}$ as $\mathfrak{o}(S)$ for a suitable set S of nonarchimedean spots on $\mathbb{Q}$. Then do the analogous thing for the polynomial ring $\mathbb{F}_q[x]$ in its quotient field $\mathbb{F}_q(x)$.
(ii) Let S be the set of spots on $\mathbb{F}_q(x)$ found in part (i) that gives $\mathbb{F}_q[x] = \mathfrak{o}(S)$, let $\mathfrak{q}$ be the $(x+1)$-adic spot, and let ∞ be as defined in the text. Describe the elements of the ring

$$R = \mathfrak{o}\big(S \cup \{\infty\} - \{\mathfrak{q}\}\big).$$

(13) (i) Let p be an odd prime number, and let $\mathfrak{q}$ be a spot on $\mathbb{F}_p(x)$ defined by a linear prime in $\mathbb{F}_p[x]$. Show that the problem of detemining whether a given element $f(x) \in \mathbb{F}_p(x)$ is square in $\mathbb{F}_p(x)_\mathfrak{q}$ can be reduced to the calculation of a Legendre symbol of the form $\left(\dfrac{a}{p}\right)$.
(ii) Demonstrate your method in part (i) by determining whether the element

$$f(x) = x^{10} - 5x^3 + 3x^9 - 15x^2 + 8x + 61$$

is square in $\mathbb{F}_{71}(x)_\mathfrak{q}$, if $\mathfrak{q}$ is the $(x+3)$-adic spot on $\mathbb{F}_{71}(x)$.

(14) With q odd, let ∞ be the "infinite" spot on $\mathbb{F}_q(x)$; that is, the spot containing the valuation $|\ \ |_\infty$ satisfying $\left|\dfrac{f}{g}\right|_\infty = q^{\partial f - \partial g}$ for $f, g \in \mathbb{F}_q[x]$.
(i) Show that the completion $\mathbb{F}_q(x)_\infty$ is a nondyadic local field. [Suggestion: Start by finding a prime element for the spot. And choose it as simply as you can!]
(ii) State and prove a necessary and sufficient condition for a polynomial $f(x) = \sum_{i=0}^{n} a_i x^i \in \mathbb{F}_q[x]$ to be a square in $\mathbb{F}_q(x)_\infty$.
(iii) Demonstrate the usefulness of your condition in part (ii) by answering this question: Is the polynomial $f(x) = 2x^6 + x^3 + x + 2$ a square in $\mathbb{F}_{27}(x)_\infty$? [Note: You may find it easier to first consider the question over $\mathbb{F}_3(x)_\infty$.]

Chapter 4

Quadratic Spaces over $\mathbb{Q}_p$

Our goal in this chapter will be to classify quadratic spaces over the fields of p-adic numbers. (In the next chapter we will use these results to classify spaces over $\mathbb{Q}$.) While therefore most of our work here will be over $\mathbb{Q}_p$, from time to time we will include results over more general local fields or even arbitrary fields (of characteristic not 2) if moving into the wider context can be done without leading us too far from our objective.

4.1. The Hilbert Symbol

Definition 4.1. Let $(F, \mathfrak{p})$ be either a complete archimedean field or a local field. For $a, b \in F^*$, define the **Hilbert symbol**

$$(a,b)_{\mathfrak{p}} = \begin{cases} 1 & \text{if } ax^2 + by^2 = 1 \text{ is solvable in } F, \\ -1 & \text{otherwise.} \end{cases}$$

If p is a known prime element of F at $\mathfrak{p}$, we may write $(a,b)_p$ instead of $(a,b)_{\mathfrak{p}}$, and in fact once $\mathfrak{p}$ and p have been fixed we will write (a,b) instead of $(a,b)_{\mathfrak{p}}$ if confusion is unlikely.

Example 4.2. (i) If $F = \mathbb{C}$ then $(a,b) = 1$ for all $a, b \in \mathbb{C}^*$.

(ii) If $F = \mathbb{R}$ then $(a,b) = \begin{cases} 1 & \text{if } a > 0 \text{ or } b > 0, \\ -1 & \text{otherwise.} \end{cases}$

Theorem 4.3. *Let $F = (F, \mathfrak{p})$ be a nondyadic local field (e.g., $\mathbb{Q}_p$ with $p \neq 2$), and let $V \cong \langle \alpha_1, \alpha_2, \alpha_3 \rangle$ be a regular quadratic F-space, with $|\alpha_1| = |\alpha_2| = |\alpha_3|$. (Here $|\cdot|$ is a fixed valuation in $\mathfrak{p}$.) Then V is isotropic.*

Proof. Without loss of generality (scale the space appropriately) we can assume $|\alpha_1| = |\alpha_2| = |\alpha_3| = 1$; that is, $\alpha_i \in \mathfrak{u}$. Since the residue class field $\overline{F}$ is a finite field of characteristic not 2, the space $\langle \overline{\alpha_1}, \overline{\alpha_2} \rangle$ is universal over $\overline{F}$; therefore there is a representation

$$\overline{-\alpha_3} \to \langle \overline{\alpha_1}, \overline{\alpha_2} \rangle.$$

That is, there exist $\lambda_1, \lambda_2 \in \mathfrak{o}$ such that $\overline{\alpha}_1 \overline{\lambda_1}^2 + \overline{\alpha}_2 \overline{\lambda_2}^2 = -\overline{\alpha}_3$; equivalently, $\alpha_1\lambda_1^2 + \alpha_2\lambda_2^2 \equiv -\alpha_3 \pmod{\mathfrak{p}}$. It then follows from the Local Square Theorem that $\alpha_1\lambda_1^2 + \alpha_2\lambda_2^2 = -\alpha_3\lambda_3^2$ for some $\lambda_3 \in \mathfrak{u}$. The conclusion follows. □

Corollary 4.4. *If F is a nondyadic local field then $(a, b) = 1$ for all $a, b \in \mathfrak{u}$. That is, in a nondyadic local field the Hilbert symbol is trivial on $\mathfrak{u} \times \mathfrak{u}$.*

Proof. The quadratic space $\langle a, b, -1 \rangle$ is isotropic, and hence $1 \to \langle a, b \rangle$. □

Proposition 4.5. *Suppose F is a complete archimedean field or a local field, $a, b \in F^*$, and $E = F(\sqrt{b})$. Then the following are equivalent:*

(i) $(a, b) = 1$.

(ii) *The space $\langle -1, a, b \rangle$ is isotropic.*

(iii) *The equation $z^2 - ax^2 - by^2 = 0$ has a nontrivial solution in $F \times F \times F$.*

(iv) $a \in N_{E/F}E$.

Proof. Equivalence of (i), (ii), and (iii) is immediate, but we will prove that (i)$\Longleftrightarrow$(iv). If $b \in F^{*2}$ the result is clear, since in this case $E = F$ and $N_{E/F}$ is the identity map on F. On the other hand, if $b \notin F^{*2}$ then $(a, b) = 1$ if and only if $ax^2 + by^2 = 1$ is solvable in F with $x \neq 0$, and this is equivalent to $a = N(\frac{1}{x} + \frac{y}{x}\sqrt{b})$. □

Proposition 4.6. (i) $(a, b) = (b, a) = (ac^2, bd^2)$ *(So the Hilbert symbol yields a well-defined map $F^*/F^{*2} \times F^*/F^{*2} \to \{\pm 1\}$.)*

(ii) $(a, -a) = (a, 1 - a) = 1$.

(iii) $(a, b) = 1 \Longrightarrow (aa', b) = (a', b)$.

Proof. Statement (i) is clear from the definition of the Hilbert symbol. In statement (ii) the claim that $(a, -a) = 1$ follows from the fact that $\langle a, -a \rangle$ is a hyperbolic plane. (The other claim is obvious.) Statement (iii) is trivial if $b \in F^{*2}$. On the other hand, if $b \notin F^{*2}$ then let $E = F(\sqrt{b})$. From the hypothesis and the preceding proposition we know that $a \in N_{E/F}E$, so the conclusion now follows from the fact that the map $N_{E/F} : E^* \to F^*$ is a homomorphism. □

Now we will focus on the theory specifically over the fields of p-adic numbers.

Definition 4.7. An element $z = (z_1, \ldots, z_n) \in \mathbb{Z}_p^n$ is **primitive** if some $z_i \in \mathfrak{u} = \mathbb{Z}_p^*$.

Lemma 4.8. *In $\mathbb{Q}_p$, let $u \in \mathfrak{u}$. If the equation $z^2 - px^2 - uy^2 = 0$ has a nontrivial solution in $\mathbb{Q}_p^3$, then it has a solution $(z, x, y) \in \mathbb{Z}_p^3$ with $y, z \in \mathfrak{u}$.*

Proof. We can suppose $(z, x, y) \in \mathbb{Z}_p^3$ is a primitive solution. (So at least one of the coordinates is in $\mathfrak{u}$.) Now if $p|z$ then $p|y$, so $p^2|y^2$ and hence $p|x$, a contradiction of primitivity. Therefore $z \in \mathfrak{u}$, and similarly $y \in \mathfrak{u}$. □

Remark 4.9. If p is an odd prime and $a \in \mathbb{Z}_p^* = \mathfrak{u}$ has canonical representation $a = a_0 + pa_1 + \cdots$, then by the Local Square Theorem we know $a \in \mathfrak{u}^2$ if and only if $a_0 \in \mathfrak{u}^2$. It is therefore consistent with our earlier definition of the Legendre symbol to extend the Legendre symbol to all of $\mathfrak{u}$ by putting $\left(\frac{a}{p}\right) = \left(\frac{a_0}{p}\right)$, and we will do this from now on.

Theorem 4.10 (Computation of Hilbert symbols, Part I). *Assume p is an odd prime and $a, b \in \mathbb{Q}_p$; say $a = p^\alpha u, b = p^\beta v$, with $u, v \in \mathfrak{u}$. Then*

$$(a, b)_p = (-1)^{\alpha\beta\frac{p-1}{2}} \left(\frac{u}{p}\right)^\beta \left(\frac{v}{p}\right)^\alpha.$$

Proof. It is enough to check the result when $\alpha, \beta \in \{0, 1\}$. And in fact we already considered the case in which $\alpha = \beta = 0$ when we showed that the Hilbert symbol is trivial on $\mathfrak{u} \times \mathfrak{u}$ in a nondyadic local field.

(i) Suppose $\alpha = 1$ and $\beta = 0$, and we must check that $(pu, v) = \left(\frac{v}{p}\right)$. But $(pu, v) = (p, v)$ by part (iii) of the preceding proposition, so it is left for us to show that $(p, v) = \left(\frac{v}{p}\right)$. This result is clear if $v \in \mathfrak{u}^2$, so we can assume $v \notin \mathfrak{u}^2$ and hence $\left(\frac{v}{p}\right) = -1$. If at the same time we had $(v, p) = 1$, then by the lemma we would have $z^2 - px^2 - vy^2 = 0$ for some $y, z \in \mathfrak{u}$ and $x \in \mathbb{Z}_p$. But this would give $v = \frac{z^2}{y^2}(1 - p\frac{x^2}{z^2}) \in \mathfrak{u}^2$, a contradiction. Therefore $(v, p) = -1$, as required.

(ii) The case $\alpha = 0$ and $\beta = 1$ is covered by part (i), so we can now suppose $\alpha = \beta = 1$; and we must show that $(pu, pv) = (-1)^{\frac{p-1}{2}} \left(\frac{u}{p}\right) \left(\frac{v}{p}\right)$.

We have

$$\begin{aligned}(pu,pv) &= (pu,pv)(pu,-pu) = (pu,-p^2uv)\\ &= (pu,-uv) = (p,-uv)\\ &= \left(\frac{-uv}{p}\right) = \left(\frac{-1}{p}\right)\left(\frac{u}{p}\right)\left(\frac{v}{p}\right) = (-1)^{\frac{p-1}{2}}\left(\frac{u}{p}\right)\left(\frac{v}{p}\right).\end{aligned}$$

This finishes the proof. □

Corollary 4.11. *In $\mathbb{Q}_p$ with $p \neq 2$, if $\varepsilon \in \mathfrak{u}$ then*

$$(p,\varepsilon)_p = \begin{cases} 1 & \text{if } \varepsilon \in \mathfrak{u}^2,\\ -1 & \text{if } \varepsilon \notin \mathfrak{u}^2.\end{cases}$$

For the computation of 2-adic Hilbert symbols, we first recall that a 2-adic unit $u \in \mathfrak{u} = \mathbb{Z}_2^*$ has a canonical representation

$$u = 1 + a_1 2 + a_2 2^2 + a_3 2^3 + \cdots$$

with $a_i \in \{0,1\}$. Following the notation in Serre [**Se**], we now define

$$\varepsilon(u) = a_1 \quad \text{and} \quad \omega(u) = a_1 + a_2 \pmod 2.$$

Thus

$$\varepsilon(u) = \begin{cases} 0 & \text{if } u \equiv 1 \pmod 4,\\ 1 & \text{if } u \equiv 3 \pmod 4,\end{cases} \quad \text{and} \quad \omega(u) = \begin{cases} 0 & \text{if } u \equiv 1,7 \pmod 8,\\ 1 & \text{if } u \equiv 3,5 \pmod 8.\end{cases}$$

It is straightforward to check that ε and ω are homomorphisms from the multiplicative group $\mathfrak{u}$ to the additive group $\mathbb{Z}/2\mathbb{Z}$.

Theorem 4.12 (Computation of Hilbert symbols, Part II). *Let $a, b \in \mathbb{Q}_2$; say $a = 2^\alpha u, b = 2^\beta v$, with $u, v \in \mathfrak{u}$. Then*

$$(a,b)_2 = (-1)^{\varepsilon(u)\varepsilon(v)+\alpha\omega(v)+\beta\omega(u)}.$$

Proof. As in part I it is enough to check the formula when $\alpha, \beta \in \{0,1\}$. (I) We begin with the case $\alpha = \beta = 0$, in which we will see that, unlike the case when $p \neq 2$, the 2-adic Hilbert symbol is no longer trivial on $\mathfrak{u} \times \mathfrak{u}$. We must check that for $u, v \in \mathfrak{u}$ we have

$$(u,v) = \begin{cases} 1 & \text{if } u \equiv 1 \pmod 4 \text{ or } v \equiv 1 \pmod 4,\\ -1 & \text{if } u \equiv v \equiv 3 \pmod 4.\end{cases}$$

First suppose $u \equiv 1 \pmod 4$. (The case $v \equiv 1 \pmod 4$ is similar.) If $u \equiv 1 \pmod 8$ (i.e., u has canonical expansion $u = 1 + a_3 2^3 + a_4 2^4 + \cdots$) then $u \in \mathfrak{u}^2$ by the Local Square Theorem, and we are done. On the other hand, if $u \equiv 5 \pmod 8$ then $u + 4v \equiv 1 \pmod 8$, and hence $u + 4v = w^2$ for some $w \in \mathfrak{u}$. But then $u\frac{1}{w^2} + v(\frac{2}{w})^2 = 1$, and hence $(u,v) = 1$.

It remains to show that if $u \equiv v \equiv 3 \pmod 4$ then $(u, v) = -1$. If to the contrary $(u, v) = 1$, then the equation $ux^2 + vy^2 = z^2$ has a nontrivial solution in $\mathbb{Q}_2^3$ and hence has a primitive solution $(x, y, z) \in \mathbb{Z}_2^3$. (So at least one of x, y, z is a 2-adic unit.) Then $x^2 + y^2 + z^2 \equiv 0 \pmod 4$ since $u \equiv v \equiv -1 \pmod 4$, an *impossibility* for a primitive (x, y, z). (Consider the possible canonical expansions of x, y, z.)

(II) Assume $\alpha = 1$ and $\beta = 0$, and the goal is to show that

$$(2u, v) = (-1)^{\varepsilon(u)\varepsilon(v)+\omega(v)}.$$

We begin with a special case, namely showing that $(2, v) = (-1)^{\omega(v)}$; equivalently, that $(2, v) = 1$ if and only if $v \equiv \pm 1 \pmod 8$.

First suppose $(2, v) = 1$. Then there are $x, y, z \in \mathbb{Z}_2$, with y, z units, such that

$$z^2 - 2x^2 - vy^2 = 0.$$

And, since $y^2 \equiv z^2 \equiv 1 \pmod 8$, this gives $v \equiv 1 - 2x^2 \pmod 8$; moreover, $2x^2 \equiv 0$ or $2 \pmod 8$, and therefore $v \equiv \pm 1 \pmod 8$, as desired.

Now for the converse. If $v \equiv 1 \pmod 8$ then v is square (by the Local Square Theorem), and hence $(2, v) = 1$; so we can suppose $v \equiv -1 \pmod 8$, and we must show that the equation $z^2 - 2x^2 - vy^2 = 0$ has a solution in $\mathbb{Q}_2$. Here the Local Square Theorem gives us that $v \in -\mathfrak{u}^2$; so we need to solve $z^2 - 2x^2 + y^2 = 0$ in $\mathbb{Q}_2$. But $(1, 1, 1)$ is an obvious solution, and this completes the proof that $(2, v) = (-1)^{\omega(v)}$.

In part (I) we showed that $(u, v) = (-1)^{\varepsilon(u)\varepsilon(v)}$ for $u, v \in \mathfrak{u}$; therefore, to complete the proof of part (II) it will be enough to show that $(2u, v) = (2, v)(u, v)$. Moveover, from the properties of Hilbert symbols proved earlier in this section, we know this to be the case if $(2, v) = 1$ or $(u, v) = 1$; so we can assume $(2, v) = (u, v) = -1$, and we must show $(2u, v) = 1$.

The special case considered at the start of (II) tells us that $v \equiv 3$ or $5 \pmod 8$, and when combined with part (I) we get that in fact $v \equiv 3 \pmod 8$. And since $(u, v) = -1$ we know $u \equiv 3$ or $-1 \pmod 8$. We want to show that the equation $2ux^2 + vy^2 = 1$ is solvable in $\mathbb{Q}_2$, and by the local square theorem the given congruence restrictions on u and v allow us to assume that

$$u = -1 \text{ and } v = 3 \qquad \text{or} \qquad u = 3 \text{ and } v = -5.$$

In other words, it suffices to check that the equations

$$-2x^2 + 3y^2 = 1 \qquad \text{and} \qquad 6x^2 - 5y^2 = 1$$

are solvable. But they have obvious solutions!

(III) The final case: $\alpha = \beta = 1$, and we must show that

$$(2u, 2v) = (-1)^{\varepsilon(u)\varepsilon(v)+\omega(u)+\omega(v)}.$$

We have

$$(2u, 2v) = (2u, 2v)(2u, -2u) = (2u, -uv) = (-1)^{\varepsilon(u)\varepsilon(-uv)+\omega(-uv)}$$

with the last equality holding by part (II). Now use the fact that ε and ω are homomorphisms along with these observations: $\varepsilon(-1) = 1$, $\omega(-1) = 0$, and $\varepsilon(u)\big(1 + \varepsilon(u)\big) = 0$ for all $u \in \mathfrak{u}$ to reach the conclusion. □

Corollary 4.13. *In $\mathbb{Q}_2$, if $\varepsilon \in \mathfrak{u}$ then $(5, \varepsilon)_2 = 1$ and $(5, 2\varepsilon)_2 = -1$.*

Example 4.14. $(-1, -1)_2 = -1$.

Our formulas for calculating Hilbert symbols lead to the following result, an important tool in calculating Hasse symbols of quadratic spaces, the subject of the next section.

Corollary 4.15. *The Hilbert symbols are bilinear mappings, in the sense that*

$$(ab, c) = (a, c)(b, c) \quad \text{and} \quad (a, bc) = (a, b)(a, c).$$

Moreover these bilinear maps are nondegenerate, in the sense that if a is not a square then there is an element b such that $(a, b) = -1$.

4.2. The Hasse Symbol (and an Alternative)

Now we can use the Hilbert symbol to define a new invariant for quadratic spaces.

Definition 4.16. Let V be a regular n-dimensional quadratic space over $\mathbb{C}, \mathbb{R}$, or some $\mathbb{Q}_p$; say $V \cong \langle a_1, \ldots, a_n \rangle$. Then we define the **Hasse symbol**

$$S_pV = \prod_{i<j} (a_i, a_j)_p$$

if $n > 1$, and define $S_pV = 1$ if $n = 1$. (Here if the field is $\mathbb{C}$ or $\mathbb{R}$, take $p = \infty$, the spot on the field containing the ordinary absolute value.)

Proposition 4.17. *The Hasse symbol S_pV is independent of the choice of orthogonal basis and hence is a well-defined invariant of V.*

Proof. The case $n = 1$ is trivial. As for the case $n = 2$, we have $S_pV = 1$ if and only if there is a representation $1 \to V$, a property independent of the orthogonalization. We will use F to denote the underlying field.

Now suppose $n \geq 3$ and suppose further that $V \cong \langle a_1, \ldots, a_n \rangle$ with respect to basis $\{v_i\}$ and $V \cong \langle b_1, \ldots, b_n \rangle$ with respect to basis $\{w_i\}$. By Witt's Chain Equivalence Theorem we can assume $v_i = w_i$ for all $i \geq 3$.

Then $\langle a_1, a_2\rangle \cong \langle b_1, b_2\rangle$ by Witt's Cancellation Theorem, and hence $a_1a_2 \in b_1b_2F^{*2}$; and by the case $n = 2$ we have $(a_1, a_2)_p = (b_1, b_2)_p$. Then

$$\prod_{i<j}(a_i, a_j)_p = (a_1, a_2)_p \prod_{2<j}(a_1a_2, a_j)_p \prod_{2<i<j}(a_i, a_j)_p = \prod_{i<j}(b_i, b_j)_p$$

as desired. □

Proof of the following proposition is left to the reader.

Proposition 4.18. $S_p(V \perp W) = S_pV \cdot S_pW \cdot (dV, dW)_p$.

Remark 4.19. We will see that dimension, discriminant, and Hasse symbol constitute a complete set of invariants for a quadratic $\mathbb{Q}_p$-space. Having said that, it is interesting to note that there are variations on the Hasse symbol in the literature. For instance, for $V \cong \langle a_1, \ldots, a_n\rangle$ suppose we had defined the Hasse symbol to be

$$S'_pV = \prod_{i\le j}(a_i, a_j)_p.$$

(This is the definition used in [**O'M1**], for example.) Then

$$S'_pV = S_pV \cdot \prod_{i=1}^{n}(a_i, a_i) = S_pV \cdot \prod_{i=1}^{n}(a_i, -1) = S_pV \cdot (dV, -1). \tag{4.1}$$

So S'_pV is again a well-defined invariant. It can happen that two spaces have the same Hasse symbol under one definition but not under the other. For instance, over $\mathbb{Q}_2$ let $V \cong \langle 1\rangle$ and $W \cong \langle -1\rangle$; then $S_pV = S_pW = 1$ while $S'_pV = 1$ and $S'_pW = -1$. In fact it follows from equation (4.1) that for any two spaces V and W we have

$$S'_pV = S'_pW \iff S_pV = S_pW \cdot (dV \cdot dW, -1).$$

Therefore if $dV = dW$ (and, in checking to see whether two given spaces are isometric, verifying that they have the same discriminant is usually the first step) then $S_pV = S_pW$ if and only if $S'_pV = S'_pW$. So, while the two "Hasse" symbols are not the same, they serve the same purpose, and our definition has the advantage of being a product of fewer Hilbert symbols.

4.3. Classification of Quadratic $\mathbb{Q}_p$-Spaces

Lemma 4.20. *Let V be a binary $\mathbb{Q}_p$-space, and let $b \in \mathbb{Q}_p^*$. Then*

$$b \to V \quad \iff \quad (b, -dV)_p = S_pV.$$

Proof. Suppose $V \cong \langle a_1, a_2\rangle$. Then there is a representation $b \to V$ if and only if the equation $a_1x^2 + a_2y^2 = b$ is solvable in $\mathbb{Q}_p$, and this in turn is equivalent to $\left(\frac{a_1}{b}, \frac{a_2}{b}\right)_p = 1$. But

$$\left(\frac{a_1}{b}, \frac{a_2}{b}\right) = (a_1 b, a_2 b) = (a_1, a_2)(b, -a_1 a_2) = S_p V \cdot (b, -dV). \qquad \square$$

Proposition 4.21. *Let V be a regular quadratic $\mathbb{Q}_p$-space of dimension 3. Then*

$$V \text{ is isotropic if and only if } S_p V = (-1, -dV).$$

Proof. Suppose $V \cong \langle a_1, a_2, a_3 \rangle$. Then V is isotropic if and only if $a_1 x^2 + a_2 y^2 = -a_3$ is solvable in $\mathbb{Q}_p$, and this is equivalent to $\left(-\frac{a_1}{a_3}, -\frac{a_2}{a_3}\right)_p = 1$ and then to $1 = (-a_3 a_1, -a_3 a_2)$. Using bilinearity it is now straightforward to see that this latter symbol is equal to $(-1, -dV) \cdot S_p V$, giving the result. $\square$

Definition 4.22. For $a \in \mathbb{Q}_p^*$ and $\varepsilon \in \{\pm 1\}$, we define

$$H_a^\varepsilon = \{x \in \mathbb{Q}_p^* \mid (a, x)_p = \varepsilon\}.$$

Lemma 4.23. *Suppose $H_a^\varepsilon \neq \emptyset$ and $H_{a'}^{\varepsilon'} \neq \emptyset$. Then*

$$H_a^\varepsilon \cap H_{a'}^{\varepsilon'} = \emptyset \iff a \in a' \mathbb{Q}_p^{*2} \text{ and } \varepsilon' = -\varepsilon.$$

Proof. "$\Longleftarrow$" is clear; we prove "$\Longrightarrow$".

If $a \in \mathbb{Q}_p^{*2}$ then $H_a^\varepsilon = \mathbb{Q}_p^*$, in which case since $\emptyset \neq H_{a'}^{\varepsilon'}$ the hypothesis $H_a^\varepsilon \cap H_{a'}^{\varepsilon'} = \emptyset$ can't happen; similarly if $a' \in \mathbb{Q}_p^{*2}$. Therefore without loss of generality we can assume $a, a' \notin \mathbb{Q}_p^2$.

View the Hilbert symbol $(a, \cdot)_p$ as a map $(a, \cdot)_p : \mathbb{Q}_p^*/\mathbb{Q}_p^{*2} \to \{\pm 1\}$. Since $a \notin \mathbb{Q}_p^{*2}$ this is a surjective homomorphism; therefore the kernel, H_a^1, is the union of half the square classes in $\mathbb{Q}_p^*/\mathbb{Q}_p^{*2}$, as is H_a^{-1} (the other coset of the kernel), and hence $H_a^1 \cup H_a^{-1} = \mathbb{Q}_p^*$, a disjoint union; and similarly $H_{a'}^1 \cup H_{a'}^{-1} = \mathbb{Q}_p^*$. Then the hypothesis $H_a^\varepsilon \cap H_{a'}^{\varepsilon'} = \emptyset$ forces the equalities $H_a^{-\varepsilon} = H_{a'}^{\varepsilon'}$ and $H_a^\varepsilon = H_{a'}^{-\varepsilon'}$.

Now suppose $\varepsilon = 1$. (The case $\varepsilon = -1$ is similar.) Then $1 \in H_a^1 = H_{a'}^{-\varepsilon'}$, and therefore $\varepsilon' = -1$ and $H_a^1 = H_{a'}^1$. From this it follows that $(a, x) = (a', x)$ for all $x \in \mathbb{Q}_p^*$, and hence $a' \in a\mathbb{Q}_p^{*2}$, as desired. $\square$

Proposition 4.24. *Let V be a regular $\mathbb{Q}_p$-space of dimension 4. Then V is anisotropic if and only if $dV = 1$ and $S_p V = -(-1, -1)_p$.*

Proof. Write $V \cong \langle a_1, a_2, a_3, a_4 \rangle$. Then V is isotropic if and only if some $x \in \mathbb{Q}_p^*$ is represented by both $\langle a_1, a_2 \rangle$ and $\langle -a_3, -a_4 \rangle$; equivalently, the spaces $\langle -x, a_1, a_2 \rangle$ and $\langle -x, -a_3, -a_4 \rangle$ are isotropic. But, from our characterization of 3-dimensional isotropic spaces in the preceding proposition, this in turn is equivalent to the existence of $x \in \mathbb{Q}_p^*$ such that

$$(x, -a_1a_2) = (a_1, a_2) \qquad \text{and} \qquad (x, -a_3a_4) = (-a_3, -a_4)$$

simultaneously. In the notation of the lemma, the set of solutions of the left equation is $H_{-a_1a_2}^{(a_1,a_2)}$ and the solution set for the right equation is $H_{-a_3a_4}^{(-a_3,-a_4)}$. Now, each of these sets is nonempty, since $a_1 \in H_{-a_1a_2}^{(a_1,a_2)}$ and $-a_3 \in H_{-a_3a_4}^{(-a_3,-a_4)}$. Moreover, anisotropy of V is equivalent to nonexistence of an element x as described above; that is,

$$H_{-a_1a_2}^{(a_1,a_2)} \cap H_{-a_3a_4}^{(-a_3,-a_4)} = \emptyset.$$

But by the lemma, this is equivalent to the following two conditions:

$$a_1a_2 \in a_3a_4\mathbb{Q}_p^{*2} \quad \text{and} \quad (a_1, a_2) = -(-a_3, -a_4).$$

The condition $a_1a_2 \in a_3a_4\mathbb{Q}_p^{*2}$ is equivalent to $dV = 1$, and in this case

$$\begin{aligned} S_pV &= \prod_{1\le i<j\le 4} (a_i, a_j) = (a_1, a_2)(a_1a_2, a_3a_4)(a_3, a_4) \\ &= (a_1, a_2)(-1, a_3a_4)(a_3, a_4) = (a_1, a_2)(-1, -1)(-a_3, -a_4); \end{aligned}$$

and from the condition $(a_1, a_2) = -(-a_3, -a_4)$ this is equivalent to $S_pV = -(-1, -1)_p$, as claimed. □

Remark and Notation 4.25. We have seen that in $\mathbb{Z}_p$ we have

$$(\mathfrak{u} : \mathfrak{u}^2) = \begin{cases} 2 & \text{if } p \neq 2, \\ 4 & \text{if } p = 2; \end{cases}$$

and that $\{1, 3, 5, 7\}$ is a representative set for $\mathfrak{u}/\mathfrak{u}^2$ if $p = 2$. So when $p \neq 2$ any two nonsquares in $\mathfrak{u} = \mathbb{Z}_p^*$ are in the same square class (i.e., coset of $\mathfrak{u}/\mathfrak{u}^2$); whereas when $p = 2$ the numbers $3, 5, 7$ represent the distinct square classes of nonsquare units. Every 2-adic unit has canonical representation $\varepsilon = 1 + \sum_{i\ge1} a_i 2^i$ with $a_i \in \{0, 1\}$. The square units all have $a_1 = a_2 = 0$; indeed, by the local square theorem the square class of any unit can be determined by comparing its coefficients a_1, a_2 with the corresponding coefficients of 1, 3, 5, and 7. Among the nonsquare units $\{3, 5, 7\}$, the number 5 is closest to being square, in the sense that it is the only one in whose 2-adic expansion $a_1 = 0$. So as ε varies through nonsquare units, the value $|1 - \varepsilon|_2$ is minimal when $\varepsilon = 5$. [In the more general treatment of quadratic forms over arbitrary local fields in O'Meara's book [**O'M1**], the notion of "quadratic defect" is developed, and in that terminology the number 5 is a nonsquare in $\mathbb{Z}_2^*$ having minimal quadratic defect.] Indeed, from the Local

Square Theorem, any nonsquare 2-adic unit ε in whose expansion $a_1 = 0$ is in $5\mathfrak{u}^2$. In what follows, we put

$$\Delta = \begin{cases} \text{any nonsquare } p\text{-adic unit if } p \neq 2, \\ 5 \text{ if } p = 2. \end{cases}$$

Corollary 4.26. *The 4-dimensional $\mathbb{Q}_p$-space $\langle 1, -\Delta, p, -p\Delta\rangle$ is anisotropic. Every regular $\mathbb{Q}_p$-space of dimension 3 represents all but at most one element of $\mathbb{Q}_p^*/\mathbb{Q}_p^{*2}$.*

Proof. The first statement follows immediately from the preceding proposition and our assorted formulas for evaluating Hilbert symbols. For the second statement, note that if $\dim V = 3$ and $b \not\to V$ then $V \perp \langle -b\rangle$ is anisotropic, and hence $-b = dV$ by the proposition. □

We will soon see that the isometry class given in the corollary is the *only* isometry class of anisotropic 4-dimensional $\mathbb{Q}_p$-spaces.

Theorem 4.27. *Every quadratic $\mathbb{Q}_p$-space of dimension $n \geq 5$ is isotropic, and hence every $\mathbb{Q}_p$-space of dimension $n \geq 4$ is universal.*

Proof. For the first part, it suffices to show that a regular $\mathbb{Q}_p$-space V of dimension 5 must be isotropic. If $p \neq 2$ this is an immediate consequence of Theorem 4.3. But in fact for all p we know from the preceding corollary that V represents some nonzero element $a \neq dV$, and therefore $V \cong \langle a\rangle \perp W$, with $\dim W = 4$ and $dW \neq 1$. Thus W is isotropic and, *a fortiori*, so is V.

The second claim now follows from Proposition 2.27. □

Remark 4.28. The preceding theorem holds over any local field of characteristic not 2 in place of $\mathbb{Q}_p$. The nondyadic case follows from Theorem 4.3, but the general dyadic case—which we will not need—requires considerably more work. See [**O'M1**], Chapter 6, for details.

Theorem 4.29. *Suppose V and W are regular quadratic spaces over $\mathbb{Q}_p$. Then*

$$V \cong W \iff \dim V = \dim W,\ dV = dW,\ \textit{and}\ S_pV = S_pW.$$

Proof. Only the sufficiency requires proof, and we argue by induction on the common dimension n. The case $n = 1$ is trivial, so assume that $n > 1$ and that the result has been demonstrated for spaces of lower dimension. We write $\delta = dV = dW$.

We claim that V and W represent some common nonzero element b. For $n = 2$ this is clear if $\delta = -1$; and if $\delta \neq -1$ then $(b, -\delta)_p = S_pV$ for all elements b in half the square classes of $\mathbb{Q}_p^*$ (i.e., half the cosets of $\mathbb{Q}_p^*/\mathbb{Q}_p^{*2}$), and hence such b are represented by both V and W, by our earlier lemma on binary spaces. And if $n \geq 3$ then the space $V \perp W^{-1}$ is isotropic, from which the claim follows.

There are orthogonal splittings $V \cong \langle b \rangle \perp V_1$ and $W \cong \langle b \rangle \perp W_1$, say. Here the $(n-1)$-dimensional subspaces V_1, W_1 satisfy $dV_1 = dW_1 = \delta b^{-1}$ and $S_pV_1 = S_pW_1 = S_pV \cdot (b, -\delta)$, so the induction hypothesis finishes the proof. □

Corollary 4.30. *The isometry class of $\langle 1, -\Delta, p, -p\Delta \rangle$ is the only isometry class of 4-dimensional anisotropic $\mathbb{Q}_p$-spaces.*

Recall from Chapter 2 that for a field F, the value $s(F)$ is the smallest number of squares required to represent -1 as a sum of squares in F.

Theorem 4.31. $s(\mathbb{Q}_p) = \begin{cases} 1 & \text{if } p \equiv 1 \pmod 4, \\ 2 & \text{if } p \equiv 3 \pmod 4, \\ 4 & \text{if } p = 2. \end{cases}$

Proof. It is immediate from Corollary 3.53 that $s(\mathbb{Q}_p) = 1$ if $p \equiv 1 \pmod 4$ and that $s(\mathbb{Q}_p) \geq 2$ if $p \equiv 3 \pmod 4$. To finish the case in which $p \equiv 3 \pmod 4$, we need only show that for these primes p the $\mathbb{Q}_p$-space $\langle 1, 1, 1 \rangle$ is isotropic; equivalently, that

$$\langle 1, 1, 1 \rangle \cong \langle 1, -1, -1 \rangle.$$

But this is clear from Theorem 4.29 in view of the triviality of the Hilbert symbol on p-adic units when p is odd.

Finally, suppose $p = 2$. We know from Theorem 4.27 that $s(\mathbb{Q}_2) \leq 4$, and by Proposition 2.30 we know that $s(\mathbb{Q}_2) \neq 3$; so to finish the proof it suffices to check that over $\mathbb{Q}_2$ we have

$$-1 \not\longrightarrow \langle 1, 1 \rangle;$$

equivalently, that the $\mathbb{Q}_2$-space $\langle 1, 1, 1 \rangle$ is anisotropic. If it were isotropic, there would be an isometry

$$\langle 1, 1, 1 \rangle \cong \langle 1, -1, -1 \rangle.$$

But these two spaces have different Hasse symbols, by Example 4.14, so there is no such isometry. □

Having shown that the isometry class of a regular quadratic $\mathbb{Q}_p$-space V is determined by its dimension, a natural number; its discriminant, the square class of some element in $\mathbb{Q}_p^*$; and its Hasse symbol, ± 1; it is now reasonable to ask whether every element of $\mathbb{N} \times \mathbb{Q}_p^* \times \{1, -1\}$ is actually an invariant triple for some $\mathbb{Q}_p$-space. In particular, to what extent are the invariants independent? The following result answers this question.

Theorem 4.32. *Let* $(n_0, d_0, s_0) \in \mathbb{N} \times \mathbb{Q}_p^* \times \{1, -1\}$. *Then there exists a regular quadratic* $\mathbb{Q}_p$*-space* V *satisfying* $(\dim V, dV, S_pV) = (n_0, d_0, s_0)$*, but for the following two exceptional cases:* $n_0 = 1$ *and* $s_0 = -1$*;* $(n_0, d_0, s_0) = (2, -1, -1)$.

Proof. The stated exceptions clearly are exceptions. Suppose $n_0 = 2$ and $d_0 \neq -1$. A regular binary space V with $dV = d_0$ has the form $V \cong \langle a, ad_0 \rangle$ for some $a \in \mathbb{Q}_p^*$. Then $S_pV = (a, -d_0)$; hence, since $-d_0 \notin \mathbb{Q}_p^{*2}$, the Hasse symbol S_pV can be made 1 or -1 by choosing a appropriately.

If $n_0 = 3$ first choose $a \in \mathbb{Q}_p^*$ such that $-ad_0 \notin \mathbb{Q}_p^{*2}$. Then take $V \cong \langle a \rangle \perp W$, with $\dim W = 2$ and $dW = ad_0$. Then $dV = d_0$ and $S_pV = (a, ad_0) \cdot S_pW$. By the case $n_0 = 2$, the value of S_pW can be chosen to make $S_pV = s_0$.

Finally, if $n_0 \geq 4$ let U be any 3-dimensional space satisfying $dU = d_0$ and $S_pU = s_0$, and then take $V \cong U \perp \langle 1, \ldots, 1 \rangle$. □

Recall from Chapter 2 that for a field F of characteristic not 2 the **Witt ring** $W(F)$ has as its elements the isometry classes of anisotropic quadratic F-spaces. Let's determine the cardinality $\#W(\mathbb{Q}_p)$ when $p \neq 2$ using our classification theory. We will count anisotropic spaces "up to isometry." For dimension $n = 0$ the zero space is the only space, and it is anisotropic. For $n = 1$ there are four regular spaces, all anisotropic. For $n = 2$ there are six regular spaces other than the hyperbolic plane. When $n = 3$ there are eight regular spaces in all, four of which have the form $\langle 1, -1, \alpha \rangle$, the other four being anisotropic. Finally, when $n = 4$ there are eight regular spaces altogether, but only one is anisotropic. In all, there are sixteen anisotropic spaces: $\#W(\mathbb{Q}_p) = 16$. The reader can easily use the same reasoning to conclude that $\#W(\mathbb{Q}_2) = 32$.

4.4. Hermitian Spaces over Quadratic Extensions of $\mathbb{Q}_p$

Let E be a quadratic extension of $\mathbb{Q}_p$, with $*$ the nontrivial automorphism of E having $\mathbb{Q}_p$ as its fixed field. Write $E = \mathbb{Q}_p(i)$ with $i^2 = -\lambda \in \mathbb{Q}_p$ and hence $N_{E/\mathbb{Q}_p}(i) = \lambda$. Because we have proved in the present chapter that $u(\mathbb{Q}_p) = 4$, it follows from Chapter 2 that the dimension and discriminant of a Hermitian E-space V constitute a full set of invariants for the isometry class of V. The space V is a $2n$-dimensional quadratic $\mathbb{Q}_p$-space with respect to the induced quadratic form, and we will denote this quadratic space by $V_{\mathbb{Q}_p}$. Consider the invariants of $V_{\mathbb{Q}_p}$. The discriminant $dV_{\mathbb{Q}_p}$ is completely determined by n and dV; therefore, from Jacobson's Theorem and the fact that dimension, discriminant, and Hasse symbol constitute the full system of isometry invariants for quadratic $\mathbb{Q}_p$-spaces, it must be the case that all the $2n$-dimensional quadratic $\mathbb{Q}_p$-spaces whose quadratic structure is induced from Hermitian E-spaces of a given discriminant have the same Hasse symbol. What is that Hasse symbol? Answering this question will provide some useful experience with Hilbert symbol manipulations.

If $V \cong \langle a_1, \ldots, a_n\rangle$ over E with respect to $\{v_1, \ldots, v_n\}$, then

$$V_{\mathbb{Q}_p} \cong \langle a_1, \ldots, a_n\rangle \perp \langle a_1, \ldots, a_n\rangle^\lambda$$

with respect to $\{v_1, \ldots, v_n, iv_1, \ldots, iv_n\}$, and therefore $dV_{\mathbb{Q}_p} = \lambda^n$. To compute the Hasse symbol, the following lemma will be useful. The proof is an exercise for the reader. (Suggestion: Use induction on n.)

Lemma 4.33. *Let W be a quadratic $\mathbb{Q}_p$-space of dimension m, and let $\beta \in \mathbb{Q}_p^*$. Then*

$$S_pW^\beta = S_pW \cdot \left(\beta, (-1)^{\binom{m}{2}}(dW)^{m-1}\right)_p.$$

This lemma in combination with our earlier formula

$$S_p(X \perp Y) = S_pX \cdot S_pY \cdot (dX, dY)_p$$

gives the $\mathbb{Q}_p$-space computation

$$\begin{aligned}
S_pV_{\mathbb{Q}_p} &= S_p\langle a_1, \ldots, a_n\rangle \cdot S_p\langle a_1, \ldots, a_n\rangle^\lambda \cdot \left(\prod a_i, \lambda^n \prod a_i\right)_p \\
&= \left(\prod_{i<j}(a_i, a_j)_p\right)^2 \cdot \left(\lambda, (-1)^{\binom{n}{2}}\left(\prod_i a_i\right)^{n-1}\right)_p \cdot \left(\prod_i a_i, -\lambda^n\right)_p \\
&= \left(\lambda, (-1)^{\binom{n}{2}}\left(\prod_i a_i\right)^{n-1}\right)_p \cdot \left(\prod_i a_i, -\lambda^n\right)_p \\
&= \left(\lambda, (-1)^{\binom{n}{2}}\left(\prod_i a_i\right)^{n-1}\right)_p \cdot \left(-1, \prod_i a_i\right)_p \cdot \left(\lambda, \prod_i a_i\right)_p^n \\
&= \left(\lambda, (-1)^{\binom{n}{2}}\prod_i a_i\right)_p \cdot \left(-1, \prod_i a_i\right)_p \\
&= \left(\lambda, (-1)^{\binom{n}{2}}\right)_p \cdot \left(-\lambda, \prod_i a_i\right)_p \\
&= \left(-1, (-1)^{\binom{n}{2}}\right)_p \cdot \left(-\lambda, (-1)^{\binom{n}{2}}\prod_i a_i\right)_p \\
&= \pm\left(-\lambda, (-1)^{\binom{n}{2}}\prod_i a_i\right)_p
\end{aligned}$$

with a leading "+" in the last line if $p \neq 2$, or if $p = 2$ and $n \equiv 0, 1 \pmod 4$; and a leading "−" if $p = 2$ and $n \equiv 2, 3 \pmod 4$.

Recall that discriminants of Hermitian spaces are *norm* classes—thus, in the present context, elements of $\mathbb{Q}_p^*/N_{E/\mathbb{Q}_p}E^*$—while discriminants of quadratic spaces are square classes. So if also $V \cong \langle b_1, \ldots, b_n \rangle$ then $\prod_i b_i = \mu \prod_i a_i$ for some $\mu \in N_{E/\mathbb{Q}_p}E^*$, in which case the value of $S_p V_{\mathbb{Q}_p}$ gets multiplied by the Hilbert symbol $(-\lambda, \mu)_p$. Thus in order for our discussion to make sense, it had better be the case that $(-\lambda, \mu)_p = 1$, so we now check that this is actually so. Suppose $\mu = N(r + si) = r^2 + \lambda s^2$, with $r, s \in \mathbb{Q}_p$. If $r = 0$ then the result is clear. And if $r \neq 0$ then $\mu \left(\frac{1}{r}\right)^2 - \lambda \left(\frac{s}{r}\right)^2 = 1$, again giving $(-\lambda, \mu)_p = 1$, as required.

Exercises

(1) Let p be an odd prime number, and let $\alpha \in \mathbb{Q}^*$. Show that the calculation of the Hilbert symbol $(\alpha, p)_p$ is equal to a product of symbols of the form $(q, p)_p$ with q prime and less than p, possibly with an additional factor of $(-1, p)_p$.

(2) Find rational integers a, b with the property that $(a, b)_3 = (a, b)_5 = -1$ while $(a, b)_p = 1$ for all primes $p \geq 7$.

(3) Begin work on this problem by checking that the material in this chapter through "Computation of Hilbert Symbols, Part I" on the Hilbert symbol $(\cdot, \cdot)_p$, with p an odd prime number, is actually true more generally in any nondyadic local field, assuming that p is now interpreted to mean a prime element of the valuation ring. Having done that, turn your attention to the local field $\mathbb{F}_{11}(x)_\mathfrak{p}$, with $\mathfrak{p}$ the $(x+2)$-adic spot, and do the following:
(i) Obtain the canonical $(x+2)$-adic expansion of the polynomial

$$f(x) = 5x^3 + 10x^2 + 9x + 10$$

using $\mathbb{F}_{11}$ as the representative set of the residue class field.
(ii) Compute the Hilbert symbol $\big(f(x), x+2\big)_\mathfrak{p}$.

(4) Show that every quadratic $\mathbb{Q}_p$-space of dimension $n \geq 4$ has an orthogonal component of dimension $n - 3$ that has an orthonormal basis.

(5) Find all the primes p (if there are any) such that the quadratic $\mathbb{Q}_p$-space $\langle 1, 1, 1, 1 \rangle$ is anisotropic.

(6) Let V be a binary quadratic space over the field $\mathbb{Q}_{13}$ of 13-adic numbers, with $dV = 26$, and let $\alpha \in \mathbb{Q}_{13}^*$. Prove that either $\alpha \to V$ or $8\alpha \to V$, but not both.

(7) (i) Verify the formula for $S_p(V \perp W)$ stated in the text.
(ii) Compute $S_p\langle\underbrace{-1,\ldots,-1}_{n}\rangle$.
(iii) Compute $S_p(H \otimes V)$ if H is a hyperbolic plane and V is a regular n-dimensional $\mathbb{Q}_p$-space.

(8) (i) Prove that if V is an n-dimensional $\mathbb{Q}_p$-space and $\lambda \in \mathbb{Q}_p^*$, then

$$S_pV^\lambda = \left(\lambda, (-1)^{\frac{n(n-1)}{2}}(dV)^{n+1}\right) \cdot S_pV.$$

(ii) Suppose $\dim V = 2$ and $\dim W = n$. Obtain a formula for the Hasse symbol $S_p(V \otimes W)$. Express your answer as a product of at most three Hilbert symbols.

(9) (i) Show that there are infinitely many primes p with the property that the Hilbert symbol is constant on similarity classes in the Witt ring. That is, $S_pU = S_pV$ whenever U and V represent the same element of the Witt ring $W(\mathbb{Q}_p)$.
(ii) Give an explicit example of spaces U and V over some $\mathbb{Q}_p$ such that $S_pU \neq S_pV$ even though $[U] = [V]$ in $W(\mathbb{Q}_p)$.

Chapter 5

Quadratic Spaces over $\mathbb{Q}$

In Chapter 4 we solved the classification and representation problems for quadratic spaces over the p-adic number fields. The central goal of the present chapter is to reduce the corresponding problems for rational quadratic spaces—problems that we will refer to as "global"—to the p-adic ("local") cases. (This is the "local–global" approach to the rational theory.) The aggressive reader will immediately complain that this is no cause for exultation, since we will have reduced one hard global problem to infinitely many easier local problems. Fortunately it will be clear that for each global problem only finitely many associated local problems need to be considered.

5.1. The Product Formula and Hilbert's Reciprocity Law

The following lemma is an immediate consequence of the definition of the p-adic valuation $|\cdot|_p$. Despite its triviality it gets a big-time name because it has a more general formulation in terms of valuations on arbitrary global fields.

Lemma 5.1 (Product Formula). *Let $a \in \mathbb{Q}^*$. Then $|a|_p = 1$ for almost all primes p, and*

$$\prod_{\substack{p \\ \text{(including } \infty)}} |a|_p = 1.$$

Like the Product Formula, the following result holds in the wider context of arbitrary global fields. Again the stated product should be understood to be over all primes, including ∞.

Theorem 5.2 (Hilbert's Reciprocity Law). *Let $a, b \in \mathbb{Q}^*$. Then $(a,b)_p = 1$ for almost all p, and*

$$\prod_p (a,b)_p = 1.$$

Proof. Without loss of generality we can assume that $a, b \in \mathbb{Z}$. Indeed, using the bilinearity of Hilbert symbols, we can assume that

$$a, b \in \{-1\} \cup \{p \mid p \text{ is prime}\}.$$

The theorem's first assertion holds because $|a|_p = |b|_p = 1$ for almost all p, and the Hilbert symbol $(\cdot,\cdot)_p$ is trivial on p-adic units when p is odd.

Case (i): $a = b = -1$. Here we have $(-1,-1)_\infty = (-1,-1)_2 = -1$, while $(-1,-1)_p = 1$ for all odd p.

Case (ii): a is prime and $b = -1$. Since trivially $(-1,2)_p = 1$ for all p, we can assume a is an odd prime. Then $(-1,a)_p = 1$ for all $p \notin \{2,a\}$, while $(-1,a)_2 = (-1,a)_a = (-1)^{\frac{a-1}{2}}$, giving the result.

Case (iii): a and b are both prime. If $a = b$ then $(a,b)_p = (a,-1)_p$ for all p, and this is covered by the previous case; so we can assume $a \neq b$. If $b = 2$ then $(a,2)_p = 1$ for all $p \notin \{2,a\}$, while

$$(a,2)_p = \begin{cases} (-1)^{\frac{a^2-1}{8}} & \text{if } p = 2, \\ \left(\dfrac{2}{a}\right) = (-1)^{\frac{a^2-1}{8}} & \text{if } p = a. \end{cases}$$

Case (iv): a and b are distinct odd primes. In this case $(a,b)_p = 1$ if $p \notin \{2,a,b\}$, while

$$(a,b)_p = \begin{cases} (-1)^{\frac{a-1}{2}\frac{b-1}{2}} & \text{if } p = 2, \\ \left(\dfrac{a}{b}\right) & \text{if } p = b, \\ \left(\dfrac{b}{a}\right) & \text{if } p = a. \end{cases}$$

The result now follows from Gauss's Law of Quadratic Reciprocity. □

Corollary 5.3. *Let V be a regular quadratic $\mathbb{Q}$-space of dimension ≥ 1. Then*

$$\prod_p S_p V = 1.$$

5.2. Extension of the Scalar Field

Given n-dimensional quadratic spaces V and W over a field F, with respective Gram matrices $A, C \in M_n(F)$, it may happen that determining whether there is a matrix congruence $A \underset{F}{\cong} C$ is difficult while for some extension field $E \supset F$ checking whether $A \underset{E}{\cong} C$ is relatively easy. In particular, since $GL_n(F)$ is a subgroup of $GL_n(E)$, if $A \underset{E}{\ncong} C$ then $A \underset{F}{\ncong} C$; while if we can show that $A \underset{E}{\cong} C$, perhaps we can use this as evidence toward proving that $A \underset{F}{\cong} C$ and hence that $V \cong W$. We need to do some work in order to make all this coherent, because V and W may not be vector spaces over E; or, if they are, they may have dimension less than n over E, and hence A and C would no longer be the relevant Gram matrices.

The formal algebraic structure that extends the F-space V to an E-space of the same dimension is the **tensor product** $V \otimes_F E$, which we will write more briefly as V_E. This is an n-dimensional vector space over E with the property that, once the vectors in V are suitably identified with vectors in V_E, every basis of V over F is also a basis of V_E over E. Readers unfamiliar with the tensor product concept can construct V_E as follows. Take any n-dimensional vector space V' over E that is disjoint from V. Fix bases $\mathbb{B} = \{v_1, \ldots, v_n\}$ for V over F and $\mathbb{B}' = \{v'_1, \ldots, v'_n\}$ for V' over E. Then, with respect to these bases, the elements of V and V' are represented by n-tuples in F^n and E^n respectively. So in a natural way we can view V as a subset of V'. (Really we are viewing the elements $\sum_i \alpha_i v_i$ and $\sum_i \alpha_i v'_i$ as identical if $(\alpha_1, \ldots, \alpha_n) \in F^n$.) Once we do this we give the resulting vector space over E the label V_E. Notice that now the basis $\mathbb{B}$ for V over F is also a basis for V_E over E, and from this it follows that the same holds for *every* basis of V.

5.3. Local to Global: The Hasse–Minkowski Theorem

Given an n-dimensional vector space V over $\mathbb{Q}$ and a prime p (possibly $p = \infty$), we will write V_p instead of $V_{\mathbb{Q}_p}$ and call V_p the **localization** of V at (or with respect to) p. If V is actually a quadratic $\mathbb{Q}$-space, we will write S_pV for the Hasse symbol S_pV_p.

The heart of this chapter's local-global strategy will be a demonstration that a given quadratic space over $\mathbb{Q}$ is simultaneously "close enough" to its localizations to imitate their behavior. And for this purpose we will first need the following two important theorems on simultaneous approximation of elements in $\mathbb{Q}$'s localizations by elements of $\mathbb{Q}$.

Theorem 5.4 (Weak Approximation Theorem). *Let T be a finite set of primes, possibly including ∞. Given elements $\alpha_p \in \mathbb{Q}_p$ for each $p \in T$ and a real number $\varepsilon > 0$, there exists a number $\alpha \in \mathbb{Q}$ such that $|\alpha - \alpha_p|_p < \varepsilon$ for all $p \in T$.*

Proof. Since $\mathbb{Q}$ is dense in each $\mathbb{Q}_p$, we can assume that $\#T \geq 2$; say $T = \{p_1, \ldots, p_t\}$.

(I) We first claim that there is an element $\beta \in \mathbb{Q}$ such that $|\beta|_{p_1} > 1$ and $|\beta|_{p_i} < 1$ for $2 \leq i \leq t$. If $\infty \notin T$, choose $\beta = \frac{p_2 \cdots p_t}{p_1}$. If $\infty \in T$ and $\infty = p_1$, put $\beta = p_2 \cdots p_t$. Finally, if $\infty \in T - \{p_1\}$, say $\infty = p_2$, put $\beta = \frac{p_3 \cdots p_t}{K p_1}$, with K any integer greater than $\frac{p_3 \cdots p_t}{p_1}$ that is relatively prime to $p_3 \cdots p_t$.

(II) By part (I), for each $j \in \{1, \ldots, t\}$ there is an element $\beta_j \in \mathbb{Q}$ such that $|\beta_j|_{p_j} > 1$ and $|\beta_j|_{p_i} < 1$ for $i \neq j$. Then

$$\lim_{n\to\infty} \frac{\beta_j^n}{1+\beta_j^n} = \begin{cases} 1 \text{ with respect to } | \quad |_{p_j}, \\ 0 \text{ with respect to } | \quad |_{p_i}, \text{ for } i \neq j. \end{cases}$$

Now define

$$c_n = \sum_{j=1}^{t} \frac{\alpha_j \beta_j^n}{1+\beta_j^n}.$$

Then for $1 \leq j \leq t$ we have $\lim_{n\to\infty} c_n = \alpha_j$ with respect to $| \quad |_{p_j}$. Therefore choosing $\alpha = c_n$ with n sufficiently large does the job. □

Theorem 5.5 (Strong Approximation Theorem). *Let T be a finite set of primes not including ∞. Given elements $z_p \in \mathbb{Z}_p$ for each $p \in T$ and a real number $\varepsilon > 0$, there exists an integer $z \in \mathbb{Z}$ such that $|z - z_p|_p < \varepsilon$ for all $p \in T$.*

Proof. Since $\mathbb{Z}$ is dense in $\mathbb{Z}_p$ with respect to $|\cdot|_p$, for every $p \in T$ there exists $z_p' \in \mathbb{Z}$ such that $|z_p' - z_p|_p < \varepsilon$. (Briefly:

$$\Bigg|\underbrace{\sum_{i=0}^{t} a_i p^i}_{z_p'} - \underbrace{\sum_{i=0}^{\infty} a_i p^i}_{z_p}\Bigg|_p < \varepsilon$$

if t is sufficiently large.) Now choose $\nu \in \mathbb{N}$ large enough so that $\frac{1}{p^\nu} < \varepsilon$ for all $p \in T$. By the Chinese Remainder Theorem there is a rational integer z such that $z \equiv z_p' \pmod{p^\nu}$ for all $p \in T$, and therefore $|z - z_p'|_p < \varepsilon$ for all $p \in T$. The result now follows by applying the strong triangle inequality for each $|\cdot|_p$. (Here we use the assumption that $\infty \notin T$.) □

Why is "strong" approximation stronger than "weak" approximation? In weak approximation a rational number α is produced that satisfies a finite collection of approximation criteria, with no bound imposed on $|\alpha|_p$ for $p \notin T$. But because the element z produced by strong approximation is an integer, we know $|z|_p \leq 1$ for all $p \notin T \cup \{\infty\}$, and therein—that is, in the control of the p-adic size of z at those primes not in the set T of primes at which the approximations take place—lies its strength. In the following lemma, while the element t produced need not be an integer, we do have *very* tight control of its p-adic size $|t|_p$ with respect to most p; in fact $|t|_p$ is out of control for at most two values of p.

At a crucial moment in the following proof (Question: Isn't *every* moment of a proof "crucial"?) we will need to invoke Dirichlet's theorem (from 1837) on primes in an arithmetic progression, which states, *If a and b are relatively prime positive integers, then the arithmetic progression $a, a+b, a+2b, \ldots$ contains infinitely many prime numbers.*

Lemma 5.6. *Let $\mathbb{P}$ be a finite set of primes, possibly including ∞. An element $t_p \in \mathbb{Q}_p$ is given for each $p \in \mathbb{P}$. Then there is an element $t \in \mathbb{Q}$ such that*

(i) *$t \in t_p\mathbb{Q}_p^{*2}$ for each $p \in \mathbb{P}$;*

(ii) *$|t|_p = 1$ for all other finite primes p, but for at most one exceptional prime p_0.*

Proof. For each finite prime $p \in \mathbb{P}$, let $\nu_p = \operatorname{ord}_p t_p$, and then define

$$\beta = \pm \prod_{p \in \mathbb{P}-\{\infty\}} p^{\nu_p} \in \mathbb{Q},$$

where the sign of β is the sign of t_∞. (If $\infty \notin \mathbb{P}$, choose the sign of β arbitrarily.) So

$$|\beta|_p = \begin{cases} |t_p|_p & \text{for all finite } p \in \mathbb{P}, \\ 1 & \text{for all finite } p \notin \mathbb{P}. \end{cases}$$

Therefore for all $p \in \mathbb{P} - \infty$ we have $\beta = \varepsilon_p t_p$ for some $\varepsilon_p \in \mathfrak{u}_p$. Now by the Strong Approximation Theorem there exists $z \in \mathbb{Z}$ such that $z \equiv \varepsilon_p \pmod{8p}$ for all $p \in \mathbb{P}-\{\infty\}$. Hence by the Local Square Theorem $z \in \varepsilon_p\mathbb{Q}_p^{*2}$ for all $p \in \mathbb{P} - \{\infty\}$.

Now, the integers z and $8\prod_{p\in\mathbb{P}-\{\infty\}} p$ are relatively prime. It therefore follows from Dirichlet's theorem on primes in an arithmetic progession that there is a prime number p_0 satisfying the congruence $p_0 \equiv z \pmod{8\prod_{p\in\mathbb{P}-\{\infty\}} p}$. Again by the Local Square Theorem, we have $p_0 \in z\mathbb{Q}_p^{*2}$ for all $p \in \mathbb{P} - \{\infty\}$. Now define $t = p_0\beta$, and check that t has the desired properties. $\square$

The following theorem is central to the theory of quadratic forms. Our guide for this presentation is the book *Rational Quadratic Forms* by Cassels [**C2**].

Theorem 5.7 (Hasse–Minkowski Theorem). *Let V be a regular quadratic space over the field $\mathbb{Q}$ of rational numbers. Then V is isotropic if and only if V_p is isotropic for all p.*

Proof. The necessity is clear, since $V \subseteq V_p$ for all p. So assume V_p is isotropic for all p, and we will show V to be isotropic by induction on $n = \dim V$.

The case $n = 1$ does not occur, since when $n = 1$ isotropy and regularity are incompatible.

Suppose $n = 2$, and let $\alpha = dV$. From the hypothesis we know $\alpha \in -\mathbb{Q}_p^{*2}$ for all p. Therefore $\alpha < 0$ and $\operatorname{ord}_p \alpha \equiv 0 \pmod 2$ for all $p \neq \infty$. Therefore $\alpha \in -\mathbb{Q}^{*2}$, and so V is a hyperbolic plane.

If $n = 3$, without loss of generality we can suppose (by scaling the space if necessary) that $V \cong \langle 1, -a, -b\rangle$, with a and b square-free integers and $|a| \leq |b|$; and within this case we will argue by induction on $m = |a| + |b|$. The first case to occur is $m = 2$, when $V \cong \langle 1, \pm 1, \pm 1\rangle$. But $V \not\cong \langle 1, 1, 1\rangle$, since V_∞ is isotropic, hence the result.

Now assume that $m \geq 3$ and that the result has been proved for spaces $\langle 1, -c, -d\rangle$ with $|c| + |d| < m$. We are considering $V \cong \langle 1, -a, -b\rangle$, with $|a| \leq |b|$ and $|b| \geq 2$. Say $b = \pm p_1 \cdots p_k$, with the p_i distinct primes.

We claim that the congruence $a \equiv c^2 \pmod b$ is solvable for some integer c. First note that it is a straightforward exercise to check that there is a ring isomorphism

$$\mathbb{Z}/b\mathbb{Z} \approx \prod_{i=1}^{k} \mathbb{Z}/p_i\mathbb{Z}$$

and from this exercise it follows that to prove the claim it suffices to show that each of the congruences $a \equiv c^2 \pmod{p_i}$ is solvable. So fix $p = p_i$ for some i, and we want to show $a \equiv c^2 \pmod p$ is solvable. If $p|a$ this is clear (take $c = 0$), so we can assume otherwise, and hence $a \in \mathfrak{u}_p = \mathbb{Z}_p^*$.

Since V_p is isotropic, there is a primitive element $(x, y, z) \in \mathbb{Z}_p^3$ such that $z^2 - ax^2 - by^2 = 0$. We are given that $p|b$ in $\mathbb{Z}_p$ and hence $p|(z^2 - ax^2)$. If we also had $p|x$, primitivity of (x, y, z) would be defied, so we must have $x \in \mathfrak{u}_p$; and then, since $ax^2 \equiv z^2 \pmod p$, we obtain $a \equiv (x^{-1}z)^2 \pmod p$. Recall (from Chapter 3) that $T = \{0, 1, \ldots, p-1\}$ is a complete set of representatives for the residue class field $\mathbb{Z}_p/p\mathbb{Z}_p$. So we can choose $c \in T$ such that $c \equiv x^{-1}z \pmod p$, and this gives $a \equiv c^2 \pmod p$, proving the claim.

It follows that we can write $a + bb' = t^2$, with $t, b' \in \mathbb{Z}$ and $|t| \leq |\frac{b}{2}|$. Thus $bb' = N(t + \sqrt{a})$, where $N : \mathbb{Q}(\sqrt{a}) \to \mathbb{Q}$ is the norm mapping. Moreover, since $b' = b \cdot \left(\frac{1}{b^2}(t^2 - a)\right)$ and every square is a norm, it follows that b' is a norm if and only b is. We leave it to the reader to check that a space $\langle 1, -a, -b\rangle$ is isotropic over a field F if and only if b is in the norm group $N_{F(\sqrt{a})/F}\big(F(\sqrt{a})\big)^*$. So in our present situation we know $V \cong \langle 1, -a, -b\rangle$ is isotropic if and only if $V' \cong \langle 1, -a, -b'\rangle$ is isotropic. But

$$|b'| = \left|\frac{t^2 - a}{b}\right| \leq \left|\frac{t^2}{b}\right| + \left|\frac{a}{b}\right| \leq \frac{|b|}{4} + 1 < |b|.$$

The local spaces V'_p are isotropic for all p since $b' \in N_{\mathbb{Q}_p(\sqrt{a})/\mathbb{Q}_p}\big(\mathbb{Q}_p(\sqrt{a})\big)^*$ at all p. Therefore V' is isotropic, by the induction hypothesis on m, and hence V is isotropic as well. This completes the case $n = 3$.

Now for the case $n = 4$. Assume $V \cong \langle a_1, a_2, a_3, a_4\rangle$, with $a_i \in \mathbb{Z}$ square-free, and let

$$\mathbb{P} = \{\infty\} \cup \{p \mid p | 2a_1a_2a_3a_4\}$$

So for all $p \notin \mathbb{P}$ we have $a_i \in \mathfrak{u}_p$ for $1 \leq i \leq 4$ and hence $(a_i, a_j)_p = 1$.

For all p, since V_p is isotropic there is an element $t_p \in \mathbb{Q}_p^*$ such that $t_p \to \langle a_1, a_2\rangle$ and $-t_p \to \langle a_3, a_4\rangle$ over $\mathbb{Q}_p$. Then, with t and p_0 as in the lemma, the $\mathbb{Q}_p$-spaces $\langle a_1, a_2, -t\rangle$ and $\langle a_3, a_4, t\rangle$ are isotropic for all $p \in \mathbb{P}$. Moreover $\langle a_1, a_2, -t\rangle$ and $\langle a_3, a_4, t\rangle$ are isotropic over $\mathbb{Q}_p$ for all $p \notin \mathbb{P} \cup \{p_0\}$ because $|a_i|_p = |t|_p = 1$ for all such p. In summary, the localizations of these two ternary (i.e., 3-dimensional) spaces are isotropic for all $p \neq p_0$. Then, from Hilbert's Reciprocity Law and our criterion (in Chapter 4) for isotropy of ternary $\mathbb{Q}_p$-spaces, it follows that these spaces are isotropic over p_0 as well, hence isotropic over $\mathbb{Q}$ by the case $n = 3$. Therefore V is isotropic, as we wanted to show.

Finally, suppose $n \geq 5$, and suppose the theorem has been proved for spaces of lower dimension. Write

$$V \cong \underbrace{\langle a_1, a_2\rangle}_{U} \perp \underbrace{\langle a_3, \ldots, a_n\rangle}_{W}.$$

Let $T = \{p \mid W_p \text{ is anisotropic }\}$. Then T is finite; and in fact if $T = \emptyset$ then W is isotropic by the induction hypothesis. Therefore we can suppose $T \neq \emptyset$.

Since V_p is isotropic, for each $p \in T$ there is an element $t_p \in \mathbb{Q}_p^*$ such that $t_p \to U_p$ and $-t_p \to W_p$; say $t_p = a_1\beta_{1p}^2 + a_2\beta_{2p}^2$, with $\beta_{1p}, \beta_{2p} \in \mathbb{Q}_p$. By the Weak Approximation Theorem in combination with the Local Square Theorem, there exist $\beta_1, \beta_2 \in \mathbb{Q}$ sufficiently close to β_{1p}, β_{2p} for all $p \in T$ so that the element $t = a_1\beta_1^2 + a_2\beta_2^2$ satisfies $t \in t_p\mathbb{Q}_p^{*2}$ for all $p \in T$. Then $V = U \perp W \cong \langle s\rangle \perp \langle t\rangle \perp W$ for some $s \in \mathbb{Q}$, and $\big(\langle t\rangle \perp W\big)_p$ is isotropic for all $p \in T$ and hence isotropic for all p (from the definition of T). Therefore $\langle t\rangle \perp W$ is isotropic by the induction hypothesis, and so V is isotropic. □

The following result is also referred to in the literature as the Hasse–Minkowski Theorem.

Corollary 5.8. *Let V and W be regular quadratic spaces over $\mathbb{Q}$. Then*

$$W \longrightarrow V \iff W_p \longrightarrow V_p \quad \textit{for all } p.$$

Proof. Only "$\Longleftarrow$" requires proof.

Suppose $W \cong \langle\alpha\rangle \perp W_1$, with $\alpha \in \mathbb{Q}^*$. Since $W_p \longrightarrow V_p$ for all p, the localizations $\big(V \perp \langle -\alpha\rangle\big)_p$ are isotropic for all p, hence $V \perp \langle -\alpha\rangle$ is isotropic over $\mathbb{Q}$. Therefore $\alpha \longrightarrow V$, and so $V \cong \langle\alpha\rangle \perp V_1$, say. But from the hypothesis, for all p we know $V_p \cong W_p \perp X_p \cong \langle\alpha\rangle \perp W_{1p} \perp X_p$ for some $\mathbb{Q}_p$-space X_p, and hence $V_{1p} \cong W_{1p} \perp X_p$ by Witt's Cancellation Theorem. Now repeat the argument, using W_1 and V_1 instead of W and V. (Of course, strictly speaking we are arguing by induction on $\dim W$.) □

Corollary 5.9. *Let V and W be regular quadratic $\mathbb{Q}$-spaces, with* $\dim V = \dim W$. *Then*

$$W \cong V \iff W_p \cong V_p \quad \textit{for all } p.$$

Proof. This is a special case of the preceding corollary, since a representation of a regular space is injective and therefore (since $\dim V = \dim W$) a surjective isometry. □

Recall that a regular quadratic space over $\mathbb{R}$ or $\mathbb{Q}$ is said to be **indefinite** if it represents both positive and negative numbers. (Over $\mathbb{R}$ this is equivalent to the space being isotropic.)

Corollary 5.10 (Meyer's Theorem, 1884). *A regular indefinite quadratic space over $\mathbb{Q}$ of dimension $n \geq 5$ is isotropic.*

Proof. Since $n \geq 5$ the given space is isotropic over all $\mathbb{Q}_p$ for p finite; and it is isotropic over $\mathbb{Q}_\infty = \mathbb{R}$ from the hypothesis. Now apply the Hasse–Minkowski Theorem. □

Remark 5.11. (i) Determining whether a given quadratic $\mathbb{Q}$-space V of dimension $n \leq 2$ is isotropic is just a matter of looking at the discriminant dV. But for $n \geq 3$ at first glance the Hasse–Minkowski Theorem and its corollaries appear to reduce the question of isotropy (as well as questions of isometry and representation) of $\mathbb{Q}$-spaces to infinitely many local computations, making "reduction" an inappropriate term. But if $V \cong \langle a_1, \ldots, a_n\rangle$, with the a_i square-free integers, and p is an odd prime not dividing $\prod_i a_i$, then the local space V_p is automatically isotropic, by Theorem 4.3. Therefore in practice the localizations V_p need inspection only for a finite number of odd primes, along with the irritating prime $p = 2$.

(ii) Having used the Hasse–Minkowski Theorem to determine that the equation $q(v) = 0$ is solvable for some nonzero $v \in V$, it is natural to ask for an explicit isotropic vector. This issue is thoroughly explored in Garbanati [**Gar**].

5.4. The Bruck–Ryser Theorem on Finite Projective Planes

If we stand on some railroad tracks and follow them with our eyes as the tracks head towards the horizon, the tracks appear to meet. [A safer experiment: Look at van Gogh's painting of his bedroom and observe that the opposite walls would intersect if they were extended sufficiently far.] If one function of mathematics is to model aspects of the universe in which we live, it is reasonable to create a system of geometry in which every two distinct lines meet in a common point, and **projective geometry** is such a system. Our purpose here is to demonstrate an application of the theory of quadratic forms over $\mathbb{Q}$ to a part of projective geometry that has a combinatorial flavor.

We start by introducing the most basic vocabulary of projective geometry. Let $\mathbb{P}$ and $\mathbb{L}$ be disjoint nonempty sets, and suppose I is a nonempty relation from $\mathbb{P}$ to $\mathbb{L}$; that is, $\emptyset \neq I \subseteq \mathbb{P} \times \mathbb{L}$. If $p \in \mathbb{P}$ and $l \in \mathbb{L}$, we will write pIl instead of the more cumbersome $(p, l) \in I$. We will call elements of $\mathbb{P}$ **points** and elements of $\mathbb{L}$ **lines**; and if pIl we will say that p and l are **incident**, or, less formally: p is *on* l, l is *on* p, or l *passes through* p. Two or more points on the same line are said to be **collinear**, while two lines on the same point are said to be **concurrent**. The **range** of a line is the set of all points on that line, and the **pencil** of a point is the set of all lines on that point.

Definition 5.12. A **projective plane** Π is a triple $(\mathbb{P}, \mathbb{L}, I)$ (in the notation of the preceding paragraph) satisfying the following three properties:

(i) Each pair of distinct points are on exactly one line.

(ii) Each pair of distinct lines are on exactly one point.

(iii) $\mathbb{P}$ contains a set of four points, no three of which are collinear.

Example 5.13 (The **real** projective plane)**.** First the general idea. Start with $\mathbb{R}^2$. If we were to let $\mathbb{P} = \mathbb{R}^2$ and let $\mathbb{L}$ be the set of lines in $\mathbb{R}^2$, this would *not* be a projective plane, since parallel lines don't intersect. The idea is to appropriately adjoin the "horizon" to get a projective plane. (Remember: Our eyes tell us that the seemingly parallel railroad tracks meet on the horizon.)

Now for the details. To each line l in $\mathbb{R}^2$ adjoin its *slope* m_l; that is, put $l' = l \cup \{m_l\}$. (If l is vertical, set $l' = l \cup \{\infty\}$.) The set $\mathbb{R} \cup \{\infty\}$ of slopes is called the **line at infinity**. Now define a projective plane $\Pi = (\mathbb{P}, \mathbb{L}, I)$ whose set $\mathbb{L}$ of lines has as its elements all the l' corresponding to ordinary lines $l \subseteq \mathbb{R}^2$ and also the line at infinity; and for $\mathbb{P}$ take the set $\mathbb{R}^2 \cup \mathbb{R} \cup \{\infty\}$. The incidence relation is ordinary membership: pIl means $p \in l$. Check that Π is actually a projective plane.

Definition 5.14. A projective plane $\Pi = (\mathbb{P}, \mathbb{L}, I)$ is said to be **finite** if $\mathbb{P}$ and $\mathbb{L}$ are finite sets.

The following result is fundamental to the subject. The proof is included in the exercises. (If you prefer, see the details in the combinatorial literature; e.g., [**Ca**] or [**H**].)

Theorem 5.15. *Let $\Pi = (\mathbb{P}, \mathbb{L}, I)$ be a finite projective plane. Then there is an integer $n \geq 2$ (called the* **order** *of Π) such that*

(i) *Every line is on exactly $n+1$ points.*

(ii) *Every point is on exactly $n+1$ lines.*

(iii) $\#\mathbb{P} = \#\mathbb{L} = n^2 + n + 1$.

Example 5.16. Let F be any field. On $F^3 - \{(0,0,0)\}$ define a relation $\sim$ by putting $(a,b,c) \sim (a',b',c')$ if and only if the two points lie on the same line through the origin; that is, there is a value $\alpha \in F^*$ such that $a' = \alpha a, b' = \alpha b, c' = \alpha c$. This is an equivalence relation on $F^3 - \{(0,0,0)\}$, and we will write $[a,b,c]$ for the $\sim$-class of (a,b,c). Let

$$\mathbb{P} = \{[a,b,c] \mid (a,b,c) \neq (0,0,0)\},$$

and let $\mathbb{L}$ be a set disjoint from $\mathbb{P}$ but having the same cardinality. Fix a bijection from $\mathbb{P}$ onto $\mathbb{L}$, and let $\langle a,b,c\rangle$ be the element of $\mathbb{L}$ corresponding to $[a,b,c]$. Finally, define $I \subseteq \mathbb{P} \times \mathbb{L}$ by

$$[a,b,c]I\langle e,f,g\rangle \iff ae + bf + cg = 0$$

and check that $\Pi_F = (\mathbb{P}, \mathbb{L}, I)$ is a projective plane. In particular, if $\#F < \infty$ then Π_F is a finite projective plane.

Now we compute the order of Π_F if $F = \mathbb{F}_q$. In this case $F^3 - \{(0,0,0)\}$ has $q^3 - 1$ elements, and the equivalence class $[a,b,c]$ has $q-1$ elements. Therefore

$$\#\mathbb{P} = \frac{q^3-1}{q-1} = q^2 + q + 1$$

and so Π_F is a finite projective plane of order q. But $q = p^n$ for some prime p. Therefore, *every prime power is the order of a finite projective plane.* This question now arises:

Is there a finite projective plane whose order is not a prime power?

This is a long-standing open question in combinatorial mathematics. The Bruck–Ryser theorem, which is based on the classification of quadratic spaces over $\mathbb{Q}$, is the only theorem to date that has successfully eliminated a large class of positive integers as possible orders of finite projective planes.

To get to the Bruck–Ryser theorem we first need to create a link between finite projective planes and quadratic spaces. So let $\Pi = (\mathbb{P}, \mathbb{L}, I)$ be a finite projective plane of order n, and put $N = n^2 + n + 1 = \#\mathbb{P} = \#\mathbb{L}$. Write $\mathbb{L} = \{l_1, \ldots, l_N\}, \mathbb{P} = \{p_1, \ldots, p_N\}$. The associated **incidence matrix** is the $N \times N$ matrix $T = (t_{ij})$ given by

$$t_{ij} = \begin{cases} 1 & \text{if } p_i \in l_j, \\ 0 & \text{otherwise.} \end{cases}$$

Thus the jth column of T indicates which points are incident with the line l_j, while the ith row describes which lines are incident with the point p_i. Notice that the symmetric matrix ${}^tTT = {}^tTI_NT = A = (a_{ij})$ is given by

$$a_{ij} = \begin{cases} n+1 & \text{if } i = j \text{ (since every line is on } n+1 \text{ points),} \\ 1 & \text{if } i \neq j \text{ (since distinct lines are on one common point).} \end{cases}$$

Thus, in the notation of quadratic spaces, ${}^tTT = A \cong I_N$ over $\mathbb{Q}$.

Example 5.17. There is a finite projective plane of order 2 having the pictorial representation shown here, and the matrix T below it is the associated incidence matrix.

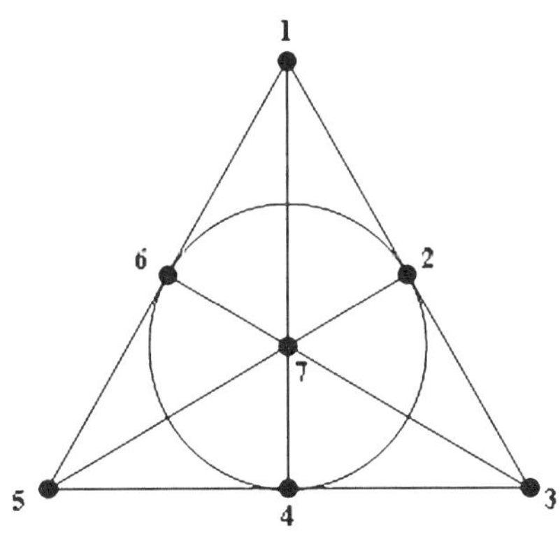

$$T = \begin{pmatrix} 1 & 0 & 1 & 1 & 0 & 0 & 0 \\ 1 & 0 & 0 & 0 & 1 & 0 & 1 \\ 1 & 1 & 0 & 0 & 0 & 1 & 0 \\ 0 & 1 & 0 & 1 & 0 & 0 & 1 \\ 0 & 1 & 1 & 0 & 1 & 0 & 0 \\ 0 & 0 & 1 & 0 & 0 & 1 & 1 \\ 0 & 0 & 0 & 1 & 1 & 1 & 0 \end{pmatrix}.$$

Theorem 5.18 (Bruck–Ryser Theorem). ([**BR**], 1949) *Let n be the order of a finite projective plane, and suppose $n \equiv 1$ or $2 \pmod 4$. Then n is a sum of two integer squares.*

Proof. We will need the following theorem of Fermat (already cited in the opening pages of Chapter 1, and to be proved in Chapter 9): A positive integer n is a sum of two squares if and only if $\operatorname{ord}_p n$ is even for all primes $p \equiv 3 \pmod 4$.

Let T be the incidence matrix of a finite projective plane of order n, and set $A = {}^tTT$ as in the preceding discussion. With $N = n^2 + n + 1$, define $S = (s_{ij}) \in GL_N(\mathbb{Q})$ by

$$s_{ij} = \begin{cases} 1 & \text{if } i = 1 \text{ and } j \geq 2, \\ n^{-1} & \text{if } i \geq 2 \text{ and } j = 1, \\ 1 & \text{if } i = j \geq 2, \\ 0 & \text{otherwise.} \end{cases}$$

That is,

$$S = \begin{pmatrix} 0 & 1 & \dots & 1 \\ n^{-1} & 1 & & \\ \vdots & & \ddots & \\ n^{-1} & & & 1 \end{pmatrix}.$$

Then it is straightforward to check that ${}^tS\langle 1, n, \dots, n\rangle S = A$ (here we have abused notation and used $\langle 1, n, \dots, n\rangle$ for a diagonal matrix) and therefore

$$\langle 1, n, \dots, n\rangle \cong A \cong \langle 1, \dots, 1\rangle \quad \text{over } \mathbb{Q},$$

an isometry of N-dimensional spaces. It follows that for all primes p we have

$$1 = S_p \underbrace{\langle 1, n, \dots, n\rangle}_{N} = (n, n)_p^{\binom{N-1}{2}} = (n, -1)_p^{\frac{(N-1)(N-2)}{2}}.$$

But $N = n^2 + n + 1 \equiv 3 \pmod 4$, since $n \equiv 1$ or $2 \pmod 4$ by hypothesis, and therefore $\frac{(N-1)(N-2)}{2}$ is odd; hence $(n, -1)_p = 1$ for all p.

Now let p be an odd prime such that $\operatorname{ord}_p n$ is odd. Say $n = p^{2k+1}u$, with $\gcd(p, u) = 1$. Then

$$1 = (n, -1)_p = (p, -1)_p(u, -1)_p = \left(\frac{-1}{p}\right) = \begin{cases} 1 & \text{if } p \equiv 1 \pmod 4, \\ -1 & \text{if } p \equiv 3 \pmod 4, \end{cases}$$

and therefore we must have $p \equiv 1 \pmod 4$, so n is a sum of two squares as claimed. □

In light of the Bruck–Ryser Theorem we conclude immediately that there are no finite projective planes of order $6, 14, 21$, and so on. (The reader should crank out some examples from the "and so on.") The first integers that are not prime powers and not covered by Bruck–Ryser are 10 and 12. In 1988 researchers at the University of Montreal used thousands of hours of computer time to show that there are no finite projective planes of order 10, but whether there exist planes of order 12 remains an open question.

5.5. Sums of Integer Squares (First Version)

We begin the move from the rational theory of quadratic forms to the integral theory by using the rational theory to obtain integral results. Again recall Fermat's classical "Two-Square" Theorem: A positive integer n is a sum of two integer squares if and only if $\operatorname{ord}_p n$ is even for all primes $p \equiv 3 \pmod 4$. (Most elementary number theory books contain a proof. And we will give a new proof in Chapter 9.) Our goals in this section are to use the rational theory of quadratic spaces to go beyond Fermat's theorem and obtain the famous theorems of Gauss and Lagrange on sums of three and four squares. Here we follow the treatment in Serre [**Se**]; in Chapter 9 we will revisit sums of squares using the structure theory of lattices.

Lemma 5.19. *If $\alpha \in \mathbb{Q}^*$, then α is a sum of three rational squares if and only if $\alpha > 0$ and $-\alpha \notin \mathbb{Q}_2^{*2}$.*

Proof. Let $V \cong \langle 1, 1, 1 \rangle$ over $\mathbb{Q}$. Then by the Hasse–Minkowski Theorem, there is a representation $\alpha \longrightarrow V$ if and only if $\alpha \longrightarrow V_p$ for all p. Now, if $p \neq 2$ the space $\langle 1, 1, 1 \rangle$ is isotropic—and hence universal—over $\mathbb{Q}_p$; so representation of α by V_p is automatic. Therefore $\alpha \longrightarrow V$ if and only if $\alpha > 0$ and $\alpha \longrightarrow V_2$. But $\alpha \longrightarrow V_2$ if and only if $-\alpha \notin \mathbb{Q}_2^{*2}$, since $\langle 1, 1, 1, 1 \rangle$ is the unique anisotropic $\mathbb{Q}_2$-space of dimension 4. $\square$

A necessary condition for a given rational number to be represented by a given $\mathbb{Z}$-lattice L spanning a rational quadratic space V is for V to represent the number. The literature on quadratic forms contains a number of results of the form "If V represents α and ......, then L represents α," and the following theorem, which Serre attributes to Davenport and Cassels, is an example.

Theorem 5.20. *Let V be a positive definite quadratic $\mathbb{Q}$-space of dimension n, and suppose $V \cong A \in M_n(\mathbb{Z})$ with respect to the basis $\mathbb{B}$. Let L be the $\mathbb{Z}$-lattice spanned by $\mathbb{B}$, and suppose for every $x \in V$ there exists $y \in L$ such that $q(x - y) < 1$. Then every integer represented by V is represented by L.*

Proof. If $m \in \mathbb{Z}$ and $m \to V$, then $t^2 m \to L$ for some $t \in \mathbb{N}$. (If $q(v) = m$, choose $t \in \mathbb{N}$ such that $tv \in L$.) Suppose t is minimal with this property, and we will be finished if we can show $t = 1$.

Suppose $q(x) = t^2 m$, with $x \in L$. From the hypothesis, there is a vector $y \in L$ such that $q\Big(\dfrac{x}{t} - y\Big) < 1$. Set $z = \dfrac{x}{t} - y$.

If $q(z) = 0$ then $z = 0$ because V is definite, and hence $q(y) = q\Big(\dfrac{x}{t}\Big) = m$, giving $t = 1$ as desired. So we can suppose $q(z) \neq 0$ and look for a contradiction.

Start by defining the following integers:

$$a = q(y) - m, \qquad b = 2\big(mt - B(x, y)\big), \qquad t' = at + b$$

and put $x' = ax + by \in L$. Then $q(x') = mt'^2$. But also $tt' = t^2 q(z)$, and therefore $t' = tq(z)$. From the definition of z we know $0 < q(z) < 1$ and hence $0 < t' < t$, contradicting the minimality of t. Therefore the case $q(z) \neq 0$ cannot occur. □

Theorem 5.21 (Gauss's Three-Square Theorem). *A positive integer n is a sum of three integer squares if and only if n is not of the form $4^\nu k$ with $k \equiv 7 \pmod 8$.*

Proof. We first claim that n is the sum of three integer squares if and only if n is the sum of three rational squares. Of course only the sufficiency requires proof. So let $V \cong \langle 1, 1, 1\rangle$ in the $\mathbb{Q}$-basis $\{v_1, v_2, v_3\}$, and suppose $n \longrightarrow V$; say $q(v) = n$, where $v = \sum_i \alpha_i v_i$ with $\alpha_i \in \mathbb{Q}$. Let $L = \perp_i \mathbb{Z} v_i$. Choose $b_i \in \mathbb{Z}$ such that $|\alpha_i - b_i| \leq \frac{1}{2}$ for all i, and let $w = \sum_i b_i v_i$. Then $w \in L$ and $q(v - w) = \sum_i (\alpha_i - b_i)^2 < 1$, hence $n \longrightarrow L$ by the preceding theorem, proving the claim.

The theorem asserts that two statements are equivalent. We show their *negations* are equivalent.

We have $n = 4^\nu k$ with $k \equiv 7 \pmod 8$ if and only if $-n = 4^\nu \cdot (1 + 8b)$ for some $b \in \mathbb{Z}$, which in turn is equivalent to $-n \in \mathbb{Q}_2^{*2}$ by the Local Square Theorem. But from the lemma preceding the previous theorem this is equivalent to n not being a sum of three rational squares. □

Corollary 5.22 (Lagrange's Four-Square Theorem). *Every positive integer is a sum of four integer squares.*

Proof. [We first note that every positive integer is a sum of four *rational* squares by the Hasse–Minkowski Theorem, so one might wish to now imitate the argument in the first paragraph of the preceding proof. But that attempt fails because the final inequality comes out "$\leq$" instead of the required "$<$".]

Write $n = 4^\nu k$ with $4 \not| k$. If $k \not\equiv 7 \pmod 8$ then n is a sum of three integer squares by Gauss's Theorem. But if $k \equiv 7 \pmod 8$ then $k-1$ is a sum of three integer squares, hence k is a sum of four integer squares, and so n is as well. □

Exercises

(1) Give a brief and very intuitive description of our approximation theorems (Weak or Strong) that a nonmathematician can understand. [Suggestion: Relate the process to buying a car.]

(2) Use the Weak Approximation Theorem (and not some other method) to find an element $\alpha \in \mathbb{Q}$ such that both $|\alpha - 7|_3$ and $|\alpha + 1|_5$ are less than $\frac{1}{30}$.

(3) Verify the isomorphism described in the proof of the Hasse–Minkowski Theorem for spaces of dimension 3. Then, assuming that the definition of Euler's φ-function is known, but nothing more about φ is known, what does that isomorphism tell you about $\varphi(m)$ if m is a square-free integer ≥ 2?

(4) Consider quadratic $\mathbb{Q}$-spaces $\langle 3, 5, 6\rangle$ and $\langle 1, 1, 10\rangle$. Are they isometric? Prove your claim.

(5) (i) Let T be a finite set of primes (possibly including ∞) in $\mathbb{Q}$. Suppose for each $p \in T$ a regular n-dimensional $\mathbb{Q}$-space U_p is given. Show that there is a $\mathbb{Q}$-space V such that $V_p \cong U_p$ for all $p \in T$.
(ii) Suppose for *every* prime p a regular n-dimensional quadratic $\mathbb{Q}$-space U_p is given; and the issue arises as to whether there is a $\mathbb{Q}$-space V such that $V_p \cong U_p$ for all $p \in T$. Give necessary conditions on the discriminants and Hasse symbols of the U_p for there to be such a V.
(iii) Give an *explicit* example of a family $\{U_p\}$ of local spaces as described in part (ii) for which *no* $\mathbb{Q}$-space V exists satisfying $V_p \cong U_p$ for all p.

(6) Show that there is an injective mapping of Witt rings

$$W(\mathbb{Q}) \longrightarrow \bigoplus_{\text{all } p} W(\mathbb{Q}_p).$$

(You need not show that the mapping is a homomorphism.)

(7) Suppose V is an n-dimensional quadratic $\mathbb{Q}$-space that has a unimodular Gram matrix. That is, $V \cong A$ for some $A \in GL_n(\mathbb{Z})$. Show that V is determined up to isometry by V_∞. In other words, if W is another n-dimensional $\mathbb{Q}$-space with a unimodular Gram matrix, then $V \cong W$ if and only if $V_\infty \cong W_\infty$.

(8) Let V be a regular quadratic $\mathbb{Q}$-space. Prove that V is universal if and only if for every prime p (including $p = \infty$) the space V_p is a universal $\mathbb{Q}_p$-space.

(9) Suppose V and W are regular n-dimensional quadratic spaces over $\mathbb{Q}$ having the same discriminant. Let q be a prime (possibly $q = \infty$), and suppose there is Hasse symbol equality $S_pV = S_pW$ for all $p \neq q$. Does it follow that $V \cong W$? [Note: There is what appears to be an immediate answer, and it is wrong.]

(10) Let V be a positive definite quadratic $\mathbb{Q}$-space of dimension $n \geq 4$. Prove that V has a 3-dimensional subspace U with the property that $dU = dV$ and $S_pU = S_pV$ for all primes p.

(11) Bruck and Ryser have described a canonical form for a positive definite $\mathbb{Q}$-space V, as follows: $V \cong \langle a_1, \dots, a_n \rangle$, with a_1 the smallest positive integer represented by V, then a_2 the smallest positive integer represented by the orthogonal complement of $\langle a_1 \rangle$, and so on. [Note: We are assuming "positive definiteness" for simplicity, though Bruck and Ryser do not restrict themselves to that case.] Describe an algorithm for producing V's canonical form, starting from a given orthogonalization $V \cong \langle b_1, \dots, b_n \rangle$. You may use the result of the preceding exercise.

(12) Here is an outline of the proof of Theorem 5.15. Fill in the details.
(i) Show that there is a line l with at least three points. Then show that if l' is another line, there is a point P not on either line. Finish this part by using lines through P to establish a bijection between the points of l and the points of l'. This produces a number $n \geq 2$ such that every line is on $n + 1$ points.
Half of (iii): Let l be a line and P a point not on l. There are $n + 1$ lines through P intersecting l; but these are *all* of the lines through P, and they intersect pairwise only at P. Deduce that $\#\mathbb{P} = n^2 + n + 1$.
Now finish the proof by proving part (ii) and the other half of (iii).

(13) Let $E/\mathbb{Q}$ be a quadratic extension, and let V be an n-dimensional regular Hermitian E-space, with H its Hermitian form. Show that if $n \geq 3$ and H represents both positive and negative numbers, then V is isotropic.

(14) Show that the $\mathbb{Z}$-lattice $L \cong \langle 1, 1, 1, 1, -1 \rangle$ represents every integer.

Chapter 6

Lattices over Principal Ideal Domains

We saw in Chapter 2 that, with R an integral domain, the representation and equivalence problems for quadratic polynomial forms in $R[x_1, \ldots, x_n]$ are equivalent to the representation and isometry problems for free quadratic R-modules of rank $\leq n$. And nearly all our work since then has been in the so-called "fractional" theory of quadratic forms: when R is a field. It is now time to begin preparation for the "integral" theory: when R is not a field. Our main goal will be the theory over $\mathbb{Z}$, though we will also have much to say about the parallel theory of quadratic forms over $\mathbb{F}_q[x]$. As in the fractional theory, there is once again a local-global approach to the subject—so along the way we will need to develop the subject over $\mathbb{Z}_p$—but we will see that the analogue of the Hasse-Minkowski Theorem fails in the integral situation. In fact the analysis of this failure provides a rich area of ongoing research in the integral theory.

Except for occasional remarks, all our rings will be principal ideal domains, and before actually getting to quadratic forms we need some foundation work on the structure of the relevant modules. **Throughout this chapter R will denote a principal ideal domain with quotient field $F \supsetneq R$.** *Unless otherwise indicated, the words "module," "submodule," "ideal," and "lattice" should be understood to have an invisible prefix "R-"* (though at first—and occasionally later on too—we will make it visible for emphasis), and *"space" will mean a finite-dimensional vector space over F.*

6.1. Lattice Basics

Definition 6.1. Let V be an n-dimensional F-space. An R-module $L \subset V$ is an **R-lattice** in V if $L = 0$ or if there exists a linearly independent subset $\{v_1, \ldots, v_r\}$ of V such that $L = Rv_1 \oplus \cdots \oplus Rv_r$. Here r is the **rank** of L (and we set rank $0 = 0$). We say L is **on** V if $r = n$; that is, L spans V. A nonzero vector in L that extends to a basis of L is called a **primitive vector** in L.

Let us recall a bit of linear algebra. Suppose V and W are spaces with respective bases $\mathbb{B} = \{v_1, \ldots, v_n\}$ and $\mathbb{B}' = \{w_1, \ldots, w_m\}$; and $\varphi : V \to W$ is linear, say $\varphi(v_j) = \sum_i a_{ij} w_i$. Then **the matrix of φ with respect to $\mathbb{B}$ and $\mathbb{B}'$** is $(a_{ij}) \in M_{m\times n}(F)$, and we will denote this matrix by $(\varphi)_{\mathbb{B}}^{\mathbb{B}'}$. If also $\mathbb{B}''$ is a basis for a space X and $\psi : W \to X$ is linear, then

$$(\psi \circ \varphi)_{\mathbb{B}}^{\mathbb{B}''} = (\psi)_{\mathbb{B}'}^{\mathbb{B}''} (\varphi)_{\mathbb{B}}^{\mathbb{B}'}.$$

Now if $\mathbb{B} = \{v_1, \ldots, v_n\}$ and $\mathbb{B}' = \{w_1, \ldots, w_n\}$ are bases for a lattice L on V and $\varphi \in \operatorname{Aut} L$ is given by $\varphi(v_j) = w_j$ for $1 \le i \le n$, then $(\varphi^{-1})_{\mathbb{B}}^{\mathbb{B}} (\varphi)_{\mathbb{B}}^{\mathbb{B}} = (\varphi^{-1} \circ \varphi)_{\mathbb{B}}^{\mathbb{B}} = I_n$ and therefore $(\varphi)_{\mathbb{B}}^{\mathbb{B}} \in GL_n(R)$. That is, $(\varphi)_{\mathbb{B}}^{\mathbb{B}}$ is a **unimodular** R-matrix. Here the matrix $T = (\varphi)_{\mathbb{B}}^{\mathbb{B}}$ is called the **transition matrix from $\mathbb{B}$ to $\mathbb{B}'$**, since its jth column consists of the coefficients of the new basis vector w_j in terms of $\mathbb{B}$. Conversely, if $T = (t_{ij}) \in GL_n(R)$ and $\varphi : L \to L$ is given by $\varphi(v_j) = \sum_i t_{ij} v_i$, then $\{\varphi(v_j)\}_{1\le j\le n}$ is a linearly independent subset of L; and if $T^{-1} = (s_{ij})$ then $v_j = \sum_i s_{ij}\varphi(v_i)$, so $\{\varphi(v_i)\}_{1\le i\le n}$ generates L over R and hence is also a basis for L.

Proposition 6.2. *Let L be a submodule of the space V. Then L is a lattice if and only if L is finitely generated.*

Proof. Only the sufficiency requires proof. Let $\mathbb{B} = \{v_1, \ldots, v_n\}$ be a basis for V, and consider the lattice $M = \bigoplus_1^n Rv_i$. Since L is finitely generated, there is a nonzero $r \in R$ such that $rL \subseteq M$. If we can show that rL has a basis $\{w_i\}_{1\le i\le k}$, then $\{r^{-1}w_i\}_{1\le i\le k}$ will be a basis for L. Therefore without loss of generality we can assume $L \subseteq M$, and we will be through if we can show that every submodule of M is a lattice. We will argue by induction.

If $n = 1$ and $L \ne 0$, let $A = \{\, \alpha \in R \mid \alpha v_1 \in L \,\}$. Then A is an ideal, say $A = (a)$, and then $L = R(av_1)$.

Now suppose that $n > 1$ and that the result has been checked for spaces of lower dimension. Every $x \in L$ has a unique expression $\sum_1^n \lambda_i v_i$ with $\lambda_i \in R$. Let A be the ideal consisting of all the coefficients of v_1 that occur in such sums for elements of L. Say $A = (a)$, and suppose we have $w_1 = av_1 + \lambda_2 v_2 + \cdots + \lambda_n v_n \in L$. Then

$$L = Rw_1 \oplus (L \cap (Rv_2 \oplus \cdots \oplus Rv_n)),$$

so the induction hypothesis finishes the job. □

From the proof of the preceding proposition we immediately have

Corollary 6.3. *Every submodule of a lattice is a lattice.*

From time to time we will find it useful to know that there is a lattice basis whose vectors are in controlled locations, and the following result is a move in this direction.

Proposition 6.4. *Let L be a lattice on V, and let $\{v_1, \ldots, v_n\}$ be a basis for V. Then there is a basis $\{w_1, \ldots, w_n\}$ for L such that $w_j \in Fv_1 + \cdots + Fv_j$ for $1 \leq j \leq n$.*

Proof. As in the proof of the previous proposition, without loss of generality we can assume that $L \subseteq M = Rv_1 + \cdots + Rv_n$; and there is a nonzero element $r \in R$ such that $rM \subseteq L$. For $1 \leq j \leq n$, define

$$L_j = L \cap (Rv_1 + \cdots + Rv_j) = \Big\{ x \in L \mid x = \sum_{i=1}^{j} \alpha_i v_i, \text{ with } \alpha_i \in R \Big\}$$

and let S_j denote the ideal of coefficients of v_j occurring in these expressions for elements of L_j. Since $rv_j \in L$ we know $S_j \neq 0$; write $S_j = (\beta_{jj})$. Then L contains an element of the form $w_j = \sum_{i=1}^{j} \beta_{ij} v_i$, with all $\beta_{ij} \in R$.

We claim that $\{w_1, \ldots, w_n\}$ is a basis for L. These vectors are clearly linearly independent, since the matrix expressing the w_j's in terms of the v_i's is upper triangular with nonzero diagonal entries $\beta_{11}, \ldots, \beta_{nn}$. So it remains only to check that the w_j's generate L over R. Let $x \in L$; say $x = \alpha_1 v_1 + \cdots + \alpha_n v_n$, with $\alpha_i \in R$. Then $\alpha_n \in S_n$, and hence $\alpha_n = \lambda_n \beta_{nn}$ for some $\lambda_n \in R$. Then the element $x - \lambda_n w_n$ is in L and has the form $\alpha'_1 v_1 + \cdots + \alpha'_{n-1} v_{n-1}$ for some $\alpha'_i \in R$. Iteration of this argument eventually leads to an expression $x = \lambda_1 w_1 + \cdots + \lambda_n w_n$ with all $\lambda_i \in R$. □

Corollary 6.5. *Let L be a lattice on V, and let v be a nonzero vector in V. Then some vector of the form λv, with $\lambda \in F^*$, is a primitive vector in L.*

In practice, how can we extend a given primitive vector in a lattice to a basis for that lattice? The following lemma, due to Hermite in 1849, is fundamental for this construction.

Lemma 6.6. *Let $a_1, \ldots, a_n \in R$, with $\gcd(a_1, \ldots, a_n) = d$. Then there is a matrix $A \in M_n(R)$ with ${}^t(a_1, \ldots, a_n)$ as its first column and $\det A = d$. In particular, if $\gcd(a_1, \ldots, a_n) = 1$ then there is a unimodular matrix with first column ${}^t(a_1, \ldots, a_n)$.*

Proof. We argue by induction on n. The result is trivial if $n = 1$ and an elementary exercise if $n = 2$. Now suppose $n \geq 3$, and assume the result has been shown for values less than n; so there is a matrix $A_0 \in M_{n-1}(R)$ with first column ${}^t(a_1, \ldots, a_{n-1})$ and $\det A_0 = d_0 = \gcd(a_1, \ldots, a_{n-1})$. Since $d = \gcd(d_0, a_n)$, there exist $x, y \in R$ such that $d_0 x - a_n y = d$. Now define A to be the matrix in $M_n(R)$ having A_0 as its upper left $(n-1) \times (n-1)$ block, $(a_n, 0, \ldots, 0, x)$ as its bottom row, and the transpose of

$$\left(\frac{a_1 y}{d_0}, \ldots, \frac{a_{n-1} y}{d_0}, x\right)$$

as its right-hand column. Calculate $\det A$ by expanding on the bottom row, getting

$$\det A = x d_0 + (-1)^{n-1} a_n \det E$$

where E is the matrix obtained from A by crossing out the first column and nth row. Notice that the matrix $d_0 E$ is obtained from A_0 by multiplying columns 2 through $n-1$ of A_0 by d_0, then shifting each column one position to the left (and the first to the nth), and then multiplying the last column by y. It follows from this that $\det E = (-1)^{n-2} y$, and hence that $\det A = d$. □

And now we can give what is essentially a constructive version of the preceding corollary.

Corollary 6.7. *Let $\mathbb{B} = \{v_1, \ldots, v_n\}$ be a basis for L, and let $v' = \sum_i t_i v_i$, with $t_i \in R$ for $1 \leq i \leq n$ and $\gcd(t_1, \ldots, t_n) = 1$. Then v' is primitive in L.*

Proof. By the preceding lemma, the transpose of the n-tuple $(t_1, \ldots, t_n)$ is the first column of a matrix $T \in GL_n(R)$, and then this T is the transition matrix from $\mathbb{B}$ to another basis for L containing v'. □

Note that the preceding corollary is algorithmic *provided* that we have an algorithm for expressing the greatest common divisor of typical elements $a, b \in R$ in the form $ax + by$ with $x, y \in R$.

6.2. Valuations and Fractional Ideals

Before continuing with R-lattices it will be helpful to bring valuation theory into the picture, and we will do this by imitating the definition of p-adic valuations, which were so useful in the material over $\mathbb{Q}$ and $\mathbb{Q}_p$ in the earlier chapters.

The principal ideal domain R is a unique factorization domain, meaning that every nonzero nonunit $\alpha \in R$ has an "essentially unique" factorization into primes (i.e., irreducible elements): If $\alpha = \varepsilon p_1 \cdots p_r = \delta q_1 \cdots q_s$ with $\varepsilon, \delta \in R^*$ and the p_i and q_i primes , then $r = s$ and, after shuffling subscripts as needed, $p_i = \lambda_i q_i$ for some $\lambda \in R^*$. Therefore, in terms of ideals, $(p_i) = (q_i)$ for $1 \le i \le r$. It follows that every $\alpha \in F^*$ has an essentially unique factorization $\alpha = \varepsilon p_1^{\nu_1} \cdots p_r^{\nu_r}$ with the p_i nonassociated primes and $\nu_i \in \mathbb{Z}$. This uniqueness allows us to associate a valuation $|\cdot|_p$ with each nonzero prime ideal $\mathfrak{p} = (p)$ of R as follows: Start by fixing a real number C satisfying $0 < C < 1$. Then define $|0|_p = 0$, and with $\alpha = \varepsilon p_1^{\nu_1} \cdots p_r^{\nu_r}$ as above, define

$$\operatorname{ord}_p \alpha = \begin{cases} 0 & \text{if no } p_i \text{ is an associate of } p, \\ \nu_j & \text{if } p_j \text{ is an associate of } p, \end{cases}$$

and put $|\alpha|_p = C^{\operatorname{ord}_p \alpha}$.

It is easily checked that $|\cdot|_p$ is a nonarchimedean valuation on F, and we call it the p**-adic** (or $\mathfrak{p}$**-adic**) **valuation.**

Notation 6.8. In what follows we will use the symbol S in two compatible ways, easily distinguished by the context: (i) as the set of nonzero prime ideals of R, with a typical prime ideal usually denoted by a gothic letter $(\mathfrak{p}, \mathfrak{q}, \dots)$; (ii) as a complete set of nonassociated prime elements of R (in other words, as a set of prime elements of R consisting of one generator of each nonzero prime ideal).

We now have a valuation-theoretic description of R:

$$R = \{\, x \in F \mid \; |x|_p \le 1 \text{ for all } p \in S \,\}.$$

Now let $\alpha \in F^*$. We define the **fractional** R**-ideal** (α) to be the submodule

$$(\alpha) = \alpha R = \{\, \alpha r \mid r \in R \,\} = \{\, x \in F \mid \; |x|_p \le |\alpha|_p \text{ for all } p \in S \,\}.$$

In particular, every nonzero ideal is a fractional ideal (these are the "integral" fractional ideals), and in general a fractional ideal is just a nonzero scalar multiple of a nonzero ideal. Notice that $(\alpha) = (\beta)$ if and only if $\alpha\beta^{-1} \in R^*$. Moreover, $(\alpha) \supseteq (\beta)$ if and only if $\beta = \alpha c$ for some $c \in R$. We will say α **divides** β—written $\alpha|\beta$—when this is the case, and this is consistent with the usual terminology inside R.

We will denote the set of fractional R-ideals by $I(R)$. If $\mathfrak{A}, \mathfrak{B} \in I(R)$, define the **product** $\mathfrak{A}\mathfrak{B}$ to be the set of all finite sums of products ab with $a \in \mathfrak{A}$ and $b \in \mathfrak{B}$. Then $\mathfrak{A}\mathfrak{B} \in I(R)$; in fact, if $\mathfrak{A} = (\alpha)$ and $\mathfrak{B} = (\beta)$ then $\mathfrak{A}\mathfrak{B} = (\alpha\beta)$. Moreover $I(R)$ is a *group* with respect to this product, with identity element $(1) = R$ and $(\alpha)^{-1} = (\alpha^{-1})$. It now follows that if $p_1, \dots, p_r$ are nonassociated primes in R and $\alpha = \varepsilon p_1^{\nu_1} \cdots p_r^{\nu_r}$, with all $\nu_i \in \mathbb{Z}$ and $\varepsilon \in R^*$, then $(\alpha) = (p_1)^{\nu_1} \cdots (p_r)^{\nu_r}$, and thus the multiplicative group $I(R)$ is generated by the set S of nonzero prime ideals.

Example 6.9. Let $(F, \mathfrak{p})$ be a local field, with $\mathfrak{o}$ its ring of integers and π a prime element. Then $\mathfrak{o}$ is a principal ideal domain, and $\mathfrak{p} = (\pi)$ is the unique prime ideal of $\mathfrak{o}$ (and therefore $S = \{\mathfrak{p}\}$). The fractional $\mathfrak{o}$-ideals are of the form $\mathfrak{p}^\nu = (\pi^\nu)$, with $\nu \in \mathbb{Z}$, and are linearly ordered by inclusion:

$$\cdots \supseteq (\pi^{\nu-1}) \supseteq (\pi^\nu) \supseteq (\pi^{\nu+1}) \supseteq \cdots .$$

The linear ordering of the fractional ideals in the local setting will turn out to lead to major simplifications in the classification of lattices on quadratic spaces when we move from the global context to the local context, for instance from lattices over $\mathbb{Z}$ to lattices over $\mathbb{Z}_p$.

6.3. Invariant factors

If two lattices reside in the same space, it can be very helpful to describe their relative positions in that space: Does one lattice contain the other? If so, is it a tight fit? If neither contains the other, can one coherently describe their "overlap"? Analyzing this kind of data is the function of invariant factor theory, and for this we need a bit of classical matrix theory.

Definition 6.10. Matrices $A, B \in M_n(F)$ are said to be **equivalent** over R if there exist unimodular matrices $P, Q \in GL_n(R)$ such that $B = PAQ$, and in this case we write $A \underset{R}{\sim} B$. (We will usually omit the subscript R.)

In seeking a set of necessary and sufficient conditions for two F-matrices A, B to be equivalent, it suffices to settle the issue for matrices in $M_n(R)$; because one can multiply the given F-matrices by a suitable nonzero element $r \in R$ to "clear denominators", find the unimodular matrices P, Q that demonstrate the equivalence of the resulting R-matrices rA, rB, then multiply through by r^{-1} to see that P, Q work for A, B as well. So unless otherwise indicated, we will assume without loss of generality that our given matrices are in $M_n(R)$.

Notice that matrix equivalence over R is weaker than matrix congruence, since the matrices P, Q used in an equivalence are not required to be related. In particular, if $A \underset{R}{\not\sim} B$ then $A \underset{R}{\not\cong} B$.

There is an extensive literature on matrix equivalence over principal ideal domains (see especially Newman's *Integral Matrices* [**N1**] or Curtis and Reiner's *Representation Theory of Finite Groups and Associative Algebras* [**CR**]), and there it is shown that every matrix $A \in M_n(R)$ is equivalent over R to a diagonal matrix $\Sigma_R(A) = \operatorname{diag}(s_1(A), \ldots, s_r(A), 0, \ldots, 0)$, where $r = \operatorname{rank} A$ and $s_i(A) | s_{i+1}(A)$ for $1 \leq i \leq r-1$. [Note: If the ring R is understood from the context, we will write $\Sigma(A)$ instead of $\Sigma_R(A)$.] The diagonal entries $s_i(A)$ are unique up to unit factors, and they are called the **invariant factors** of A. In fact if we define $d_0(A) = 1$ and for $1 \leq k \leq n$

define $d_k(A)$ to be the greatest common divisor of the $k \times k$ subdeterminants of A, then $s_k(A) = d_k(A)/d_{k-1}(A)$. (In particular, $s_k(A) = 0$ for all $k >$ rank A. The $d_k(A)$ are the **determinantal divisors** of A and the prime-power divisors of the invariant factors, listed as often as they occur, are the **elementary divisors** of A. It follows from the uniqueness that $A \underset{R}{\sim} B$ if and only if $\Sigma(A) = \Sigma(B)$; and, since the $s_i(A)$ are unique only up to unit factors, in fact $A \underset{R}{\sim} B$ if and only if A and B have the same elementary divisors. Equivalently, $\operatorname{ord}_p s_k(A) = \operatorname{ord}_p s_k(B)$ for all k.

The matrix $\Sigma(A)$ is the **Smith normal form** of A, named for the distinguished Nineteenth Century British mathematician Henry John Stephen Smith. (Note: When $A \in M_n(F) - M_n(R)$ the matrix $\Sigma(A)$ is sometimes referred to as the **Smith–McMillan form**. This is the custom in the literature of control theory, for example.) Computing $\Sigma(A)$ via determinantal divisors would in general be an extremely tedious process, and fortunately there are usually alternatives.

If R is a Euclidean ring or the ring of integers of a local field, $\Sigma(A)$ can be achieved by performing a sequence of elementary row and column operations; and the transforming matrices $P, Q \in GL_n(R)$ are then the product of the corresponding elementary R-matrices. (So in these cases—which are our main cases of interest—invariance of the invariant factors and elementary divisors of a matrix can be deduced from invariance of the determinantal divisors under elementary row and column operations.) In brief, proceed as follows. Use elementary row and column operations to move the smallest (with respect to the Euclidean function) nonzero entry of A into the upper left corner. Subtract suitable multiples of the first column from all other columns, so that now the resulting matrix either has zeros for the other entries in the first row or else some nonzero element smaller than the upper left entry of the matrix appears, in which case move that smaller element into the upper left. Iteration of this eventually replaces all entries of the top row after the first entry by zeros. Now use row operations in the same way to replace entries 2 through n of the first column by 0s. During the course of this activity, if at any time an element smaller than the upper left element appears in the matrix, move it to the upper left and continue. Having cleared out the first row and column (after the first entry), with the new a_{11} the smallest entry in the matrix, if a_{11} is a divisor of every other matrix entry, then $a_{11} = s_1$, the first invariant factor. On the other hand, if for some $i, j \geq 2$ it happens that $a_{11} \nmid a_{ij}$, then add row i to row 1 and continue the process. Eventually the most recent a_{11} divides every other element in the matrix and is the only nonzero element in the first row or column. Then $s_1 = a_{11}$. Now repeat the process on the lower right $(n-1) \times (n-1)$ matrix to eventually get $\Sigma(A)$.

If R is a discrete valuation ring, with p a prime element, the procedure can go much more quickly. Namely, look for a matrix entry a_{ij} with $\text{ord}_p\, a_{ij}$ minimal and move it to the upper left corner using elementary row and column operations. Then the new a_{11} is now a divisor of all other entries of the matrix, so we can use it to clear out all other entries of the first row and column, declare that $s_1 = a_{11}$, and then proceed with the lower $(n-1) \times (n-1)$ matrix.

The preceding paragraph suggests a "local-global" approach to finding $\Sigma(A)$ for $A \in GL_n(F)$. For a nonzero prime ideal $\mathfrak{p}$ of R, define

$$R_{(\mathfrak{p})} = \{\alpha \in F \mid \text{ord}_{\mathfrak{p}}\alpha \geq 0\}.$$

Then $R_{(\mathfrak{p})}$ is the subring of F generated by R and the inverses in F of all elements of R that are outside $\mathfrak{p}$. [Notice in particular that we are remaining in F here, with no need to pass to a completion.] If $\mathfrak{p} = (p)$, then $\varepsilon \in F$ is in $R^*_{(\mathfrak{p})}$ if and only if $\text{ord}_p\, \varepsilon = 0$. Moreover, every element of $R_{(\mathfrak{p})}$ has the form $\alpha = \varepsilon p^\nu$ for some $\varepsilon \in R^*_{(\mathfrak{p})}$ and $\nu = \text{ord}_p\alpha \geq 0$. The divisibility relation $\alpha \mid \beta$ holds in $R_{(\mathfrak{p})}$ if and only if $\text{ord}_p\alpha \leq \text{ord}_p\beta$. The ring $R_{(\mathfrak{p})}$ is a principal ideal domain, and it is a local ring with $\mathfrak{p} = (p)$ its unique maximal ideal. In fact, the proper ideals of $R_{(\mathfrak{p})}$ are $(p) \supset (p^2) \supset (p^3) \supset \cdots \supset 0$. Write $\Sigma_{\mathfrak{p}}(X)$ or $\Sigma_p(X)$ for the Smith form of an R-matrix X, using $R_{\mathfrak{p}}$ as the relevant principal ideal domain, and write $\sim_p$ for matrix equivalence over $R_{\mathfrak{p}}$. Observe that if $\Sigma(A) = \text{diag}(s_1, \ldots, s_n)$, with $\text{ord}_p\, s_i = \nu_i(p)$, then $\Sigma_p(A) = \text{diag}(p^{\nu_1(p)}, \ldots, p^{\nu_n(p)})$, and it follows from this that

$$s_i = \prod_{p \mid |A|} p^{\nu_i(p)}.$$

In other words, the ith invariant factor of an F-matrix A is the product of ith local invariant factors of A viewed as an R_p-matrix as p ranges through S. We therefore have the following theorem.

Theorem 6.11 (Local-Global Principle for Smith Normal Form). *Let $A, B \in M_n(R)$. Then $A \sim_R B$ if and only if $A \sim_p B$ for all prime elements p of R. Moreover,*

$$\Sigma(A) = \prod_{p \mid \det A} \Sigma_p(A);$$

where it is understood that the primes involved in the product come from a complete set of nonassociated primes of R.

Examples 6.12.

(i) If $A \in GL_n(R)$ then $\Sigma(A) = I_n$.

(ii) If $\det A$ is square-free, then for all primes $p \mid \det A$ we have $\Sigma_p(A) = \mathrm{diag}(1,\ldots,1,p)$, and therefore $\Sigma(A) = \mathrm{diag}(1,\ldots,1,\det A)$.

(iii) Over $\mathbb{Z}$, consider the matrix

$$A = \begin{pmatrix} 9 & 2 & 16 \\ 11 & 25 & 13 \\ 5 & 4 & 6 \end{pmatrix}.$$

Since $\det A = -416 = -2^5 \cdot 13$, we have $\Sigma(A) = \Sigma_2(A) \cdot \Sigma_{13}(A)$. Now $\Sigma_{13}(A) = \mathrm{diag}(1,1,13)$, since $\mathrm{ord}_{13} 416 = 1$. And as for $\Sigma_2(A)$, using the fact that 9 is a unit in the ring $\mathbb{Z}_{((2))}$ we obtain

$$\begin{aligned} A \quad & \sim_2 \quad \begin{pmatrix} 1 & \frac{2}{9} & \frac{16}{9} \\ 11 & 25 & 13 \\ 5 & 4 & 6 \end{pmatrix} \sim_2 \begin{pmatrix} 1 & 0 & 0 \\ 11 & \frac{203}{9} & \frac{101}{9} \\ 5 & \frac{26}{9} & \frac{-59}{9} \end{pmatrix} \sim_2 \begin{pmatrix} 1 & 0 & 0 \\ 0 & 203 & 101 \\ 0 & 26 & -59 \end{pmatrix} \\ & \sim_2 \quad \mathrm{diag}(1,1,*). \end{aligned}$$

(Here notice that the very last "$\sim_2$" is immediate from the fact that $\mathrm{ord}_2 203 = 0$, so we need not explicitly compute the last entry in the diagonal matrix.) Therefore $\Sigma(A) = \mathrm{diag}(1,1,416)$.

What does all this have to do with lattices? Suppose L and M are R-lattices on V with respective bases $\mathbb{B} = \{v_1,\ldots,v_n\}$ and $\mathbb{C} = \{w_1,\ldots,w_n\}$; say $v_j = \sum_i \alpha_{ij} w_i$ for $1 \le j \le n$. Writing ι for the identity map on V, this gives $A = (\alpha_{ij}) = (\iota)^{\mathbb{C}}_{\mathbb{B}}$. Now suppose $\Sigma(A) = PAQ = \mathrm{diag}(s_1,\ldots,s_n)$, with $P, Q \in GL_n(R)$. Define new bases $\mathbb{B}' = \{v'_1,\ldots,v'_n\}$, $\mathbb{C}' = \{w'_1,\ldots,w'_n\}$ for L, M resp. by $(\iota)^{\mathbb{B}}_{\mathbb{B}'} = Q = (q_{ij})$ (that is, $v'_j = \sum_i q_{ij} v_i$) and $(\iota)^{\mathbb{C}}_{\mathbb{C}'} = P^{-1}$. Then

$$(\iota)^{\mathbb{C}'}_{\mathbb{B}'} = (\iota)^{\mathbb{C}'}_{\mathbb{C}} (\iota)^{\mathbb{C}}_{\mathbb{B}} (\iota)^{\mathbb{B}}_{\mathbb{B}'} = PAQ = \mathrm{diag}(s_1,\ldots,s_n)$$

and therefore $v'_j = s_j w'_j$ for $1 \le j \le n$. It follows that

$$\begin{aligned} M &= Rw'_1 \oplus Rw'_2 \oplus \cdots \oplus Rw'_n \\ \text{while} \qquad L &= R(s_1 w'_1) \oplus R(s_2 w'_2) \oplus \cdots \oplus R(s_n w'_n) \\ &= (s_1)w'_1 \oplus (s_2)w'_2 \oplus \cdots \oplus (s_n)w'_n \end{aligned}$$

with $s_i | s_{i+1}$ for $1 \le i \le n-1$; equivalently, the fractional ideals (s_i) satisfy $(s_1) \supseteq (s_2) \supseteq \cdots \supseteq (s_n)$. The fractional ideals $(s_1) \supseteq \cdots \supseteq (s_n)$ are the **invariant factors** of L in M. In summary, we have now proved the following important structure theorem.

Theorem 6.13 (Invariant Factor Theorem). *Let L and M be R-lattices on the n-dimensional vector space V over F. Then there is a basis $\{v_1, \dots, v_n\}$ for V and there are fractional ideals $(s_1) \supseteq (s_2) \supseteq \cdots \supseteq (s_n)$ such that*

$$\begin{aligned} M &= Rv_1 \oplus Rv_2 \oplus \cdots \oplus Rv_n \\ \textit{and} \qquad L &= R(s_1v_1) \oplus R(s_2v_2) \oplus \cdots \oplus R(s_nv_n) \\ &= (s_1)v_1 \oplus (s_2)v_2 \oplus \cdots \oplus (s_n)v_n. \end{aligned}$$

The fractional ideals (s_i) are uniquely determined by L and M.

Given the situation described in the theorem, here are some immediate observations:

(i) The invariant factors of M in L are $(s_n^{-1}) \supseteq (s_{n-1}^{-1}) \supseteq \cdots \supseteq (s_1^{-1})$.

(ii) $L \subseteq M$ if and only if $s_1 \in R$ (and therefore $s_i \in R$ for all i); and $L = M$ if and only if $s_i \in R^*$ for all i.

(iii) If $L \subseteq M$ then there is an isomorphism $M/L \approx \bigoplus_i (R/(s_i))$, and therefore $(M : L) = \prod_i (R : (s_i))$ if $\big(R : (s_i)\big) < \infty$ for all i.

Corollary 6.14. *Let M be an R-lattice of rank n, and let $\varphi \in$ $\mathrm{Hom}(M, M)$. Suppose $\varphi \sim A \in M_n(R)$ with respect to some basis of M, with $\sum(A) =$ $\mathrm{diag}(s_1, \dots, s_n)$. Then $M/\varphi(M) \approx \bigoplus_i (R/(s_i))$.*

Proof. In the argument leading to the Invariant Factor Theorem, take $L = \varphi(M)$ and $v_j = \varphi(w_j)$ for $1 \le j \le n$. □

Notice that if the map φ in the corollary has rank $k < n$, then

$$s_{k+1} = \cdots = s_n = 0.$$

Example 6.15. In the notation of the preceding corollary, if $R = \mathbb{Z}$ and $\det \varphi \neq 0$, then $\big(M : \varphi(M)\big) = |\det \varphi|$.

6.4. Lattices on Quadratic Spaces

Suppose L and M are R-lattices on regular quadratic F-spaces V and W of dimensions m and n, respectively. Recall from Chapter 2 that a **representation** from L into M is a homomorphism $\varphi : L \to M$ such that $B(x, y) = B(\varphi x, \varphi y)$ for all $x, y \in L$. (Here we have abused notation and used B for the bilinear forms on both V and W.) And the representation φ is an **isometry** onto M if it is an isomorphism. By restating the issue in terms of relations between the associated Gram matrices, we see that for the existence of a representation $L \to M$ it is necessary for there to be a representation $V \to W$. Therefore, to determine whether a representation $L \to M$ exists, there is no loss of generality in assuming that L and M

are sublattices of the same quadratic space. Isometric lattices on the same space are said to be in the same **isometry class** (or, more briefly, the same **class**) .

We would like to determine a complete set of computable invariants for an R-lattice up to isometry; that is, a set of invariants with the property that two lattices are isometric if and only if their invariant sets coincide. Up to this point the only invariants we have for a lattice on a quadratic space are its rank and discriminant. Now it is time to explore invariants that more fully characterize the quadratic form on the lattice.

Definition 6.16. Let L be an R-lattice on a quadratic F-space V.

(i) The **scale** of L is the fractional R-ideal sL generated by the set $\{B(x,y) \mid x, y \in L\}$.

(ii) The **norm** of L is the fractional R-ideal nL generated by $\{q(x) \mid x \in L\}$.

(iii) The **volume** of L is the fractional R-ideal $vL = (dL)$ generated by the discriminant.

Proposition 6.17. *Suppose L and M are lattices on the same n-dimensional quadratic space. Then*

(i) $2(sL) \subseteq nL \subseteq sL$. *Therefore if* $2 \in R^*$ *then* $nL = sL$.

(ii) *If* $L \subseteq M$ *then* $vL \subseteq vM$ *and* $dM|dL$. *Moreover, there are only finitely many lattices* K *satisfying* $L \subseteq K \subseteq M$.

(iii) $L \subsetneq M \Longrightarrow vL \subsetneq vM$.

(iv) $$L = J \perp M \Longrightarrow \begin{cases} sL = sJ + sM, \\ nL = nJ + nM, \\ vL = vJ \cdot vM. \end{cases}$$

(v) *If* $L = Rv_1 + \cdots + Rv_n$ *then*
$$sL = \sum_{i,j} B(v_i, v_j)R \quad \text{and} \quad nL = \sum_i q(v_i)R + 2sL.$$

(vi) $vL \subseteq (sL)^n$.

(vii) $sL^\alpha = \alpha(sL), nL^\alpha = \alpha(nL), vL^\alpha = \alpha^n(vL)$. *(Here* L^α *denotes the "scaled" lattice: the lattice* L *with the quadratic form* q^α *given by* $q^\alpha(x) = \alpha q(x)$ *for all* $x \in L$.*)*

Proof sketch. (Fill in the details!)

(i) $2B(x,y) = q(x+y) - q(x) - q(y) \in nL$.

(ii) and (iii) Use the Invariant Factor Theorem.

(iv) Clear.

(v) The result for sL is clear. As for nL, if $x = \sum_i \alpha_i v_i$ it follows that $q(x) = \sum_i \alpha_i^2 q(v_i) + 2\sum_{i<j} \alpha_i\alpha_j B(v_i, v_j)$ and hence $nL \subseteq \sum_i Rq(v_i) + 2sL$, so the result follows from (i).

(vi) Suppose $L \cong A = (a_{ij}) \in GL_n(F)$. Then

$$\det A = \sum_{\sigma \in S_n} \pm a_{1\sigma(1)} \cdots a_{n\sigma(n)} \in (sL)^n.$$

(vii) Clear. □

Remark 6.18. Clearly if lattices L and M are isometric then they represent the same elements, and an isometry $L \to M$ carries the vectors in L representing α onto the vectors in M doing so. In current terminology, two lattices representing exactly the same elements and representing each of those the same number of times are said to be **isospectral**. It is natural to wonder whether, conversely, isospectral lattices must be isometric. In fact the answer in general is *no*, though examples—for instance when $R = \mathbb{Z}$—can be very difficult to construct. See Conway ([**Con2**], Chapter 2) or the earlier paper by Earnest and Nipp [**EN**] for discussion of this issue.

6.5. Orthogonal Splitting and Triple Diagonalization

In the study of complex phenomena it is common practice to consider the decomposition of complicated objects into simpler ones, the "fundamental particles" of the subject, and then to consider how those simpler objects interact. For instance, physicists and chemists tell us about the wonders of molecules, atoms, and miscellaneous levels of subatomic structures before moving on to describe the world that confronts our senses. In our study of quadratic spaces over $\mathbb{Q}$ a key step was diagonalization: the decomposition of a space into the orthogonal sum of one-dimensional subspaces; and this was followed by a long trail of arguments, involving Hilbert symbols and then Hasse symbols, that described how those spaces interacted, and then a local-to-global argument to finish the job. But it all started with orthogonal decomposition à la Gram–Schmidt.

Why should an orthogonal splitting simplify matters? Because the form on a regular space or a lattice on such a space is described by a symmetric Gram matrix $A \in GL_n(F)$, hence by $1 + 2 + \cdots + n = \frac{n(n+1)}{2}$ elements of F; while an orthogonal decomposition produces zeros in the Gram matrix, effectively reducing the number of elements describing the form; and in its most extreme case—when there is an orthogonal basis—to just n elements.

As we begin to contemplate the classification and representation problems for lattices, it is reasonable to ask whether every lattice of rank $n \geq 2$

has an orthogonal splitting into sublattices of smaller rank, and the following examples show that this is not generally the case.

Example 6.19. (i) With $R = \mathbb{Z}$ and $n \geq 2$, consider the $\mathbb{Z}$-lattice

$$A_n \cong \begin{pmatrix} 2 & 1 & & & \\ 1 & 2 & 1 & & \\ & \ddots & \ddots & \ddots & \\ & & \ddots & \ddots & 1 \\ & & & 1 & 2 \end{pmatrix}$$

in the basis $\{v_1, \ldots, v_n\}$. (Note: It should be understood that except for the main diagonal and the sub- and super-diagonal all the other entries in this "triple diagonal" matrix are 0's.) We claim A_n is **indecomposable**; that is, there do not exist nonzero sublattices J_1, J_2 such that $A_n = J_1 \perp J_2$.

An induction argument shows that for all n we have $dA_n = n + 1$, and from this it follows that the underlying $\mathbb{Q}$-space is positive definite. Next observe that if $0 \neq (x_1, \ldots, x_n) \in \mathbb{Z}^n$ then $q(\sum_i x_i v_i) = 2(\sum_i x_i^2) + 2\sum_{i<j} x_i x_j$, a positive even integer. Therefore, if there were a splitting $A_n = J_1 \perp J_2$, each v_i would have to be an element of component J_1 or component J_2. But since $B(v_i, v_{i+1}) \neq 0$, in fact each v_i would have to belong to the component containing v_1. Therefore the other component would be 0, proving the claim.

Incidentally, if $n = 2$ we could have used a discriminant argument, as follows: If $A_2 \cong \langle a, b\rangle$ then, since $dA_2 = 3$, a and b would be odd, contradicting the fact that $q(v) \equiv 0 \pmod 2$ for all v in the lattice. This discriminant argument when $n = 2$ also holds when A_2 is considered as a lattice over the ring $\mathbb{Z}_2$ of 2-adic integers. (That is, any $\mathbb{Z}_2$-lattice having the given matrix as Gram matrix is indecomposable.) But we will see in the next chapter that for every prime p, every $\mathbb{Z}_p$-lattice of rank ≥ 3 has a nontrivial orthogonal splitting.

(ii) The number 2 often provides special difficulties in the theory of quadratic forms over $\mathbb{Z}$, in part because $(\mathbb{Q}_2^* : \mathbb{Q}_2^{*2}) = 8$ while $(\mathbb{Q}_p^* : \mathbb{Q}_p^{*2}) = 4$ when $p \neq 2$, and also because for a $\mathbb{Z}$-lattice L the norm ideal nL need not be equal to the scale ideal sL. For example, $nA_n = 2\mathbb{Z} \subsetneq sA_n = \mathbb{Z}$. So it is of interest to see an instance of indecomposability over a ring in which 2 is a unit, when neither of these difficulties occurs. So consider a lattice

$$L \cong \begin{pmatrix} x & 1 \\ 1 & x+2 \end{pmatrix}$$

in a basis $\{v_1, v_2\}$ over the polynomial ring $\mathbb{F}_3[x]$. We will use "∂" for the degree function. Suppose there is a splitting $L \cong \langle f, g\rangle$, with $f, g \in \mathbb{F}_3[x] - \{0\}$ and $\partial f \leq \partial g$. Then, since $dL = x^2 + 2x + 2$, either $\partial f = \partial g = 1$ or

$\partial f = 0$. But the equation $\partial f = \partial g = 1$ is impossible, since dL is irreducible. Now suppose $\partial f = 0$ and $f = q(av_1 + bv_2) = a^2x + b^2(x+2) + 2ab$, with $a, b \in \mathbb{F}_3[x]$. Here $\partial(ab) < \max\{\partial(a^2x), \partial b^2(x+2)\}$; so to get $\partial f = 0$ it is necessary that $\partial a = \partial b = m$, say, and hence

$$f = (\alpha^2 + \beta^2)x^{2m+1} + \{\text{lower degree terms}\}$$

where α, β denote the leading coefficients of a, b respectively. But the space $\langle 1, 1\rangle$ is anisotropic over $\mathbb{F}_3$, since $-1 \notin \mathbb{F}_3^{*2}$, and therefore $\alpha = \beta = 0$, contradicting the fact that $f \neq 0$. Therefore L is indecomposable, as claimed.

(iii) By way of contrast with (ii), and as an indicator that caution is needed, consider a lattice

$$M \cong \begin{pmatrix} x & 1 \\ 1 & x+2 \end{pmatrix}$$

in a basis $\{w_1, w_2\}$ over $\mathbb{F}_5[x]$. Here $q(2w_1 + w_2) = 1$, and hence there is an orthogonal splitting

$$M \cong \langle 1, x^2 + 2x + 4\rangle.$$

The following theorem, published in 1848, shows that while a lattice cannot be expected to have an orthogonal splitting, it does have a basis with a substantial dose of orthogonality.

Theorem 6.20 (Jacobi). *Let L be an R-lattice of rank n on a quadratic F-space. Then L has a basis $\{v_1, \ldots, v_n\}$ such that $B(v_i, v_j) = 0$ whenever $|i - j| \geq 2$, and hence a "triple-diagonal" Gram matrix:*

$$L \cong \begin{pmatrix} a_{11} & a_{12} & & & \\ a_{12} & a_{22} & a_{23} & & \\ & \ddots & \ddots & \ddots & \\ & & \ddots & \ddots & a_{n-1,n} \\ & & & a_{n-1,n} & a_{nn} \end{pmatrix}.$$

Proof. Suppose $L \cong A = (a_{ij})$ in the basis $\{v_1, \ldots, v_n\}$. Without loss of generality we can assume that $A \in M_n(R)$. (Scale as needed.) If $a_{1j} \neq 0$ for some $j \geq 3$, let $\delta = \gcd(a_{12}, \ldots, a_{1n})$; then there exist $x_2, \ldots, x_n \in R$ such that $\delta = \sum_{j=2}^n x_j a_{1j}$. Since $\gcd(x_2, \ldots, x_n) = 1$, there is a unimodular matrix $U = (u_{ij})_{2 \leq i,j \leq n} \in GL_{n-1}(R)$ having ${}^t(x_2, \ldots, x_n)$ as its first column. For $2 \leq j \leq n$ replace each v_j by $v_j' = \sum_{i=2}^n u_{ij} v_i$. Then the associated Gram matrix for L in the new basis $\{v_1, v_2' \ldots, v_n'\}$ has first row of the form $(a_{11}, \delta, a_{13}', \ldots, a_{1n}')$ with $\delta | a_{1j}'$ for $3 \leq j \leq n$. Subtraction of suitable multiples of v_2' from $v_3', \ldots, v_n'$ yields a new basis $\{v_j''\}$ with respect to which L has a Gram matrix with first row $(a_{11}, \delta, 0, \ldots, 0)$. Now repeat the argument on the sublattice spanned by $\{v_2'', \ldots, v_n''\}$. □

In applications of integral quadratic forms to other areas of mathematics—notably topology, group theory, and coding theory—unimodular lattices over $\mathbb{Z}$ have been especially prominent. We call an R-lattice L **unimodular** if its Gram matrix is unimodular; that is, $L \cong A = (a_{ij}) \in GL_n(R)$. (So $a_{ij} \in R$ for all i, j, and $|A| \in R^*$.) In particular, a $\mathbb{Z}$-lattice L is unimodular if and only if $sL \subseteq \mathbb{Z}$ and $dL = \pm 1$. The classification of unimodular lattices, especially over $\mathbb{Z}$, is a major open problem, and the following supplement (from [**N2**], in 1994) to Jacobi's Theorem may turn out to be useful for this purpose.

Theorem 6.21 (Newman). *Let L be a unimodular R-lattice of rank $n \geq 3$. Then there exist $a_1, a_2, \ldots, a_n, c \in R$ such that*

$$L \cong \begin{pmatrix} a_1 & 1 & & & \\ 1 & a_2 & \ddots & & \\ & \ddots & \ddots & 1 & \\ & & 1 & a_{n-1} & c \\ & & & c & a_n \end{pmatrix}.$$

Proof. By Jacobi's theorem there is a basis $\mathbb{B} = \{v_1, \ldots, v_n\}$ for L such that the associated Gram matrix $A = (a_{ij}) \in GL_n(R)$ is in triple-diagonal form. So $B(v_i, v_j) = 0$ whenever $|i-j| \geq 2$. Then $\gcd(a_{n-1,n}, a_{nn}) = 1$, since $|A| \in R^*$, and so $\lambda a_{n-1,n} + \varepsilon a_{nn} = 1$ for some $\lambda, \varepsilon \in R$. Define $v_1' = v_1 + \lambda v_{n-1} + \varepsilon v_n$. Then $\{v_1', v_2, \ldots, v_n\}$ is also a basis for L, and $B(v_1', v_n) = 1$. Interchange v_n with v_2 in the basis, and relabel v_n as v_2', getting a new Gram matrix with first row of the form $\begin{pmatrix} a_{11} & 1 & * \cdots * \end{pmatrix}$. Subtracting suitable multiples of v_2' from the subsequent basis vectors changes the first row to $\begin{pmatrix} a_{11} & 1 & 0 \ldots 0 \end{pmatrix}$. Now repeat the argument as needed on the sublattice spanned by v_2' and its followers in the most current basis. $\square$

Remark 6.22. When $R = \mathbb{Z}$ the classification of lattices up to isometry is an outstanding open problem, and we will see evidence in Chapter 9 that it is fiendishly difficult, even in the unimodular case. For two $\mathbb{Z}$-lattices to be isometric, it is necessary that the lattices they generate over the rings of p-adic integers be isometric, and the p-adic classification of lattices—the so-called local theory—has been settled. (We will cover most of that in Chapter 8.) But there is no Hasse–Minkowski Theorem for $\mathbb{Z}$-lattices; that is, two lattices on the same quadratic $\mathbb{Q}$-space can be "locally isometric" over all the rings of p-adic integers without being isometric over $\mathbb{Z}$. Extracting invariants from a $\mathbb{Z}$-lattice's Gram matrix that go beyond the invariants of the local theory remains a challenge. One might hope that the very special triple-diagonalization produced in Newman's Theorem (6.21) might lead to invariants of this kind in the unimodular case, but so far they have not been forthcoming.

6.6. The Dual of a Lattice

Now let V be a regular space, with dual space $V^* = \mathrm{Hom}(V, F)$, and $f_B : V \to V^*$ given by $v \mapsto B(v, \cdot)$ the associated isomorphism; and let $\{v_i\}_{1\le i\le n}$ be a basis of V, with $\{v_i^*\}_{1\le i\le n}$ the dual basis of V^*. Define $v_i^\sharp = f_B^{-1}(v_i^*)$ for $1 \le i \le n$. Then $\{v_1^\sharp, \ldots, v_n^\sharp\}$ is a basis for V, and

$$B(v_i^\sharp, v_j) = \left(f_B(v_i^\sharp)\right)(v_j) = v_i^*(v_j) = \delta_{ij} \quad \text{for } 1 \le j \le n.$$

Moreover, since f_B is an isomorphism, $\{v_i^\sharp\}$ is the *only* basis of V with this property. (Each $f_B^{-1}(v_i^*)$ is a singleton.) We call $\{v_i^\sharp\}_{1\le i\le n}$ the basis of V **dual** to $\{v_i\}_{1\le i\le n}$. Thus for a given quadratic space V with basis $\{v_i\}$, an informal mention of the "dual basis" may refer to the basis $\{v_i^*\}$ of V^*, as in linear algebra, or to the unique basis $\{v_i^\sharp\}$ of V satisfying $B(v_i^\sharp, v_j) = \delta_{ij}$. Which of these is intended must be determined from the context. We have now proved the first statement of the following proposition; the second statement is an exercise for the reader.

Proposition 6.23. *Let V be a regular space, with $V \cong A = (a_{ij})$ in basis $\{v_i\}_{1\le i\le n}$. Then there is a unique basis $\{v_i^\sharp\}_{1\le i\le n}$ for V such that $B(v_i^\sharp, v_j) = \delta_{ij}$ for $1 \le i, j \le n$. Moreover, if $A^{-1} = (\alpha_{ij})$ then $v_j^\sharp = \sum_i \alpha_{ij} v_i$, and $V \cong A^{-1}$ in $\{v_1^\sharp, \ldots, v_n^\sharp\}$.*

Clearly V has a basis that is equal to its dual if and only if $V \cong \langle 1, \ldots, 1\rangle$.

Now let us apply the dual basis concept to lattices. First some intuition. Imagine that a mirror in the center of a darkened room is behind a curtain just thick enough to hide the mirror from you, and your job is to use a powerful flashlight to determine the mirror's orientation without getting close to the mirror. Your method: shine your light through the curtain onto the mirror and observe the reflection. When your position is such that the reflection comes directly back to you, then you will know that your light beam is orthogonal to the plane of the mirror. If $\mathbb{B} = \{v_1, v_2, v_3\}$ is a basis for $\mathbb{R}^3$, with v_1 and v_2 in the plane of the mirror, then the vector $v_3^\sharp$ in the dual basis $\mathbb{B}^\sharp$ is in the line orthogonal to the mirror.

The preceding paragraph only hints at the sophisticated approaches used by solid-state physicists to determine the crystalline structure of materials through the study of the reflection and refraction of assorted forms of radiation by those materials. (See a book on solid state physics or Senechal's *Quasicrystals and Geometry* [**Sen**] for more details.) Such a structure is modeled by a $\mathbb{Z}$-lattice $L \in \mathbb{R}^3$ with basis $\mathbb{B} = \{v_1, v_2, v_3\}$. (Envision the crystal structure by imagining identical clusters of atoms situated at each lattice point.) Associated with $\mathbb{B}$ is a parallelpiped called the **fundamental region**

$$\{\,\alpha_1 v_1 + \alpha_2 v_2 + \alpha_3 v_3 \mid 0 \leq \alpha_i < 1\,\}$$

and $\mathbb{R}^3$ is filled with nonoverlapping translates of this region. It is the shape of the fundamental region that the physicists are after with their beams, and the preceding paragraph suggests that the dual basis $\mathbb{B}^\sharp$ is likely to be relevant to the mathematical model. The **dual lattice** of L is the lattice generated by $\mathbb{B}^\sharp$. (In fact the definition we will give is independent of the choice of basis.) As it turns out, the dual lattice concept is useful in higher dimensions as well, and also in the study of lattices over other domains R, so we will introduce it in the general setting of this chapter: with R a principal ideal domain and F its quotient field.

Definition 6.24. Let L be an R-lattice on the quadratic F-space V. Then the **dual lattice** of L is the R-lattice

$$L^\sharp = \{\,x \in V \mid B(x, L) \subseteq R\,\}.$$

(The physicists also call $L^\sharp$ the **reciprocal** or **polar** lattice.)

The following proposition, whose proof (except for part (vii)) is left as an exercise for the reader, gives the elementary properties of the dual lattice.

Proposition 6.25. *Let L be an R-lattice on the n-dimensional quadratic F-space V. Then*

(i) *If L has basis $\mathbb{B} = \{v_1, \ldots, v_n\}$, then $L^\sharp$ has basis the dual basis $\mathbb{B}^\sharp = \{v_1^\sharp, \ldots, v_n^\sharp\}$. (Hence the name "dual lattice.")*

(ii) *If L is "integral" in the sense that $sL \subseteq R$, then $L \subseteq L^\sharp$.*

(iii) $(L^\sharp)^\sharp = L$, $\quad (L_1 \perp L_2)^\sharp = L_1^\sharp \perp L_2^\sharp$.

(iv) $dL^\sharp = (dL)^{-1}$ *and hence* $vL^\sharp = (vL)^{-1}$.

(v) $(\alpha L)^\sharp = \alpha^{-1} L^\sharp$ *for all* $\alpha \in F^*$.

(vi) *If K is also a lattice on V then*
 (a) $L \subseteq K \Longrightarrow L^\sharp \supseteq K^\sharp$;
 (b) $(L + K)^\sharp = L^\sharp \cap K^\sharp$.

(vii) *L is unimodular if and only if $L = L^\sharp$.*

Proof of (vii). ("$\Longrightarrow$") If L is unimodular then L is integral and hence $L \subseteq L^\sharp$. But if $L \cong A \in GL_n(R)$ then $L^\sharp \cong A^{-1} \in GL_n(R)$, and hence $vL^\sharp = (vL) = R$. Therefore $L = L^\sharp$.

("$\Longleftarrow$") If $L = L^\sharp$ then L is integral; also $vL = (vL)^{-1}$, and hence $vL = R$; therefore $dL \in R^*$, and so L is unimodular. □

Exercises

Note: In the following exercises, wherever R and F are unspecified it should be understood that R is a principal ideal domain with quotient field F, a field of characteristic not 2.

(1) (i) Over the ring $\mathbb{Z}_2$ of 2-adic integers, construct a unimodular matrix with first row $(3, 6, 5)$. [Note: Lemma 6.6 should not be needed for this part, and in fact the answer should be virtually instantaneous.]
(ii) Do the same over $\mathbb{Z}$.

(2) Let L be an $\mathbb{F}_5[x]$-lattice with basis $\{v_1, v_2, v_3\}$, and let

$$w_1 = (x^2 + 3x + 2)v_1 + (x^2 + 1)v_2 + (x^2 + 2x + 2)v_3.$$

Show that w_1 is a primitive vector in L, and construct vectors w_2, w_3 such that $\{w_1, w_2, w_3\}$ is a basis for L.

(3) The product of fractional ideals (α) and (β) is the fractional ideal $(\alpha\beta)$. (Here α, β are elements of F^*.) In other words, multiplication in $I(R)$ is like multiplication in F^*, but with parentheses inserted appropriately. Is the mapping $F^* \to I(R)$ given by $\alpha \mapsto (\alpha)$ a group isomorphism? If not, what is its kernel?

(4) Prove that the following conditions are equivalent for a principal ideal domain R:
(i) The group $I(R)$ of fractional R-ideals is linearly ordered by inclusion.
(ii) R has a unique nonzero prime ideal.
(iii) $I(R)$ is cyclic.

(5) This exercise involves matrices over the ring $\mathbb{Z}_p$ of p-adic integers, with p a fixed prime number.
(i) How many equivalence classes (i.e., classes with respect to matrix equivalence) are there of matrices $A \in M_4(\mathbb{Z}_p)$ with $\mathrm{ord}_p(\det A) = 4$?
(ii) Would the answer to part (i) change if the involved matrices were required to be symmetric?

(6) Use the Local-Global Principle for Smith normal form to compute the Smith form of the $\mathbb{Z}$-matrix

$$\begin{pmatrix} 0 & 2 & 6 \\ 6 & 4 & 24 \\ 18 & 4 & 78 \end{pmatrix}.$$

(7) Consider the 3×3 $\mathbb{Z}$-matrix J whose entries are all 1's, and let A be its characteristic matrix:

$$A = XI_3 - J = \begin{pmatrix} x-1 & -1 & -1 \\ -1 & x-1 & -1 \\ -1 & -1 & x-1 \end{pmatrix}.$$

Use the Local-Global Principle to compute the Smith form $\Sigma_{\mathbb{Q}[x]}(A)$.

(8) Let $A, B \in M_n(R)$.
(i) Show that if $B \in GL_n(R)$ then $\Sigma(AB) = \Sigma(A)$.
(ii) Use the Local-Global Principle to prove that if $\det A$ and $\det B$ are relatively prime, then

$$\Sigma(AB) = \Sigma(A) \cdot \Sigma(B).$$

(9) If $k, n \in \mathbb{N}$, define $p_n(k)$ to be the number of ways to express k as a sum of at most n positive integers, not counting reorderings of the summands. For example, $p_3(4) = 4$ because 4 can be written as

$$4, \quad 3+1, \quad 2+2, \quad 2+1+1.$$

Use functions of the form $p_n(\cdot)$ to obtain a formula for the number of equivalence classes of matrices in $M_n(R)$ having nonzero determinant $d = q_1^{\nu_1} \cdots q_r^{\nu_r}$, where the q_i are nonassociated prime elements of R.

(10) As every linear algebra student knows, for an $n \times n$ matrix $A = (a_{ij})$ the expression for the determinant

$$\det A = \sum_{\sigma \in S_n} (-1)^{\mathrm{sgn}(\sigma)} a_{1\sigma(1)} \cdots a_{n\sigma(n)}$$

has $n!$ terms. If A is a triple-diagonal matrix, what is the maximum number $T(n)$ of those terms that can be nonzero? For the answer, it is sufficient to relate the sequence $T(1), T(2), \ldots$ to a more famous sequence.

(11) Suppose a set X of necessary and sufficient conditions has been found on sequences $a_1, \ldots, a_n$ in R for there to be an isometry of an R-lattice

$$L \cong A = \begin{pmatrix} a_1 & 1 & & & \\ 1 & a_2 & 1 & & \\ & \ddots & \ddots & \ddots & \\ & & \ddots & \ddots & 1 \\ & & & 1 & a_n \end{pmatrix}$$

onto a given R-lattice M. Obtain a set of necessary and sufficient conditions on $a_1, \ldots, a_n$ for there to be an isometry of a lattice with a Gram matrix of this form onto the scaled lattice M^{-1}.

(12) Let $R = \mathbb{Z}$, and suppose A is a matrix of the kind displayed in the preceding exercise, with $\det A = 24$. Determine A's invariant factors and elementary divisors.

(13) Suppose L and M are $\mathbb{Z}$-lattices on the 3-dimensional $\mathbb{Q}$-space V, with $L \supseteq M$ and $(L : M) = 36$. What are the possibilities for the chain $(s_1) \supseteq (s_2) \supseteq (s_3)$ of invariant factors of M in L; and if instead $\dim V = 4$, how would this affect the answer?

(14) Let L be a $\mathbb{Z}_p$-lattice on a quadratic $\mathbb{Q}_p$-space, and suppose $L \cong A = (a_{ij}) \in M_n(\mathbb{Q}_p)$. The scale sL and the norm nL are then determined by the integers $\operatorname{ord}_p a_{ij}$. How?

(15) Let L be an R-lattice on a quadratic F-space, and let $\mathfrak{A}$ be a fractional R-ideal. Show that if L represents every element of $\mathfrak{A}$ then $nL \supseteq \mathfrak{A}$, but the converse need not be the case.

(16) Let L be a $\mathbb{Z}$-lattice on an 8-dimensional quadratic $\mathbb{Q}$-space V, and suppose $L \cong \langle 1, \ldots, 1 \rangle$ in the basis $\mathbb{B} = \{v_1, \ldots, v_8\}$.
(i) Use L and its given basis to construct a lattice M on V and a basis $\mathbb{B}' = \{w_1, \ldots, w_8\}$ for M such that

$$M \cong \begin{pmatrix} 2 & 1 & & & & & & \\ 1 & 2 & 1 & & & & & \\ & 1 & 2 & 1 & & & & \\ & & 1 & 2 & 1 & & & \\ & & & 1 & 2 & 1 & & \\ & & & & 1 & 2 & 2 & \\ & & & & & 2 & 4 & 1 \\ & & & & & & 1 & 2 \end{pmatrix}$$

in the basis $\mathbb{B}'$. [Hint: Such a basis $\mathbb{B}'$ can be constructed using the vector $w_8 = \frac{1}{2}(v_1 + \cdots + v_8)$.]
(ii) List the invariant factors of M in L in descending order by inclusion, as in the Invariant Factor Theorem.

(17) Let L be an R-lattice on the regular quadratic F-space V; and assume $sL \supseteq R$. Show that if L represents all of R, then V is universal. Is the converse true?

(18) (i) Obtain a triple-diagonal matrix congruent over $\mathbb{Z}$ to

$$\begin{pmatrix} 7 & 4 & 6 \\ 4 & 2 & 9 \\ 6 & 9 & 5 \end{pmatrix}.$$

(ii) Show that part (i) could have been carried out in fewer steps over the ring $\mathbb{Z}_3$ of 3-adic integers.

(19) Consider the quadratic space $\mathbb{R}^2$ with the usual dot product as the symmetric bilinear form, and suppose $\mathbb{R}^2 \cong \begin{pmatrix} 5 & 3 \\ 3 & 2 \end{pmatrix}$ with respect to basis $\mathbb{B} = \{v_1, v_2\}$. Determine the dual basis $\mathbb{B}^\sharp = \{v_1^\sharp, v_2^\sharp\}$, and then draw a sketch of $\mathbb{R}^2$ showing both bases. (Suggestion: Represent the four vectors as arrows starting at the origin, and use different colors for $\mathbb{B}$ and $\mathbb{B}^\sharp$.)

(20) Suppose V is a quadratic $\mathbb{Q}$-space with basis $\mathbb{B} = \{v_1, \ldots, v_n\}$, and suppose $V \cong A \in GL_n(\mathbb{Z})$ with respect to $\mathbb{B}$. Show that $\mathbb{B}$ and the dual basis $\mathbb{B}^\sharp = \{v_1^\sharp, \ldots, v_n^\sharp\}$ span the same $\mathbb{Z}$-lattice on V.

(21) Let L be an R-lattice on a quadratic F-space V.
(i) Show that all the Gram matrices of L have the same invariant factors.
(ii) How are the invariant factors of the Gram matrices of L and its dual lattice $L^\sharp$ related?

6.7. Modular Lattices

We have already defined *unimodular* lattices: lattices with unimodular Gram matrices. Now we give a useful extension of this idea to *modular* lattices: lattices obtained by scaling unimodular lattices. This idea will be especially important in the local classification theory (Chapter 8), where we will see that every lattice over the integers of a local field is an orthogonal sum of modular components, and such a decomposition will provide the key invariants of the isometry class of the lattice. Here is the formal definition.

Definition 6.26. Let $\mathfrak{A} = (\alpha) \in I(R)$. An R-lattice L is said to be **$\mathfrak{A}$-modular** if $L \cong \alpha U$ for some $U \in GL_n(R)$. We say L is **modular** if L is $\mathfrak{A}$-modular for some $\mathfrak{A} \in I(R)$.

An elementary observation: the definition of $\mathfrak{A}$-modularity is independent of the particular generator α chosen for $\mathfrak{A}$.

Once again there are some properties to check, mostly left to the reader. We will prove one part here to give the flavor of the arguments.

Proposition 6.27. *Let L be a lattice on V, with* $\dim V = n$*, and let $\mathfrak{A} \in I(R)$. Then*

(i) *L is $\mathfrak{A}$-modular if and only if $sL = \mathfrak{A}$ and $vL = \mathfrak{A}^n$.*

(ii) *L is R-modular if and only if L is unimodular.*

(iii) *Let $\beta \in F^*$. Then L is $\mathfrak{A}$-modular $\Longleftrightarrow$ βL is $\beta^2\mathfrak{A}$-modular $\Longleftrightarrow$ L^β is $\beta\mathfrak{A}$-modular.*

(iv) *If x is a nonzero anisotropic vector in V then the lattice Rx is $(q(x))$-modular.*

(v) *The following three conditions are equivalent:*
(a) $L = \{ x \in V \mid B(x, L) \subseteq \mathfrak{A} \}$;
(b) $L = \mathfrak{A}L^\sharp$;
(c) L *is* $\mathfrak{A}$*-modular.*

Proof of (b)$\Longleftrightarrow$(c) in part (v). If L is (α)-modular then $B(x, L) \subseteq (\alpha)$ for all $x \in L$ and hence $L \subseteq \alpha L^\sharp$. But $L \cong \alpha U$ for some $U \in GL_n(R)$, and hence $\alpha L^\sharp \cong \alpha U^{-1}$, so $v(\alpha L^\sharp) = vL$. Therefore $L = \alpha L^\sharp$.

Conversely, if $L = \alpha L^\sharp$ then $B(x, L) \subseteq (\alpha)$ for all $x \in L$, giving $sL \subseteq (\alpha)$ and hence

$$vL \subseteq (sL)^n \subseteq (\alpha^n).$$

But

$$vL = v(\alpha L^\sharp) = \alpha^{2n}(vL^\sharp) = \alpha^{2n}(vL)^{-1},$$

which gives $vL = (\alpha^n) = (sL)^n$, so $sL = (\alpha)$; hence L is (α)-modular. □

Suppose M is a unimodular sublattice of an integral lattice L; lsay $\operatorname{rank} M = m$ and $\operatorname{rank} L = n$. It is not hard to see that M is a direct summand of L as a module. For we can choose a basis $\{v_1, \dots, v_n\}$ of L such that $\{v_1, \dots, v_m\}$ spans the subspace FM spanned by M. Then $M \subseteq K = Rv_1 + \cdots + Rv_m$. Hence $R = vM \subseteq vK \subseteq R$, and so $M = K$. (Incidentally, a module direct summand of a lattice—independent of any quadratic form that may be on the lattice—is said to be a **primitive** sublattice.) The following extremely useful theorem goes farther and tells us that M is actually an orthogonal component of L. We will give two proofs. The first proof has the advantage of being without the swarm of subscripts found in the second; moreover, the argument carries over directly to lattices over more general domains, such as the rings of integers of algebraic number fields. (These rings need not be principal ideal domains.) The second proof gives an actual algorithm for constructing the asserted splitting in our situation: when the coefficient ring is a principal ideal domain. (But also see the remark following the proofs.)

Theorem 6.28. *Let L be a lattice on the regular space V, and let $M \subseteq L$ be an $\mathfrak{A}$-modular sublattice. Then M splits L (i.e., $L = M \perp L'$ for some sublattice L') if and only if $B(M, L) \subseteq \mathfrak{A}$. In particular, every unimodular sublattice of an integral lattice L splits L.*

Proof 1. Implication "$\Longrightarrow$" is clear; we will prove the converse.

There is a splitting $V = FM \perp W$ for some subspace W. We will show that $L = M \perp (L \cap W)$, and for this only "$\subseteq$" requires proof.

Assume $B(M, L) \subseteq \mathfrak{A}$, and let $x \in L$. Say $x = y + w$ with $y \in FM$ and $w \in W$. Then

$$B(y, M) = B(x - w, M) = B(x, M) \subseteq B(L, M) \subseteq \mathfrak{A}.$$

But $M = \{ x \in FM \mid B(x, M) \subseteq \mathfrak{A} \}$ because M is $\mathfrak{A}$-modular. Therefore $y \in M$, *a fortiori* $y \in L$, and hence $w \in L \cap W$. □

Proof 2. By scaling as needed, without loss of generality we can suppose that $sL = R$ and M is unimodular. Let $\mathbb{B} = \{v_1, \ldots, v_m, \ldots, v_n\}$ be a basis for L, with $\{v_1, \ldots, v_m\}$ a basis for M, and suppose $M \cong A = (a_{ij})$ with respect to this basis, with $A \in GL_m(R)$. We claim that we can replace each v_j for which $j \geq m + 1$ by a vector of the form

$$v'_j = v_j + \sum_{k=1}^{m} \lambda_{kj} v_k$$

with $\lambda_{kj} \in R$, in such a way that $B(v'_j, v_i) = 0$ for $1 \leq i \leq m$. Once we have shown this it will follow that the set $\mathbb{B}' = \{v_1, \ldots, v_m, v'_{m+1}, \ldots, v'_n\}$ is a basis for L, because the transition matrix from $\mathbb{B}$ to $\mathbb{B}'$ is unimodular; and hence

$$L = M \perp (Rv'_{m+1} + \cdots + Rv'_n\}.$$

Fix j satisfying $m + 1 \leq j \leq n$. Then for a vector of the form given above, the condition that $B(v'_j, v_i) = 0$ for all i satisfying $1 \leq i \leq m$ is equivalent to the m-tuple $(\lambda_{1j}, \lambda_{2j}, \ldots, \lambda_{mj})$ being a simultaneous solution to the following system of m linear equations:

$$\begin{aligned}
a_{11}x_1 + a_{12}x_2 + \cdots + a_{1m}x_m &= -a_{1j}, \\
a_{21}x_1 + a_{22}x_2 + \cdots + a_{2m}x_m &= -a_{2j}, \\
&\vdots \\
a_{m1}x_1 + a_{m2}x_2 + \cdots + a_{mm}x_m &= -a_{mj}.
\end{aligned}$$

Now, since M is unimodular we have $\det A \in R^*$, hence by Cramer's rule the system has a unique solution in R. □

Remark 6.29. The uniqueness of the soution in F^m to the system of equations in Proof 2 of the preceding theorem tells us that there is only one vector in V of the form $v'_j = v_j + \sum_{k=1}^{m} \lambda_{kj} v_k$, with $\lambda_{kj} \in F$, that satisfies $B(v'_j, M) = 0$, and from the modularity of M it follows that the coefficients λ_{kj} are in R and hence $v'_j \in L$. This suggests an algorithm for computing v'_j, for $m + 1 \leq j \leq n$, that avoids solving the system of equations. Namely, first find an orthogonal basis $\{v_1^*, \ldots, v_m^*\}$ for the subspace FM; then, à la

Gram–Schmidt, define

$$v_j' = v_j - \sum_{k=1}^{m} \Big(\frac{B(v_j, v_k^*)}{q(v_k^*)}\Big) v_k^*.$$

The following concept and the subsequent lemma will be especially useful in applications of modular lattices.

Definition 6.30. Let (α) be a fractional R-ideal, and let L be an R-lattice on a quadratic F-space. Define

$$L^{(\alpha)} = \{\, x \in L \mid B(x, L) \subseteq (\alpha)\,\} = L \cap \alpha L^{\sharp}.$$

Lemma 6.31. *If $(\alpha) \subseteq (\beta)$ and L is (β)-modular, then $L^{(\alpha)} = \alpha\beta^{-1}L$.*

Proof. Since $\alpha\beta^{-1} \in R$, we have $L^{(\alpha)} = L \cap \alpha L^{\sharp} = L \cap \alpha\beta^{-1} L = \alpha\beta^{-1} L$. □

6.8. Maximal Lattices

Definition 6.32. Let $\mathfrak{A}$ be a fractional R-ideal. A lattice M on a regular quadratic F-space V is **$\mathfrak{A}$-maximal** if $nM \subseteq \mathfrak{A}$ and M is maximal with respect to this property. Equivalently, $q(v) \in \mathfrak{A}$ for all $v \in M$, and no lattice properly containing M has this property.

Remark 6.33. (1) If $L = L_1 \perp L_2$ and L is $\mathfrak{A}$-maximal, then L_1 and L_2 are $\mathfrak{A}$-maximal on their respective spaces. (The converse is not true in general. For instance see Exercise 8 at the end of Section 9.4.)

(2) If the lattice L satisfies $nL \subseteq \mathfrak{A}$ then, since $2sL \subseteq nL$, we have

$$sL \subseteq \frac{1}{2} nL \subseteq \frac{1}{2}\mathfrak{A} \quad \text{and therefore} \quad vL \subseteq (sL)^n \subseteq \left(\frac{1}{2}\mathfrak{A}\right)^n.$$

Recall that a principal ideal domain is a Noetherian ring: Every ascending chain of ideals is finite. It follows that if $\mathfrak{A}_1$ and $\mathfrak{A}_2$ are fractional R-ideals satisfying $\mathfrak{A}_1 \subset \mathfrak{A}_2$, then every ascending fractional ideal chain

$$\mathfrak{A}_1 \subset \mathfrak{B}_1 \subset \mathfrak{B}_2 \subset \cdots \subset \mathfrak{A}_2$$

is finite. Returning now to our lattice L, we see that every ascending chain $L \subset L_1 \subset L_2 \subset \cdots$ in which $nL_i \subseteq \mathfrak{A}$ for all i must have finite length; because the associated volumes satisfy $vL \subset vL_1 \subset \cdots \subset (\frac{1}{2}\mathfrak{A})^n$. Therefore every lattice with $nL \subseteq \mathfrak{A}$ is contained in an $\mathfrak{A}$-maximal lattice.

(3) If L is $\mathfrak{A}$-modular and $nL = 2\mathfrak{A}$, then L is $2\mathfrak{A}$-maximal. Therefore, if $2 \in R^*$ then every $\mathfrak{A}$-modular lattice is $\mathfrak{A}$-maximal; and in particular every unimodular R-lattice is R-maximal.

The reader should check that if L is an $\mathfrak{A}$-maximal lattice on an F-space V and $\lambda \in F^*$, then L^{λ} is $\lambda\mathfrak{A}$-maximal on V^{λ}.

Lemma 6.34. *An isotropic vector x in a maximal R-lattice L is contained in a binary orthogonal component of L.*

Proof. Without loss of generality (scale as needed) we can suppose that L is R-maximal and that x is primitive in L. Extend x to a basis

$$x = x_1, x_2, \ldots, x_n$$

of L, and let $A = (a_{ij})$ be the associated Gram matrix. So $a_{11} = 0$ and

$$a_{ij} \in \begin{cases} (\frac{1}{2}) & \text{if } i \neq j, \\ R & \text{if } i = j. \end{cases}$$

If for some prime $p \in R$ we had $\operatorname{ord}_p a_{1j} > \operatorname{ord}_p \frac{1}{2}$ for all j, $2 \leq j \leq n$, then the lattice $M = (p^{-1})x_1 + L$ would satisfy $nM \subseteq R$, contradicting the maximality of L. Therefore $(a_{12}, \ldots, a_{1n}) = (\frac{1}{2})$, and hence $sL = (\frac{1}{2})$. It follows that $2x_1$ is primitive in the dual lattice $L^\sharp$. (Otherwise we would have $p^{-1}(2x_1) \in L^\sharp$ for some prime p, from which it would follow that $\operatorname{ord}_p a_{1j} > \frac{1}{2}$.) And since $L = (L^\sharp)^\sharp$, there is therefore a vector $y \in L$ such that $B(y, 2x_1) = 1$. Now consider the sublattice

$$J = Rx_1 + Ry \cong \begin{pmatrix} 0 & \frac{1}{2} \\ \frac{1}{2} & * \end{pmatrix}$$

of L. This is $(\frac{1}{2})$-modular and hence splits L. □

Lemma 6.35. *Let L be a lattice on a hyperbolic plane, and let $\mathfrak{A} = (\lambda)$ be a fractional R-ideal. Then L is $2\mathfrak{A}$-maximal if and only if L is $\mathfrak{A}$-modular with $nL \subseteq 2\mathfrak{A}$. Moreover, if these conditions hold then $L \cong \begin{pmatrix} 0 & \lambda \\ \lambda & 0 \end{pmatrix}$.*

Proof. As in the preceding lemma, we can assume throughout that $\mathfrak{A} = R$.

If L is unimodular with $nL \subseteq (2)$, then $2sL \subseteq nL \subseteq (2)$, and so $sL \subseteq R$. But any lattice with integral scale has integral volume as well, and since $vL = R$ it follows that L is maximal; more precisely, L is (2)-maximal, as desired.

Conversely, suppose L is (2)-maximal, and we must check that L is unimodular. Extend a primitive isotropic vector x to a basis for L, getting $L \cong \begin{pmatrix} 0 & \alpha \\ \alpha & 2\beta \end{pmatrix}$ with $\alpha, \beta \in R$.

An inequality $|\alpha|_p < 1$ for some prime p would lead (as in the proof of Lemma 6.34) to a failure of (2)-maximality; hence we must have $\alpha \in R^*$, giving the required result.

Finally, if $L \cong \begin{pmatrix} 0 & \alpha \\ \alpha & 2\beta \end{pmatrix}$ in $\{x, y\}$, then $L \cong \begin{pmatrix} 0 & 1 \\ 1 & 0 \end{pmatrix}$ in $\{\alpha^{-1}x, y - \frac{\beta}{\alpha}x\}$. □

Lemma 6.36. *Let L and M be $\mathfrak{A}$-maximal lattices on a regular isotropic quadratic space V. Then there is a hyperbolic plane H such that*

$$L = (L \cap H) \perp \cdots \quad \text{and} \quad M = (M \cap H) \perp \cdots .$$

Proof. Again (!) we can assume that $\mathfrak{A} = R$.

For $0 \neq x \in V$, fix scalars $\lambda_x, \mu_x \in F^*$ such that $\lambda_x x$ is primitive in L and $\mu_x x$ is primitive in M; and then define $\rho_x = \dfrac{\mu_x}{\lambda_x}$. Because lattices are finitely generated, we know $\alpha M \subseteq L$ for suitable $\alpha \in F^*$, and in particular $\alpha(\mu_x x) \in L$; so $\alpha\mu_x \in (\lambda_x)$, and hence $\rho_x \in (\alpha^{-1})$ for all $x \neq 0$. There is therefore an isotropic vector $x \in V$ such that the fractional ideal (ρ_x) is maximal for isotropic vectors in V. With such an isotropic vector x in hand, by Lemma 6.34 there is a hyperbolic plane $H = Fx + Fy$, with $\{x, y\}$ a hyperbolic pair, such that the sublattice $L_1 \stackrel{\text{def}}{=} L \cap H$ splits L. We claim that also $M \cap H$ splits M.

Since L is R-maximal, so is L_1; hence by Lemma 6.35 the lattice L_1 is $(\frac{1}{2})$-modular. But

$$L_1 = R(\lambda_x x) + R(\lambda_y y) \cong \begin{pmatrix} 0 & \lambda_x\lambda_y \\ \lambda_x\lambda_y & 0 \end{pmatrix}$$

and so $(\lambda_x\lambda_y) = (\frac{1}{2})$.

Now $M_1 \stackrel{\text{def}}{=} M \cap H = R(\mu_x x) + R(\mu_y y)$ is a sublattice of the R-maximal lattice M, so $sM_1 \subseteq (\frac{1}{2})$ and in particular $\mu_x\mu_y \in (\frac{1}{2})$.

Reasoning with M and y as we did with L and x, we see that there is an isotropic vector z, with $\{y, z\}$ a hyperbolic pair, such that $R(\mu_z z) + R(\mu_y y)$ splits M. Here we have $(\mu_z\mu_y) = (\frac{1}{2})$ while $\lambda_z\lambda_y \in (\frac{1}{2})$. Thus

$$(\rho_x\rho_y) \subseteq R \subseteq (\rho_z\rho_y)$$

so by the maximality of (ρ_x) we actually have $(\rho_x\rho_y) = R$ and hence $(\mu_x\mu_y) = (\lambda_x\lambda_y) = (\frac{1}{2})$. Therefore M_1 is $(\frac{1}{2})$-modular and hence splits M, as desired. □

6.9. Unimodular Lattices and Pythagorean Triples

Let R be a principal ideal domain with quotient field F. A triple $(a, b, c) \in R^3$ is a **Pythagorean triple (over R)** if $a^2 + b^2 = c^2$; and the Pythagorean triple (a, b, c) is said to be **primitive** if $\gcd(a, b, c) = 1$. Our goal in this section is to explore the production of Pythagorean triples through the medium of R-lattices on quadratic F-spaces. And clearly it will suffice to consider the production of *primitive* Pythagorean triples.

If char $F = 2$ then the Pythagorean triples over R are just the triples $(a, b, a+b)$. So without loss of generality from now on we will assume that char $F \neq 2$.

Let V be a 3-dimensional quadratic F-space with basis $\mathbb{B} = \{v_1, v_2, v_3\}$, and suppose $V \cong \langle 1, 1, -1\rangle$ with respect to $\mathbb{B}$. Now consider the lattice $L = \perp_{i=1}^{3} Rv_i$ on V, and suppose $v = av_1 + bv_2 + cv_3 \in L$. Then (a, b, c) is a Pythagorean triple if and only if v is isotropic; and (a, b, c) is primitive if and only if v is a primitive vector in L. Thus an algorithm for the generation of the primitive Pythagorean triples over R is equivalent to an algorithm for generating the primitive isotropic vectors in L.

Define a new basis $\mathbb{B}' = \{w_1, w_2, w_3\}$ for L by

$$w_1 = v_1 + v_3, \qquad w_2 = v_2 + v_3, \qquad w_3 = v_1 + v_2 + v_3.$$

Then

$$L \cong \begin{pmatrix} 0 & -1 \\ -1 & 0 \end{pmatrix} \perp \langle 1\rangle \qquad \text{with respect to } \{w_1, w_2, w_3\}.$$

Let us consider the primitive isotropic vectors and their associated primitive Pythagorean triples. If $v = rw_1 + sw_2 + tw_3$ then $q(v) = -2rs + t^2$, hence

$$v \text{ is isotropic} \iff t^2 = 2rs.$$

It follows that v is *primitive* and isotropic if and only if $t^2 = 2rs$ and $\gcd(r, s) = 1$. Assuming these conditions to hold, we have

$$\begin{aligned} v &= rw_1 + sw_2 + tw_3 \\ &= (r+t)v_1 + (s+t)v_2 + (r+s+t)v_3, \end{aligned}$$

and therefore $(r+t,\ s+t,\ r+s+t)$ is the associated primitive Pythagorean triple. We have now proved the following theorem, originally proved over $\mathbb{Z}$ by Leonard Eugene Dickson in 1894.

Theorem 6.37. *The triple $(a, b, c) \in R^3$ is a primitive Pythagorean triple over R if and only if*

$$(a, b, c) = (r+t,\ s+t,\ r+s+t)$$

for some $r, s, t \in R$ satisfying $\gcd(r, s) = 1$ and $t^2 = 2rs$. Moreover, the correspondence $(a, b, c) \longleftrightarrow (r, s, t)$ is a bijection.

Examples 6.38. (i) Let us list some primitive Pythagorean triples over $\mathbb{Z}$ using the scheme suggested by the theorem. We will list triples

$$(a, b, c) = (r+t,\ s+t,\ r+s+t)$$

for which a, b, c are all positive, and we leave it to the interested reader to wade through the inequalities and check that this is equivalent to r, s, t being positive. Also, notice that the conditions $\gcd(r, s) = 1$ and $t^2 = 2rs$ in the theorem force t to be even, and one of r, s to be twice a square and the other an odd square. In the accompanying table all these conditions are in place, with t growing as we move from top to bottom. Interchanging given values of r and s would just interchange the associated a and b, so we will assume throughout that $r < s$, hence $a < b$. Except for that, for each value of t all the possibilities for r and s are displayed. If a given t has k distinct prime divisors, then there are 2^k ways to choose positive r and s satisfying the conditions of the theorem, and therefore 2^{k-1} lines associated with t in the table.

(i) Primitive Pythagorean triples over $\mathbb{Z}$:

t	$\frac{t^2}{2}$	r	s	(a, b, c)
2	2	1	2	$(3, 4, 5)$
4	8	1	8	$(5, 12, 13)$
6	18	1	18	$(7, 24, 25)$
		2	9	$(8, 15, 17)$
8	32	1	32	$(9, 40, 41)$
10	50	1	50	$(11, 60, 61)$
		2	25	$(12, 35, 37)$
12	72	1	72	$(13, 84, 85)$
		8	9	$(20, 21, 29)$
14	98	1	98	$(15, 112, 113)$
		2	49	$(16, 63, 65)$
16	128	1	128	$(17, 144, 145)$
18	162	1	162	$(19, 180, 181)$
		2	81	$(20, 99, 101)$
20	200	1	200	$(21, 220, 221)$
		8	25	$(28, 45, 53)$
22	242	1	242	$(23, 264, 265)$
		2	121	$(24, 143, 145)$
24	288	1	288	$(25, 312, 313)$
		9	32	$(33, 56, 65)$
26	338	1	338	$(27, 364, 365)$
		2	169	$(28, 195, 197)$
28	392	1	392	$(29, 420, 421)$
		8	49	$(36, 77, 85)$
30	450	1	450	$(31, 480, 481)$
		2	225	$(32, 255, 257)$
		9	50	$(39, 80, 89)$
		18	25	$(48, 55, 73)$

(ii) Now consider triples over $\mathbb{Z}[i]$, the ring of Gaussian integers. As was the case over $\mathbb{Z}$, we must have $2|t^2$. This is equivalent to the condition $(1+i)|t$, since 2 is an associate of $(1+i)^2$ and $1+i$ is prime in $\mathbb{Z}[i]$. (See the book by Pollard and Diamond [**PD**], Chapter 2.) In the accompanying little table of primitive Pythagorean triples over $\mathbb{Z}[i]$ with $|t|$ small, we will omit triples in $\mathbb{Z}^3$ and conjugates of listed triples. Also, if (a, b, c) has been listed, then (a, ci, bi) and (c, ai, bi) and other similarly derived triples are automatically qualified for listing as well, hence are omitted from the table. Finally, we will skip "trivial" triples—that is, triples in which some entry is 0.

(ii) Primitive Pythagorean triples over $\mathbb{Z}[i]$:

t	$\frac{t^2}{2}$	r	s	(a,b,c)
$1+i$	i	i	1	$(1+2i,\ 2+i,\ 2+2i)$
2	2	$2i$	$-i$	$(2+2i,\ 2-i,\ 2+i)$
$2+2i$	$4i$	$4i$	1	$(2+6i,\ 3+2i,\ 3+6i)$
$3+i$	$4+3i$	$4+3i$	1	$(7+4i,\ 4+i,\ 8+4i)$
$3+9i$	$-36+27i$	$-36+27i$	1	$(-33+36i,\ 4+9i,\ -32+36i)$
		$3+4i$	$9i$	$(6+13i,\ 3+18i,\ 6+22i)$

6.10. Remarks on Lattices over More General Rings

Our treatment of the integral theory of quadratic forms in this book will be restricted to lattices over principal ideal domains, with $\mathbb{Z}$ and $\mathbb{F}_q[x]$ the main domains of interest. In pursuit of a local-global theory à la Hasse–Minkowski, we will also need to consider the theory over the rings of p-adic integers and their function field analogues, and again these are principal ideal domains. So for us all lattices are free; that is, they have bases—a fact that in part motivates the title of this book—and this allows us to translate questions on isometry and representation directly into matrix problems.

With much more extensive number-theoretic background, one can pursue the theory of quadratic forms on lattices over the important subrings of arbitrary global fields—the finite extensions of $\mathbb{Q}$ and $\mathbb{F}_q(x)$—and their completions. Valuation theory again plays an important role in this more general setting. For example, if F is an algebraic number field and S is the set of all the discrete spots on F, the ring R_F of algebraic integers in F (the subring of all elements in F satisfying a monic polynomial in $\mathbb{Z}[x]$) and the fractional R_F-ideals again have descriptions in terms of valuation theory. In particular,

$$R_F = \{\, \alpha \in F \mid |\alpha|_{\mathfrak{p}} \leq 1 \text{ for all } \mathfrak{p} \in S \,\}.$$

(Here $|\cdot|_{\mathfrak{p}}$ denotes a valuation in $\mathfrak{p}$.) But $R = R_F$ need not be a principal ideal domain, and in that case an R-lattice L—a finitely generated R-module in a vector space over F—need not be free; so L's properties can no longer be completely captured by matrices without serious attention to the ideal theory of R. As over $\mathbb{Z}$ and $\mathbb{F}_q[x]$, one passes to the local setting for help, where happily the localized coefficient rings $R_\mathfrak{p}$ are once again principal ideal domains. In fact, when $\mathfrak{p}$ is nondyadic, the local theory is essentially unchanged from the theory over $\mathbb{Z}_p$ with p odd that we will develop in Chapter 8; but in the dyadic case the complications that we will experience in the 2-adic theory become magnified. (Lattices over the rings of integers of arbitrary local fields of characteristic not 2 were classified by O'Meara—see [**O'M1**]—in the 1950s.) Most of these remarks apply to lattices over the important subrings of global function fields as well. In fact, in that setting all the spots are nondyadic, making the theory considerably more tractable than that over the rings of integers of number fields. We will get a taste of the function field theory in Chapter 10 by solving a classification problem for $\mathbb{F}_q[x]$-lattices, while meanwhile the analogous problem for $\mathbb{Z}$-lattices remains unsolved.

We refer the reader to O'Meara [**O'M1**] for the number theory background required to pursue quadratic forms over arbitrary local and global fields. For wide-ranging explorations of computational issues in algebraic number fields, along with some explicit numerical algorithms, see Cohen [**Coh**] and Pohst–Zassenhaus [**PZ**].

Exercises

(1) Give an example of fractional ideals $I \supset J$ and a lattice L that is simultaneously J-maximal and I-maximal.

(2) Give an example of a maximal lattice that is not modular.

(3) Suppose

$$L \cong A = \begin{pmatrix} 10 & 1 & 0 & 0 \\ 1 & 3 & 2 & 0 \\ 0 & 2 & 2 & 0 \\ 0 & 0 & 0 & 50 \end{pmatrix}$$

with respect to the basis $\{v_1, \ldots, v_4\}$ over the ring $\mathbb{Z}_2$ of 2-adic integers.
(i) Determine the invariant factors of the $\mathbb{Z}_2$-matrix A.
(ii) Show that L is the orthogonal sum of two modular sublattices, and exhibit a basis for each of them.
(iii) Explain the connection between the multiplicity of the invariant factors in part (i) and the decomposition in part (ii).
(iv) With A as above, suppose M is a $\mathbb{Z}_5$-lattice satisfying $M \cong A$. It is a fact that $\det A = 900$, and the upper left 3×3 submatrix of A has

determinant 18. Without further calculation (aside from factoring these two determinants) one can conclude from this determinant data that there is a splitting $M = M_1 \perp M_2$ with the M_i modular $\mathbb{Z}_5$-lattices. How?

(4) Let L be an R-lattice on the n-dimensional regular quadratic F-space V; say $L \cong A \in GL_n(F)$ with respect to $\{v_1, \ldots, v_n\}$. From §2.3, recall the linear transformation $f_B : V \to V^*$ given by $f_B(v) = B(v, \cdot)$. Consider the restriction of that map to L.
(i) Show that the dual basis $\{v_1^*, \ldots, v_n^*\}$ of V^* is a basis for the R-module $\text{Hom}(L, R)$.
(ii) Under what condition on the scale ideal sL is $f_B(L) \subseteq \text{Hom}(L, R)$?
(iii) Show that if L is unimodular then the restriction of f_B to L is an isomorphism from L onto $\text{Hom}(L, R)$.

(5) Suppose the R-lattice L has a splitting $L = L_1 \perp \cdots \perp L_t$ in which each component L_i is (α_i)-modular for some $\alpha_i \in F^*$. What can be said about the structure of the dual lattice $L^\sharp$?

(6) Find relatively prime quadratic polynomials $f, g, h \in \mathbb{Q}[x]$ such that $f^2 + g^2 = h^2$.

Chapter 7

Initial Integral Results

Let R be a principal ideal domain that is not a field, with quotient field F and, as usual, with characteristic not 2. Given two R-lattices on n-dimensional quadratic F-spaces, the classification problem asks for a set of necessary and sufficient effectively checkable conditions for the lattices to be isometric; and this is equivalent to determining when the lattices' associated Gram matrices are congruent over R. A necessary condition for lattice isometry is that the underlying spaces be isometric, and so there is no loss of generality in assuming that the given lattices are on the same space.

Briefly recall the strategy for the matrix *similarity* problem over the complex numbers. One learns that every square matrix is similar to a unique matrix in Jordan canonical form, so to check whether two matrices are similar we compute their Jordan forms and compare the results. For the isometry problem viewed as a matrix problem, it would be nice to have some kind of canonical form with respect to matrix congruence—paralleling the reduction to Jordan form under similarity—and an algorithm for reduction of a given matrix to that canonical form. (From the lattice perspective, this would amount to production of a so-called **reduced basis** for the lattice.) And then we could just compare the canonical forms. We did make moves in this direction in the preceding chapter with Jacobi's triple-diagonal theorem and Newman's supplement, but these results don't go far enough: a lattice on a regular quadratic F-space of dimension $n \geq 2$ has infinitely many triple diagonalizations.

We will see in the next chapter that there is a canonical reduced form for lattices over the rings $\mathbb{Z}_p$ of p-adic integers. This is helpful, because for $\mathbb{Z}$-lattices L and M to be isometric it is necessary that L_p and M_p be isometric for all primes p. But it turns out that local lattice isometry for all p does not imply global isometry, except in very special circumstances. In other words, the local-global approach à la Hasse–Minkowski does not extend in its full glory to the integral setting.

The goal of the current chapter is to present some results on lattices, mostly over $\mathbb{Z}$, and mostly quite classical, that don't rely on the local theory.

7.1. The Minimum of a Lattice; Definite Binary $\mathbb{Z}$-Lattices

Recall that a quadratic $\mathbb{Q}$-space V is said to be **indefinite** if its localization V_∞ over $\mathbb{R} = \mathbb{Q}_\infty$ is isotropic; equivalently, V represents both positive and negative numbers. In particular, every isotropic $\mathbb{Q}$-space is indefinite. But not every indefinite space is isotropic: $\langle 1, -2\rangle$ is an example of an anisotropic indefinite $\mathbb{Q}$-space. A $\mathbb{Q}$-space V is **definite** if V_∞ is anisotropic, and it is **positive** (resp., **negative**) **definite** if $q(v) > 0$ (resp., $q(v) < 0$) for all $v \neq 0$ in V. A $\mathbb{Z}$-lattice L is called definite or indefinite according to whether the $\mathbb{Q}$-space it spans is definite or indefinite.

Consider a $\mathbb{Z}$-lattice L of rank n. First note that if $T \in GL_n(\mathbb{Z})$ then $(\det T)^2 = 1$, and from this it follows that the discriminant dL is a well-defined integer; that is, all Gram matrices for L have the same determinant. Recall from the preceding chapter that matrix congruence is an instance of matrix equivalence, and hence the invariant factors of the Gram matrix are preserved under congruence, so they form another set of invariants for L under isometry. Moreover, for a $\mathbb{Z}$-lattice the discriminant is $\pm$ the product of those invariant factors.

Another important invariant of an integral $\mathbb{Z}$-lattice L is its **minimum**:

$$\min L = \min\{\, |q(x)| \mid 0 \neq x \in L\,\}.$$

Not surprisingly, a nonzero vector $v \in L$ such that $|q(v)| = \min L$ is a **minimal vector** of L. If the lattice is anisotropic then a minimal vector v is primitive, because when v is expressed as a linear combination of basis vectors its coefficients must be relatively prime. On the other hand, if v is an isotropic vector in L, then $L \cap \mathbb{Q}v$ contains a primitive isotropic vector.

If L is positive definite of rank n, then L spans a copy of $\mathbb{R}^n$, since a basis of L is a basis of the n-dimensional $\mathbb{Q}$-space V, and hence of $V_\infty \cong \langle 1, \dots, 1\rangle \cong \mathbb{R}^n$. Thus in the positive definite case we can view the underlying symmetric bilinear form B as the standard dot product in ordinary Euclidean space, familiar to every linear algebra student; and in this case a minimal vector is just a shortest nonzero vector in L. (Similarly, if L is negative definite, then B is the ordinary dot product scaled by -1.)

If $n = 1$ the discriminant of a lattice completely determines its isometry class: $L \cong \langle dL\rangle$. Now consider a positive definite integral $\mathbb{Z}$-lattice L of rank

2 and discriminant d; say $L \cong \begin{pmatrix} a & b \\ b & c \end{pmatrix} \in M_2(\mathbb{Z})$ in the basis $\{v_1, v_2\}$. Here $a, c \in \mathbb{N}$ by the "positive definite" condition. Without loss of generality we can assume that $a \le c$ and also that $b \ge 0$. (Replace v_2 by $-v_2$ in the basis if necessary.)

Now if $b > \frac{a}{2}$ use the division algorithm to write $b = ak + r$, with $0 \le |r| \le \frac{a}{2}$, and replace v_2 by $v_2' = v_2 - kv_1$, getting $L \cong \begin{pmatrix} a & r \\ r & c' \end{pmatrix}$ in the basis $\{v_1, v_2'\}$, with $q(v_2') = c'$. As before, we can suppose $r \ge 0$. If $c' \ge a$ stop; otherwise interchange v_1 and v_2' and return to the initial stage. After a finite number (why only a finite number?) of iterations of this process we eventually reach a new basis $\{v_1, v_2\}$ with respect to which $L \cong \begin{pmatrix} a & b \\ b & c \end{pmatrix}$ with $a \le c$ and $0 \le b \le \frac{a}{2}$. We say that this Gram matrix for L is in **reduced form**, and the associated basis $\{v_1, v_2\}$ is a **reduced basis** for L.

Remark 7.1. Were we discussing a quadratic $\mathbb{Q}$-space $V \cong \begin{pmatrix} a & b \\ b & c \end{pmatrix}$ in $\{v_1, v_2\}$, the Gram–Schmidt procedure would dictate that we replace v_2 by $v_2 - \frac{b}{a} v_1$. But in our present situation we instead replace v_2 by

$$v_2 - kv_1 = v_2 - \left\lfloor \frac{b}{a} \right\rceil v_1$$

where $\lfloor \ \rceil$ is the **nearest integer** function, making the transition matrix

$$\begin{pmatrix} 1 & \lfloor \frac{b}{a} \rceil \\ 0 & 1 \end{pmatrix} \text{ instead of } \begin{pmatrix} 1 & \frac{b}{a} \\ 0 & 1 \end{pmatrix}.$$

In other words, *we are approximating the rational transition matrix by a unimodular transition matrix in order to retain a basis of* L, and we cannot expect that this will produce an orthogonal decomposition of L.

Notice that

$$a^2 \le ac = b^2 + d \le \frac{a^2}{4} + d, \quad \text{and hence} \quad a \le \sqrt{\frac{4}{3} d}$$

and therefore

$$\min L \le \sqrt{\frac{4}{3} d}.$$

From these inequalities we see that in the reduced form there are only finitely many possibilities for a and hence only finitely many possibilities for b. Finally, since $c = \frac{b^2 + d}{a}$, there are only finitely many possibilites for c in the reduced form. Since we have shown that every lattice under consideration has at least one reduced basis, we have proved the following.

Proposition 7.2. *There are only finitely many isometry classes of positive definite integral $\mathbb{Z}$-lattices of rank 2 and given discriminant.*

Example 7.3. Suppose $L \cong \begin{pmatrix} 47 & 112 \\ 112 & 267 \end{pmatrix}$ in $\{v_1, v_2\}$, and our goal is to find a reduced form for L. Incidentally, this will be a warm-up for our discussion of the **LLL-algorithm** in Chapter 11.

Since $112 = 2 \cdot 47 + 18$, we replace v_2 by $v_2' = v_2 - 2v_1$, getting $L \cong \begin{pmatrix} 47 & 18 \\ 18 & 7 \end{pmatrix}$ in $\{v_1, v_2'\}$. Interchange the basis vectors and relabel them as $\{v_1, v_2\}$. So now $L \cong \begin{pmatrix} 7 & 18 \\ 18 & 47 \end{pmatrix}$ in $\{v_1, v_2\}$. Replace v_2 by $v_2 - 2v_1$ to obtain $L \cong \begin{pmatrix} 7 & 4 \\ 4 & 3 \end{pmatrix}$. Again interchange the basis vectors, relabel, apply the division algorithm, and make a final vector exchange to obtain $L \cong \begin{pmatrix} 2 & 1 \\ 1 & 3 \end{pmatrix}$, the desired reduced form. Upon keeping track of the sequence of basis changes, we see that in terms of the original basis the reduced basis we have found is $\{-7v_1 + 3v_2, 5v_1 - 2v_2\}$.

Note that there was no serious need to do the vector interchanges—except in the very last step—in order to obtain the reduced form. But doing so allowed us to perform each basis change with the same ritual, namely replacing v_2 by a vector of the form $v_2 - ev_1$, with $e \in \mathbb{Z}$.

Next we claim that if a lattice L has reduced form $L \cong \begin{pmatrix} a & b \\ b & c \end{pmatrix}$ in the basis $\{v_1, v_2\}$ then $a = \min L$. This is clear if $b = 0$, so suppose $b > 0$, and let $M = L^{b^{-1}}$. Then

$$M \cong \begin{pmatrix} \alpha & 1 \\ 1 & \gamma \end{pmatrix} \quad \text{in the basis } \{v_1, v_2\}$$

with $\alpha, \gamma \in \mathbb{Q}$ and $2 \leq \alpha \leq \gamma$. Then the quadratic form on M can be expressed as $q_0 + q_1$, where

$$(M, q_0) \cong \begin{pmatrix} 2 & 1 \\ 1 & 2 \end{pmatrix} \quad \text{and} \quad (M, q_1) \cong \langle \alpha - 2, \gamma - 2 \rangle.$$

Then $\min(M, q_0) = 2$ and $\min(M, q_1) = \alpha - 2$, and for both forms v_1 is a minimal vector. Therefore v_1 is a minimal vector for M, and hence $\min M = \alpha = \frac{a}{b}$. It follows that $\min L = \min(M^b) = a$, as claimed. Similar reasoning shows that c is minimal among all values $q(x)$ for vectors $x \in L$

linearly independent of v_1. Therefore the entries a and c in L's reduced form are unique for L; and, since $d = ac - b^2$, the entry b is unique as well.

One more comment before stating a summary theorem. If the Gram matrix of our given lattice L has rational, but not integral, entries (equivalently, $sL \not\subseteq \mathbb{Z}$), then scaling L by a common denominator of the Gram matrix entries yields an integral lattice, to which the above discussion applies. In fact, if one considers lattices with scale ideal contained in the fractional ideal (λ), then scaling by λ's denominator will yield integral lattices. Now we can sum up our results.

Theorem 7.4. *Let A be a fractional $\mathbb{Z}$-ideal, and let d be a positive rational number. Then there are only finitely many isometry classes of positive definite $\mathbb{Z}$-lattices L satisfying $\operatorname{rank} L = 2$, $sL \subseteq A$, and $dL = d$. Every such lattice has a unique reduced form, and two such lattices are isometric if and only if they have the same reduced form.*

7.2. Hermite's Bound on $\min L$, with a Supplement for $k[x]$-Lattices

Except under very special circumstances, the exact determination of the minimum of a lattice is a challenging open problem. But the following theorem (from 1850), which generalizes to arbitrary ranks the bound on the minimum that we obtained earlier in the binary case, is a highlight from the classical literature of quadratic forms.

Theorem 7.5 (Hermite's Inequality). *Let V be an anisotropic $\mathbb{Q}$-space of dimension n, and let L be an integral $\mathbb{Z}$-lattice on V. Then*

$$\min L \leq \left(\frac{4}{3}\right)^{\frac{n-1}{2}} |dL|^{\frac{1}{n}}.$$

Proof. Suppose $v_1 \in L$ is a minimal vector. Then v_1 extends to a basis $\{v_1, \ldots, v_n\}$ for L, and hence without loss of generality we can suppose $L \cong A = (a_{ij}) \in M_n(\mathbb{Z})$, with $|a_{11}| = \min L$. Write $d = dL$. The theorem is trivial if $n = 1$, so assume that $n \geq 2$ and that the result has been checked for lattices on spaces of dimension $< n$.

Define a linear transformation $\varphi : V \to V$ by setting

$$\varphi(v_1) = v_1 \quad \text{and} \quad \varphi(v_j) = v_j' = v_j - \frac{a_{1j}}{a_{11}} v_1 \text{ if } j > 1,$$

and set $L' = \varphi L$. Then

$$L' = \mathbb{Z}v_1 \perp \underbrace{(\mathbb{Z}v_2' + \cdots + \mathbb{Z}v_n')}_{J} \cong \begin{pmatrix} a_{11} & 0 & \dots & 0 \\ 0 & & & \\ \vdots & & \frac{1}{a_{11}}C & \\ 0 & & & \end{pmatrix}$$

with matrix $C = (c_{ij})_{2 \le i,j \le n} \in M_{n-1}(\mathbb{Z})$ given by $c_{ij} = a_{11}a_{ij} - a_{i1}a_{1j}$ for $2 \le i, j \le n$. Notice that since $\det \varphi = 1$ we have

$$d = dL = dL' = a_{11}\left(\frac{1}{a_{11}}\right)^{n-1} |C|$$

and therefore $|C| = a_{11}^{n-2} \cdot d$.

Let $w \in L$, say $w = \sum_{j=1}^{n} \lambda_j v_j$, with all $\lambda_j \in \mathbb{Z}$. Since $v_j = v_j' + \dfrac{a_{1j}}{a_{11}} v_1$ if $j > 1$, we have

$$w = \underbrace{\left(\lambda_1 + \frac{a_{12}}{a_{11}}\lambda_2 + \cdots + \frac{a_{1n}}{a_{11}}\lambda_n\right)}_{\gamma \in \mathbb{Q}} v_1 + \underbrace{\lambda_2 v_2' + \cdots + \lambda_n v_n'}_{z \in J}. \tag{7.1}$$

Now suppose that, in specifying w, the cofficients $\lambda_2, \dots, \lambda_n$ have been chosen so that z is a minimal vector in J. Notice that $J \cong \frac{1}{a_{11}} C$ in the basis $\{v_2', \dots, v_n'\}$, and so by the induction hypothesis applied to J we have

$$|q(z)| \le \left(\frac{4}{3}\right)^{\frac{n-2}{2}} \left|\frac{1}{a_{11}}\right| |C|^{\frac{1}{n-1}}.$$

(Strictly speaking, in order to apply the induction hypothesis, first scale J by a_{11} to make the form on J integral, then apply the induction hypothesis to $J^{a_{11}}$, then rescale by $\frac{1}{a_{11}}$.) Having chosen $\lambda_2, \dots, \lambda_n$, next choose $\lambda_1 \in \mathbb{Z}$ so that $|\gamma| \le \frac{1}{2}$. Then

$$\begin{aligned} \min L = |a_{11}| \le |q(w)| &= |\gamma^2 a_{11} + q(z)| \\ &\le \frac{|a_{11}|}{4} + \left(\frac{4}{3}\right)^{\frac{n-2}{2}} \left|\frac{1}{a_{11}}\right| \left|a_{11}^{n-2} d\right|^{\frac{1}{n-1}} \\ &= \frac{|a_{11}|}{4} + \left(\frac{4}{3}\right)^{\frac{n-2}{2}} |a_{11}|^{-\frac{1}{n-1}} \cdot |d|^{\frac{1}{n-1}}. \end{aligned}$$

Therefore

$$|a_{11}|^{\frac{n}{n-1}} \le \left(\frac{4}{3}\right)^{\frac{n}{2}} \cdot |d|^{\frac{1}{n-1}}$$

and from this inequality the theorem follows immediately. □

Remark 7.6. (i) Because the lattice L in Hermite's theorem is integral, minimal vectors clearly exist, but in general no effective algorithm is known for actually *finding* an explicit minimal vector or determining the minimum exactly. This is an important area of research.

(ii) Given that the argument is by induction on n, one might have expected the induction hypothesis to be applied to a sublattice of L of rank $n-1$. Part of the genius of Hermite's proof is in his use of the transformation φ that does a Gram–Schmidt sort of maneuver to base the induction on a lattice J that might not be a sublattice of L, or even an integral lattice. In fact $J \subseteq L$ if and only if $a_{11}|a_{1j}$ for all $j \geq 2$.

Corollary 7.7. *Let L be an anisotropic unimodular $\mathbb{Z}$-lattice of rank $n \leq 5$. Then*

$$L \cong \langle 1, \ldots, 1 \rangle \quad \text{or} \quad L \cong \langle -1, \ldots, -1 \rangle.$$

[Note: We will see later that the result actually holds whenever $n \leq 7$.]

Proof. Since $|dL| = 1$, from Hermite's theorem we have $\min L = 1$. So $L \cong (a_{ij}) \in GL_n(\mathbb{Z})$, with $a_{11} = \pm 1$, in some basis $\{v_1, \ldots, v_n\}$. Replacing v_j by $v_j' = v_j - \dfrac{a_{1j}}{a_{11}} v_1$ for $j \geq 2$ yields a splitting $L \cong \langle \pm 1 \rangle \perp L'$ with $dL' = \pm 1$. Iteration eventually gives $L \cong \langle \pm 1, \ldots, \pm 1 \rangle$, and since L is anisotropic all the signs must be the same. □

We remark that the assumption of anisotropy is essential here; for example, recall from previous material that a unimodular $\mathbb{Z}$-lattice with Gram matrix $\begin{pmatrix} 0 & 1 \\ 1 & 0 \end{pmatrix}$ is indecomposable.

Example 7.8. Not all anisotropic unimodular $\mathbb{Z}$-lattices "diagonalize" or, indeed, have *any* orthogonal splitting. For $n \equiv 0 \pmod 8$, with $n \geq 8$, consider the $\mathbb{Z}$-lattice

$$K_n \cong \begin{pmatrix} \frac{n}{4} & 1 & & & & & \\ 1 & 4 & 2 & & & & \\ & 2 & 2 & 1 & & & \\ & & 1 & 2 & \ddots & & \\ & & & \ddots & \ddots & 1 \\ & & & & 1 & 2 \end{pmatrix}$$

in the basis $\{v_1, \ldots, v_n\}$. Determinant calculations show that K_n is positive definite and unimodular. Moreover the lattice is **even**, in the sense that $q(v) \equiv 0 \pmod 2$ for all $v \in K_n$. Notice that the sublattice spanned by $v_3, \ldots, v_n$ is just the lattice A_n discussed in the previous chapter; and from that discussion we know that if there were a splitting $K_n = J \perp L$, then all of $v_3, \ldots, v_n$ would belong to the same component, say J. But then

$\operatorname{rank} L \leq 2$, and if $L \neq 0$ this would force a representation $1 \to L$, by Hermite's inequality, contradicting the fact that K_n is even. Therefore $L = 0$, so K_n is indecomposable.

The lattices K_n were first discovered by Kneser [**K2**] in 1957. In fact Kneser required only that $n \equiv 0 \pmod 4$ (with $n \geq 8$ as before), and he showed that all such lattices are indecomposable, though more work is required to show this when $8 \nmid n$, since then the form on the lattice is **odd** (that is, the lattice represents at least one odd number).

It is always interesting to see instances in which mathematical ideas developed in one context can be used—perhaps with minor modifications—to give results in another. We will now adapt Hermite's $\mathbb{Z}$-lattice argument to lattices over a polynomial ring $k[x]$, with k any field of characteristic not 2. Here the measure of "size" is the degree function ∂ rather than the absolute value; and for an integral $k[x]$-lattice L on a quadratic $k(x)$-space we define

$$\min L = \min\{\, \partial q(v) \mid 0 \neq v \in L \,\}.$$

For a rational function $r(x) = \dfrac{f(x)}{g(x)}$, with $f(x), g(x) \in k[x]$, define $\partial r = \partial f - \partial g$, with the convention that $\partial 0 = -\infty < m$ for all $m \in \mathbb{Z}$. And now for the result.

Theorem 7.9 (à la Hermite). *Let k be a field of characteristic not 2, and L be an integral $k[x]$-lattice on an anisotropic quadratic $k(x)$-space V of dimension n. Then*

$$\min L \leq \frac{1}{n}\partial(dL).$$

Proof. Return to the proof of Hermite's theorem. Our present proof is identical to that proof through the selection of $\lambda_2, \ldots, \lambda_n$ just below the display marked (7.1), except that now $k[x], k(x)$, and ∂ replace $\mathbb{Z}, \mathbb{Q}$, and the absolute value function, respectively. In particular, $|C| = a_{11}^{n-2} \cdot dL$ and the vector $z = \sum_{i=2}^{n} \lambda_i v_i'$ is a minimal vector in J.

The induction hypothesis applied to J gives

$$\min J \leq \left(\frac{1}{n-1}\partial|C|\right) - \partial a_{11}.$$

Now use the division algorithm in $k[x]$ to choose λ_1 so that $\partial\gamma < 0$. Then

$$\min L = \partial a_{11} \leq \partial q(w) = \partial\left(\gamma^2 a_{11} + q(z)\right) \leq \max\left\{\partial\left(\gamma^2 a_{11}\right), \partial q(z)\right\}.$$

Therefore

$$\begin{aligned}\partial a_{11} \leq \partial q(z) &\leq \left(\frac{1}{n-1}\partial|C|\right) - \partial a_{11} \\ &= \frac{1}{n-1}\big((n-2)\partial a_{11} + \partial(dL)\big) - \partial a_{11}.\end{aligned}$$

from which the conclusion follows. □

It is worth noting the exact moment where the proof of the preceding theorem over $k[x]$ differs from Hermite's original proof over $\mathbb{Z}$. Namely, in Hermite's proof there is a triangle inequality of the form $|\gamma^2 a_{11} + q(z)| \leq |\gamma^2 a_{11}| + |q(z)|$, and in the proof over $k[x]$ this is replaced by the strong triangle inequality $\partial\left(\gamma^2 a_{11} + q(z)\right) \leq \max\left\{\partial\left(\gamma^2 a_{11}\right), \partial q(z)\right\}$. In other words, it is the nonarchimedean property of the degree function, as opposed to the archimedean property of the ordinary absolute value, that accounts for the difference between the two results.

7.3. Djokovič's Reduction of $k[x]$-Lattices; Harder's Theorem

There is another approach to $k[x]$-lattices via "reduction theory": the pursuit of canonical forms.

Definition 7.10. A symmetric matrix $A = (a_{ij}) \in M_n(k(x))$ is **reduced** if

(i) A has a **dominant diagonal**, in the sense that $\partial a_{ii} > \partial a_{ij}$ whenever $i \neq j$; and

(ii) $\partial a_{11} \leq \cdots \leq \partial a_{nn}$.

A basis $\{v_1, \ldots, v_n\}$ for a $k[x]$-lattice L on a quadratic $k(x)$-space V is a **reduced basis** for L if the associate Gram matrix $(B(v_i, v_j))$ is reduced. An indexed subset S of V is **reduced** if S is a reduced basis for the $k[x]$-lattice that it spans.

The following theorem was published in 1976 [**Dj**].

Theorem 7.11 (Djokovič's Lattice Reduction). *Every anisotropic $k[x]$-lattice has a reduced basis.*

Proof. Suppose $L \cong A = (a_{ij}) \in M_n(k(x))$ with respect to the basis $\{v_1, \ldots, v_n\}$. By scaling if necessary, without loss of generality we can assume $A \in M_n(k[x])$. We now give the proof in the form of an algorithm for achieving a basis of the desired kind.

Step (I) Reorder (and relabel) the basis vectors so that

$$\partial a_{11} \leq \cdots \leq \partial a_{nn}.$$

Let $t \geq 1$ be the largest subscript with the property that $\{v_1, \ldots, v_t\}$ is reduced. If $t = n$ we are done; so we can suppose $t < n$.

Define $d = \max_{1\le i\le t} \partial a_{i,t+1} - \partial a_{ii}$. Then $d \ge 0$ since $\{v_1, \dots, v_{t+1}\}$ is not reduced. For all i, put $\nu_i = \partial a_{ii}$, and let λ_i denote the leading coefficient of a_{ii}. Note that each λ_i is nonzero, since L is anisotropic. For each i satisfying $1 \le i \le t$ we can write

$$a_{i,t+1} = c_i x^{\nu_i+d} + \{\text{lower degree terms}\}$$

for some $c_i \in k$. (Note: Possibly $c_i = 0$ for some values of i.) Now define

$$v'_{t+1} = v_{t+1} - \frac{c_i}{\lambda_i} x^d a_{ii} - \sum_{j=1}^{t} \frac{c_j}{\lambda_j} x^d v_j.$$

Then for $1 \le i \le t$ we have

$$B(v_i, v'_{t+1}) = a_{i,t+1} - \sum_{\substack{1\le j\le t\\ j\ne i}} \frac{c_j}{\lambda_j} x^d a_{ij},$$

a polynomial of degree at most $\nu_i + d - 1$. (Each term in the sum on the right has degree at most $\nu_i + d - 1$ because $\{v_1, \dots, v_t\}$ is reduced.) If $\partial q(v'_{t+1}) \ge \nu_t$ then replace v_{t+1} by v'_{t+1} in the basis and return to the start of Step (I). Otherwise proceed to Step (II).

Step (II). Let m be the smallest index such that $\partial q(v'_{t+1}) < \nu_m$, and insert v'_{t+1} immediately before v_m in the ordered basis for L; that is, update the basis to

$$\{v_1, \dots, v_{m-1}, v'_{t+1}, v_m, \dots, v_t, v_{t+2}, \dots, v_n\}.$$

Relabel these vectors as $\{v_1, \dots, v_n\}$, respectively, and return to Step (I).

The procedure eventually halts because each pass through Step (II) reduces a term in the sequence $\{\nu_1, \dots, \nu_n\}$ of nonnegative integers, so there can be only finitely many such passes. And if the set $\{v_1, \dots, v_t\}$ is reduced, while $\{v_1, \dots, v_{t+1}\}$ is not—because the associated value d is nonnegative—then at most $d + 1$ passes through Step (I) will be needed before either a reduced set $\{v_1, \dots, v_{t+1}\}$ is achieved or a pass through Step (II) is required. □

Remark 7.12. When $n = 2$ the above reduction can be done more directly. As in the preceding argument, we can suppose $\partial a_{11} \le \partial a_{22}$. If $\partial a_{12} \ge \partial a_{11}$, then $a_{12} = a_{11}\sigma + \rho$ for some $\sigma, \rho \in k[x]$, with $\partial\rho < \partial a_{11}$. Replacing v_2 by $v'_2 = v_2 - \sigma v_1$ yields a new Gram matrix $A = (a_{ij})$ in which $\partial a_{12} < \partial a_{11}$. Then either we are finished, or interchange v_1 and v'_1 and repeat the process as needed.

Notice that our theorem "à la Hermite" over $k[x]$ is a consequence of Djoković's Theorem, because if $A = (a_{ij})$ is a Gram matrix for L with respect to a reduced basis $\{v_1, \dots, v_n\}$, then $\partial(dL) = \sum_i \partial a_{ii}$, and hence $q(v_1) \le \frac{1}{n}\partial(dL)$.

We have seen examples of indecomposable unimodular $\mathbb{Z}$-lattices. But the following result, essentially a corollary of each of the theorems—the Hermite-style inequality and Djoković's reduction—just presented, shows that every unimodular $k[x]$-lattice splits completely into components of rank 1. The theorem was originally due to Günter Harder, and it appeared in a 1969 paper [**Kb**] by Manfred Knebusch.

Theorem 7.13 (Harder). *Let L be a unimodular $k[x]$-lattice of rank n. Then there exist $\alpha_1, \ldots, \alpha_n \in k^*$ such that $L \cong \langle \alpha_1, \ldots, \alpha_n \rangle$.*

Proof. If L is isotropic and v is a primitive isotropic vector, then, since $L = L^\sharp$, there is a vector $w \in L$ such that $B(v, w) = 1$. [To see this, extend v to a basis $\mathbb{B} = \{v, v_2, \ldots, v_n\}$ for L, and then let $w = v^\sharp$ in the dual basis. Alternatively, let $A = (a_{ij})$ be the Gram matrix associated with $\mathbb{B}$, and then use the fact that since L is unimodular there exist $x_2, \ldots, x_n \in R$ such that $\sum_{i=2}^n x_i a_{1i} = 1$, and so $B(v, \sum_{i=2}^n x_i v_i) = 1$.] Then the sublattice M spanned by v and w is unimodular and hence splits L. Now if $M \cong \begin{pmatrix} 0 & 1 \\ 1 & a \end{pmatrix}$ in the basis $\{v, w\}$ then

$$q\left(\frac{1-a}{2}v + w\right) = 1$$

and hence $M \cong \langle 1, -1 \rangle$. Iterating the argument thus far as needed eventually gives a splitting

$$L \cong \langle 1, -1 \rangle \perp \cdots \perp \langle 1, -1 \rangle \perp L'$$

with L' anisotropic.

If the lattice L' is nonzero, then it is unimodular, and then $\partial(dL') = 0$. Repeated application of the preceding theorem (in combination with the Gram–Schmidt process) then gives a splitting $L' \cong \langle \beta_1, \ldots, \beta_t \rangle$, with $\beta_i \in k^*$ for all i, and we are done. $\square$

Recall (from Chapter 2) that the **u-invariant** $u(k)$ of a field k is the maximum dimension of an anisotropic quadratic k-space. (And $u(k) = \infty$ if no such maximum exists.)

Corollary 7.14. *If L is a unimodular $k[x]$-lattice of rank $n > u(k)$, then L is isotropic.*

Remark 7.15. (i) In the vernacular of the subject, Harder's Theorem is often stated informally as "every unimodular $k[x]$-lattice extends from a k-space." And a more formal algebraic equivalant of this is to say that for every unimodular $k[x]$-lattice L there is a regular quadratic k-space V such that $L \cong V \otimes_k k[x]$. The matrix equivalent is that for every symmetric matrix $A \in GL_n(k[x])$ there exists $T \in GL_n(k[x])$ such that ${}^tTAT \in GL_n(k)$.

(ii) With extension of our definitions and some additional algebraic machinery, one can consider quadratic forms on lattices over a wider class of rings than in our present treatment, for instance over polynomial rings $k[x_1, \dots, x_t]$ in several variables. In 1976 Parimala [**P**] proved that Harder's Theorem does not extend to this more general context; in particular, she proved that there exist unimodular lattices of rank 4 over $\mathbb{R}[x_1, x_2]$ that are not extended from quadratic $\mathbb{R}$-spaces. Another proof, using extensions of ideas in the present section, can be found in Gerstein [**G3**].

(iii) For a matrix $C \in M_n(k[x])$, denote by $C_0 \in M_n(k)$ the matrix obtained by substituting 0 for x in C. In other words, C_0 is the matrix of C's constant terms. Given that ${}^tTAT = \langle \alpha_1, \dots, \alpha_n \rangle \in GL_n(k)$, with $T \in GL_n(k[x])$ as we have been discussing, it is now clear that ${}^tT_0A_0T_0 = \langle \alpha_1, \dots, \alpha_n \rangle \in GL_n(k)$. So if L and M are unimodular $k[x]$-lattices, with $L \cong A \in GL_n(k[x])$ and $M \cong C \in GL_n(k[x])$, then $L \cong M$ if and only if $A_0 \cong C_0$ over k. Moreover, there is at most one isometry class of unimodular $k[x]$-lattices on a given quadratic $k(x)$-space.

Example 7.16. (i) If L is a unimodular $\mathbb{F}_q[x]$-lattice then $L \cong \langle 1, \dots, 1, dL \rangle$; and L is isotropic if $\operatorname{rank} L \geq 3$. If M is another unimodular $\mathbb{F}_q[x]$-lattice of the same rank, then $L \cong M$ if and only if $dL = dM$. It follows that on a given quadratic $\mathbb{F}_q(x)$-space there is at most one isometry class of unimodular $\mathbb{F}_q[x]$-lattices.

(ii) A unimodular $k[x]$-lattice $L \cong A \in GL_n(k[x])$ is isotropic if the quadratic k-space $\langle A_0 \rangle$ is isotropic. For example, if $k = \mathbb{Q}$ then by Meyer's Theorem L is isotropic if $n \geq 5$ and $\langle A_0 \rangle$ is indefinite. (Here A_0 is as in the preceding remarks.)

7.4. Finiteness of Class Numbers (The Anisotropic Case)

In the literature the following theorem is attributed to Hermite and Eisenstein when $R = \mathbb{Z}$.

Theorem 7.17. *Let R denote either $\mathbb{Z}$ or $\mathbb{F}_q[x]$, with F its quotient field, and let I be a fractional R-ideal. Suppose $n \in \mathbb{N}$ and $d \in F^*$. Then there are only finitely many isometry classes of anisotropic R-lattices L having rank n and discriminant d and satisfying $sL \subseteq I$.*

Proof. The case $n = 1$ is trivial; so assume that $n > 1$ and that the result has been checked for lattices of rank less than n.

By scaling, there is no loss in generality in assuming that $I = R$ and hence that $d \in R$. (For example, if $I = \left(\frac{a}{b}\right)$ with $a, b \in R$, scale the quadratic forms on all lattices under consideration by b.)

Part 1. For now assume $R = \mathbb{Z}$. In Part 2 of the proof we will give the appropriate modifications for the case $R = \mathbb{F}_q[x]$.

(i) Suppose L is a lattice with the properties in the theorem's statement, and let $v \in L$ be a minimal vector; say $q(v) = \mu$. There is no loss of generality in assuming that $\mu > 0$, which we now do. (Scale the space by -1 if necessary.) By Hermite's inequality on the minimum there are only finitely many possibilities for μ. Define

$$L_0 = \{\, y \in L \mid B(y, v) \in \mu\mathbb{Z} \,\}.$$

Clearly L_0 is a sublattice of L spanning V and containing v. The proof now proceeds in several stages.

(ii) We claim that

$$L_0 = \mathbb{Z}v \perp \left(L \cap (\mathbb{Q}v)^{\perp}\right).$$

To see this, first note that $\mathbb{Z}v$ is a (μ)-modular sublattice of L_0 satisfying $B(\mathbb{Z}v, L_0) \subseteq (\mu)$. Therefore $L_0 = \mathbb{Z}v \perp J$ for some sublattice J, from which the claim quickly follows.

(iii) Since L and L_0 span V and $L_0 \subseteq L$, we know from the Invariant Factor Theorem (and observation (iii) immediately following that theorem) that $(L : L_0) < \infty$. We now claim that $(L : L_0) \leq \mu$. For this purpose, let $w_1, \ldots, w_{\mu+1} \in L$; we willl be done if we can show that two of these vectors must be in the same coset of L/L_0. Since L is integral, we know that $B(v, w_i) \in \mathbb{Z}$ for all i; hence by the pigeon-hole principle it follows that $B(v, w_i) \equiv B(v, w_j) \pmod{\mu}$ for some $i \neq j$, and therefore $w_i - w_j \in L_0$, proving the claim.

(iv) We claim that there are only finitely many possibilities for the discriminant dL_0. By the Invariant Factor Theorem there is a basis $\{v_1, \ldots, v_n\}$ for V and positive integers $r_1, \ldots, r_n$ such that

$$L = \mathbb{Z}v_1 + \cdots + \mathbb{Z}v_n \quad \text{and} \quad L_0 = \mathbb{Z}r_1v_1 + \cdots + \mathbb{Z}r_nv_n$$

and therefore $\prod_i r_i = (L : L_0) \leq \mu$ by (iii). Thus there are only finitely many possibilities for the r_i. But $dL_0 = \left(\prod_i r_i^2\right) \cdot dL$, and so the claim is proved.

(v) Write $L_0 \cong \langle \mu \rangle \perp J$, with $J = L \cap (\mathbb{Q}v)^{\perp}$ as in (ii). From (iv) there are only finitely many possibilities for dJ, and hence by the induction hypothesis there are only finitely many possibilities for the isometry class of J, and therefore for the isometry class of L_0 as well.

(vi) Since $L \supseteq L_0$, we know that $L^{\sharp} \subseteq L_0^{\sharp}$; moreover $L \subseteq L^{\sharp}$ because L is integral. Therefore

$$L_0 \subseteq L \subseteq L_0^{\sharp};$$

and, since there are only finitely many lattices between L_0 and $L^\sharp$, there are only finitely many possibilities for the isometry class of L.

Part 2. Now suppose $R = \mathbb{F}_q[x]$. The sections of the proof in this part correspond to those in Part 1. We will just briefly indicate the essential changes beyond changing $\mathbb{Z}$ to $\mathbb{F}_q[x]$ and $\mathbb{Q}$ to $\mathbb{F}_q(x)$.

(i) Here $\mu \in \mathbb{F}_q[x] - \{0\}$. The theorem over $\mathbb{F}_q[x]$ "à la Hermite" earlier in this chapter bounding the minimum of L gives $\partial\mu \leq \dfrac{\partial d}{n}$; and hence, since $\mathbb{F}_q$ is finite, there are only finitely many possibilities for μ.

(ii) No changes.

(iii) Here the inequality $(L : L_0) \leq \mu$ of Part 1 is replaced by $(L : L_0) \leq q^{\partial\mu}$, since there are $q^{\partial\mu}$ polynomials of degree $\partial\mu$. The important point is that the index $(L : L_0)$ is bounded by a function of d and n (namely $q^{\frac{\partial d}{n}}$).

(iv) No changes beyond noting that here the r_i are polynomials, so the relevant inequality is

$$\sum_i \partial r_i = (L : L_0) \leq \partial\mu$$

so finiteness of $\mathbb{F}_q$ tells us that there are only finitely many possibilities for the r_i.

(v) and (vi) No changes are required. □

Exercises

In the first four exercises, all given lattices should be understood to be $\mathbb{Z}$-lattices.

(1) Suppose $L \cong \begin{pmatrix} 5 & -8 \\ -8 & 13 \end{pmatrix}$. Determine $\min L$ without going through a reduction process.

(2) Suppose $L \cong \begin{pmatrix} 66 & 178 \\ 178 & 481 \end{pmatrix}$. Determine $\min L$.

(3) Suppose

$$L \cong A = \begin{pmatrix} a_1 & 1 & & & \\ 1 & a_2 & 1 & & \\ & \ddots & \ddots & \ddots & \\ & & \ddots & \ddots & 1 \\ & & & 1 & a_n \end{pmatrix}$$

with $a_i \geq 2$ for all i. Show that L is positive definite, and then show that $\min L = \min_i a_i$. [Suggestion: First consider the case in which $a_i = 2$ for all i.]

(4) Suppose

$$L \cong \begin{pmatrix} 38 & 14 & 22 \\ 14 & 10 & 7 \\ 22 & 7 & 13 \end{pmatrix}.$$

(i) Show that L has an orthogonal basis.
(ii) Conway and Schneeberger showed in 1993 that if a positive definite $\mathbb{Z}$-lattice represents all of the integers

$$1, 2, 3, 5, 6, 7, 10, 14, 15,$$

then it is "universal" in the sense that it represents *every* positive integer. (See [**Con2**] and [**Bh**].) This remarkable result is called the "Fifteen Theorem." Is the lattice L in the present exercise universal in this sense?

(5) Let k be a field of characteristic not 2, and let $R = k[x]$. The goal of this exercise is to extend Harder's Theorem and prove this: *Every integral R-lattice L with $\partial(dL) \leq 1$ has an orthogonal basis.*
(i) First observe that we can assume that $\partial(dL) = 1$ and that L is isotropic.
(ii) Show that L is R-maximal.
(iii) Let v_1 be primitive and isotropic in L. Show that v_1 is primitive in the dual lattice $L^\sharp$, and deduce that there is a vector $v_2 \in L$ such that $B(v_1, v_2) = 1$.
(iv) Show that the $Rv_1 + Rv_2$ is an orthogonal component of L and has the form $\langle 1, -1 \rangle$.
(v) Finish the proof.

(6) Let L and M be integral $\mathbb{F}_q[x]$-lattices on the n-dimensional quadratic $\mathbb{F}_q(x)$-space V, and suppose $\partial(dL) = \partial(dM) \leq 1$. Use the preceding exercise to prove that $L \cong M$.

(7) Consider an $\mathbb{F}_3[x]$-lattice

$$L \cong \begin{pmatrix} x & 1 & x^2+2 \\ 1 & x^2+2x+1 & x^2+x+2 \\ x^2+2 & x^2+x+2 & x^3+x^2+2x+1 \end{pmatrix}.$$

Use Djokoviè's algorithm to find a reduced Gram matrix for this lattice. (Note: The lattice is anisotropic. You need not check this, but in fact you can proceed with confidence that isotropic vectors will not appear as you work to construct a reduced basis.)

(8) Consider an $\mathbb{F}_q[x]$-lattice

$$L \cong \begin{pmatrix} x+1 & 2 \\ 2 & x+2 \end{pmatrix}.$$

Denote the quadratic form on L by Q to distinguish it from the finite field cardinality q. For what values of q is there a nonzero vector $v \in L$ such that $Q(v) \in \mathbb{F}_q$?

(9) Construct an anisotropic unimodular lattice of rank 4 over the polynomial ring $\mathbb{Q}_3[x]$.

Chapter 8

Local Classification of Lattices

The problem of determining whether two given rational quadratic spaces are isometric is reduced by the Hasse–Minkowski Theorem to classifying extensions of those spaces over a finite collection of p-adic number fields, where Hasse symbol computations settle the matter. It is therefore reasonable to expect that an analogous strategy will work for $\mathbb{Z}$-lattices. In fact, a $\mathbb{Z}$-lattice extends in a natural way to a lattice over $\mathbb{Z}_p$ having the same Gram matrix. The implementation of this extension is the **localization** process, to be discussed in detail in the next chapter. Because isometry of $\mathbb{Z}$-lattices L and M is equivalent to congruence of their Gram matrices over $\mathbb{Z}$, clearly a *necessary* condition for $L \cong M$ is that $L_p \cong M_p$ for all primes p. So whether or not the Hasse–Minkowski Theorem has an integral analogue—and, as we will see, it does not—we may be able to give a negative answer to an instance of the lattice isometry problem by locating a prime p for which $L_p \ncong M_p$. And we will also see that there are kinds of lattices for which local isometry for all p *will* force global isometry. With these issues as motivation, the main goal of this chapter is the classification of $\mathbb{Z}_p$-lattices up to isometry.

As usual, all given lattices should be understood to be on regular spaces, whether or not that is stated explicitly.

8.1. Jordan Splittings

Jacobi's Theorem (from Chapter 7) shows that every lattice on a quadratic space has a basis whose corresponding Gram matrix is triple-diagonal. The following theorem gives a sharpening of Jacobi's theorem in the local context.

Theorem 8.1. *Let $(F, \mathfrak{p})$ be a local field, with $\mathfrak{o} = \mathfrak{o}(\mathfrak{p})$ its ring of integers, and let L be an $\mathfrak{o}$-lattice on an n-dimensional quadratic F-space V. Then L is the orthogonal sum of modular components of rank* 1 *and/or* 2. *Moreover if F is nondyadic then L has an orthogonal basis.*

Proof. The result is trivially true if $n = 1$; so we can assume that $n \geq 2$ and that the result has been checked in lower dimensions. Suppose $L \cong A = (a_{ij})$ with respect to $\{v_1, \ldots, v_n\}$. The scale sL is generated by the a_{ij}, and if actually $(a_{ii}) = sL$ for some i then the sublattice $M = \mathfrak{o}v_i$ is sL-modular, and hence $L = M \perp (\quad)$. (In fact this splitting can be carried out à la Gram–Schmidt.) On the other hand, if $|a_{ii}| < |sL|$ for all i, then, since the valuation is nonarchimedean, $|a_{ij}| = |sL|$ for some i, j, and

$$M = \mathfrak{o}v_i + \mathfrak{o}v_j \cong \begin{pmatrix} a_{ii} & a_{ij} \\ a_{ij} & a_{jj} \end{pmatrix}$$

is an sL-modular sublattice and therefore splits L. Let's see the explicit construction of this latter splitting. Without loss of generality we can assume (by scaling if needed) that $sL = \mathfrak{o}$, that M is unimodular, and that $M = \mathfrak{o}v_1 + \mathfrak{o}v_2$. We want to replace each v_j with $j \geq 3$ by a vector of the form $v_j' = v_j + x_j v_1 + y_j v_2$, where the elements $x_j, y_j \in \mathfrak{o}$ are obtained by solving the equations $B(v_j', v_1) = 0$ and $B(v_j', v_2) = 0$ simultaneously. This yields

$$v_j' = v_j + \Big(\frac{-a_{1j}a_{22} + a_{2j}a_{12}}{dM}\Big)v_1 + \Big(\frac{-a_{2j}a_{11} + a_{1j}a_{12}}{dM}\Big)v_2.$$

Whether M has rank 1 or 2, now apply the induction hypothesis to $M^{\perp}$ to complete the proof of the first statement.

Now suppose F is nondyadic. If in the preceding argument we are in the case where

$$|a_{ii}|, |a_{jj}| < |a_{ij}| = |sL|,$$

then

$$|q(v_i + v_j)| = |a_{ii} + a_{jj} + 2a_{ij}| = |a_{ij}| = |sL|$$

by the principle of domination and the fact that $|2| = 1$. In this case set $v_i' = v_i + v_j$ and $M = \mathfrak{o}v_i'$ and then again we have $L = M \perp (\quad)$. □

Corollary 8.2. *Over the ring of integers of a local field, every lattice L can be expressed as an orthogonal sum $L = L_1 \perp \cdots \perp L_t$, with each component L_i modular and $sL_1 \supset \cdots \supset sL_t$. If $L = M_1 \perp \cdots \perp M_k$ is another such splitting, then $t = k$,* $\operatorname{rank} L_i = \operatorname{rank} M_i$, *and $sL_i = sM_i$ for all i.*

Proof. Start with a splitting of L into components of rank 1 and 2 as produced in the preceding theorem. Each of those components is a modular lattice. "Perp" together (more formally, take the orthogonal sum of) all those with a common scale $\mathfrak{p}^{\nu_i}$; the result will be $\mathfrak{p}^{\nu_i}$-modular and will be one of the components L_i in the statement of the theorem. Assign the subscripts to the L_i so that $sL_1 \supset \cdots \supset sL_t$. Thus L has a splitting of the indicated form.

The original Gram matrix A is congruent to the Gram matrix C corresponding to a basis for L obtained by taking the union of bases for the components $L_1, \ldots, L_t$. Therefore A and C have the same invariant factors, since matrix congruence implies matrix equivalence. But the invariant factors of C are clearly

$$\underbrace{\mathfrak{p}^{\nu_1}, \ldots, \mathfrak{p}^{\nu_1}}_{n_1}, \underbrace{\mathfrak{p}^{\nu_2}, \ldots, \mathfrak{p}^{\nu_2}}_{n_2}, \ldots, \underbrace{\mathfrak{p}^{\nu_t}, \ldots, \mathfrak{p}^{\nu_t}}_{n_t}$$

where $n_i = \operatorname{rank} L_i$. The Gram matrix associated with the splitting $M_1 \perp \cdots \perp M_k$ must have the same invariant factors with the same multiplicities, from which the result follows. □

Definition 8.3. Over the ring of integers of a local field, a lattice splitting $L = L_1 \perp \cdots \perp L_t$ into modular components satisfying $sL_1 \supset \cdots \supset sL_t$ is a **Jordan splitting** for L.

At this point it is reasonable to wonder whether two lattices (on the same space) that have Jordan splittings of the same general contours must be isometric, and the following example addresses this issue.

Example 8.4. Consider two $\mathbb{Z}_5$-lattices

$$L \cong \langle 1, 25 \rangle \quad \text{and} \quad M \cong \begin{pmatrix} 3 & 6 \\ 6 & 87 \end{pmatrix}$$

on respective quadratic $\mathbb{Q}_5$-spaces V and W. Is there an isometry $L \cong M$? Here L is given via a Jordan splitting with two components: $L \cong \langle 1 \rangle \perp \langle 5^2 \rangle$. We begin our search for the answer by computing a Jordan splitting for M. Assume the given Gram matrix for M is with respect to the basis $\{w_1, w_2\}$. Since $3 \mid 6$, we stay in the lattice when we apply Gram–Schmidt (in other words, the transition matrix for the basis change is unimodular) and replace w_2 in the basis by $w_2' = w_2 - \dfrac{B(w_1, w_2)}{q(w_1)} = w_2 - 2w_1$, getting that

$$M \cong \langle 3, 75 \rangle \cong \langle 3 \rangle \perp \langle 3 \cdot 5^2 \rangle,$$

a Jordan splitting. [Here is a variation of this little computation. The observation that $|3|_5 = 1$ tells us that $\mathbb{Z}_5 w_1$ is unimodular and hence an orthogonal component of M; so, since $dM = 225 = 3 \cdot 75$, the complementary component must be of the form $\langle 75 \rangle$.]

Notice that $V \cong W$ (certainly a *necessary* condition for an isometry $L \cong M$), because $dV = dW = 1$ and $S_5 V = S_5 W = 1$; and the ranks and modularities of the Jordan components are the same for the two lattices. But that is no guarantee of isometry, as we will now see. Obviously there is a representation $1 \longrightarrow L$; but is there a representation $1 \longrightarrow M$? Equivalently, is the equation $3x^2 + 75y^2 = 1$ solvable in $\mathbb{Z}_5$? Since $|75|_5 < 1$, solvability of the equation would require that $|3x^2|_5 = 1$. Were this the case, then for all $y \in \mathbb{Z}_5$ we would have

$$3x^2 + 75y^2 = 3x^2\left(1 + 75\frac{y^2}{3x^2}\right) \in 3 \cdot \mathbb{Q}_5^{*2}$$

by the Local Square Theorem. But $3 \notin \mathbb{Q}_5^{*2}$, by our criteria for p-adic representation of squares developed in Chapter 3. Therefore $1 \not\longrightarrow M$, and hence $L \ncong M$.

Incidentally, the Hilbert symbol calculation $(3, 75)_5 = (3, 3)_5 = 1$ shows that the equation $3x^2 + 75y^2 = 1$ *is* solvable in $\mathbb{Q}_5$. Had this symbol's value turned out to be -1 instead, we would have immediately known that $1 \not\longrightarrow M$.

At the end of the preceding example, if 3 had turned out to be a 5-adic square, the lattices would have been isometric. So while the example shows that the ranks and modularities of a lattice's Jordan components are not a full set of invariants for isometry of two lattices on a given space, it also suggests that the discriminants of the Jordan components need to be taken into account. In fact, in the nondyadic case these discriminants are the only additional invariant that we will need to guarantee isometry. One of the morals of this story is that congruence is a finer relation than equivalence for matrices over the ring of integers of a local field.

8.2. Nondyadic Classification

Theorem 8.5. *Let $\mathfrak{o}$ be the ring of integers of a nondyadic local field F, and let L and M be $\mathfrak{o}$-lattices on an n-dimensional quadratic F-space V. Assume L and M have Jordan splittings*

$$L = L_1 \perp \cdots \perp L_t \quad \text{and} \quad M = M_1 \perp \cdots \perp M_t,$$

with $\operatorname{rank} L_i = \operatorname{rank} M_i$ *and* $sL_i = sM_i$ *for* $1 \leq i \leq t$. *Then*

$$L \cong M \quad \text{if and only if} \quad dL_i = dM_i \quad \text{for } 1 \leq i \leq t.$$

Proof. By scaling, without loss of generality we can suppose

$$sL = sM = sL_1 = sM_1 = \mathfrak{o}.$$

(i) First consider the case in which L and M are unimodular. (So $L = L_1$ and $M = M_1$.) Then $L \cong \langle a_1, \ldots, a_n \rangle$, with $a_i \in \mathfrak{u}(\mathfrak{p})$; in particular, the residue classes $\overline{a_i}$ are nonzero elements of the residue class field $F(\mathfrak{p})$. If $n \geq 2$ then the equation $\overline{a_1}x^2 + \overline{a_2}y^2 = \overline{1}$ is solvable in $F(\mathfrak{p})$, since every binary quadratic space over a finite field is universal. Say $\overline{a_1}\,\overline{\alpha_1}^2+\overline{a_2}\,\overline{\alpha_2}^2=\overline{1}$. But then

$$\begin{aligned} a_1\alpha_1^2+a_2\alpha_2^2 &= 1+\pi\lambda \quad \text{for some } \lambda\in\mathfrak{o} \\ &= \beta^2 \text{ for some } \beta\in\mathfrak{u} \quad \text{by the Local Square Theorem;} \end{aligned}$$

and so $a_1\left(\frac{\alpha_1}{\beta}\right)^2+a_2\left(\frac{\alpha_2}{\beta}\right)^2=1$. Therefore $L\cong\langle 1,a_2',\ldots,a_n'\rangle$ for some $a_i'\in\mathfrak{u}$, and upon iterating this argument we get $L\cong\langle 1,1,\ldots,1,dL\rangle$. Similarly $M\cong\langle 1,1,\ldots,1,dM\rangle$. Hence the result holds in this case.

(ii) From part (i), direction "$\Longleftarrow$" in the theorem's conclusion is now clear in all cases; that is, without the assumption of unimodularity. It remains to prove "$\Longrightarrow$".

(iii) Now assume $L\cong M$; so in fact without loss of generality we can actually assume $L=M$. Thus $L=L_1\perp\cdots\perp L_t=M_1\perp\cdots\perp M_t$ are both Jordan splittings for L; and from the argument of part (i) we can suppose $L_1\cong\langle 1,1,\ldots,1,dL_1\rangle$ in a basis $\{v_1,\ldots,v_{n_1}\}$ and $M_1\cong\langle 1,1,\ldots,1,dM_1\rangle$ in a basis $\{w_1,\ldots,w_{n_1}\}$, with $dL_1,dM_1\in\mathfrak{u}(\mathfrak{p})$.

If $n_1=1$ and $dL_1=1$, then $1\longrightarrow L$ and therefore also $dM_1=1$ by the argument in the example preceding this theorem. Therefore we can assume that $n_1\geq 2$ and that the theorem is true for lattices whose leading Jordan components have rank $<n_1$.

Since $q(v_1)=q(w_1)=1$ and $\mathfrak{p}$ is nondyadic, we have

$$q(v_1+w_1)+q(v_1-\ w_1)\in\mathfrak{u}.$$

Hence, by replacing w_1 with $-w_1$ in the basis if necessary, we can assume $q(v_1-w_1)\in\mathfrak{u}$. Then the symmetry $\tau_{v_1-w_1}$ is in the orthogonal group $O(L)$, and $\tau_{v_1-w_1}(v_1)=w_1$. Moreover

$$(\mathfrak{o}v_1)^\perp\cong\tau_{v_1-w_1}\left((\mathfrak{o}v_1)^\perp\right)=(\mathfrak{o}w_1)^\perp,$$

and applying the induction hypothesis to this orthogonal complement now finishes the proof. □

Corollary 8.6. *Let L and M be unimodular lattices over the ring of integers of a nondyadic local field. If* $\operatorname{rank}L\leq\operatorname{rank}M$ *then there is a representation* $L\longrightarrow M$. *And if* $\operatorname{rank}L=\operatorname{rank}M$ *then $L\cong M$ if and only if $dL=dM$.*

8.3. Towards 2-adic Classification

The classification of lattices over the ring of integers of a *dyadic* local field is much more complicated than the nondyadic classification. We will be content here to take some initial steps over the 2-adic integers, by first refining our analysis of Jordan splittings and then focusing on unimodular lattices. (But to give a sense of the complete picture, we will state the main 2-adic classification theorem.)

Using invariant factor theory, we saw earlier in this chapter that the ranks and scales of the Jordan components of a lattice are well-defined invariants of the lattice; that is, they are independent of the particular Jordan splitting that one might have produced. Recall that for any lattice J we have $sJ \supseteq nJ \supseteq 2sJ$. So if the coefficient ring is nondyadic then $sJ = nJ$; but if the coefficient ring is $\mathbb{Z}_2$ then, since 2 is prime in $\mathbb{Z}_2$, either $nJ = sJ$ or $nJ = 2sJ$. So in the search for 2-adic invariants we need to refine our earlier results on Jordan splittings and consider the *norms* of the Jordan components, and this is the point of the following theorem.

Theorem 8.7. *Let L and L' be isometric $\mathbb{Z}_2$-lattices on the quadratic $\mathbb{Q}_2$-space V, with respective Jordan splittings*

$$L = L_1 \perp \cdots \perp L_t \quad \text{and} \quad L' = L'_1 \perp \cdots \perp L'_k.$$

Then $t = k$; also, $\operatorname{rank} L_i = \operatorname{rank} L'_i$, $sL_i = sL'_i$, and $nL_i = nL'_i$ for all i.

Proof. By the hypothesis, there is an isometry $\sigma \in O(V)$ such that $\sigma L' = L$; so without loss of generality we can assume $L' = L$. In view of Corollary 8.2, it remains only to check the norms of the Jordan components.

First note that for $i \geq 2$ we have $nL_i, nL'_i \subseteq sL_i \subseteq 2sL_1$, and it follows from this that the leading Jordan components have the desired property: $nL_1 = nL = nL'_1$. Consequently, to prove that $nL_i = nL'_i$ for $i \geq 2$ it will suffice to produce a lattice for which L_i and L'_i are the leading components of two Jordan splittings.

We use the construction $L^{(\alpha)}$ from the end of Section 6.7. For $1 \leq i \leq t$, let $sL_i = (\alpha_i)$. Then

$$L^{(\alpha_i)} = L_1^{(\alpha_i)} \perp \cdots \perp L_t^{(\alpha_i)}.$$

Clearly $L_j^{(\alpha_i)} = L_j$ if $j \geq i$. Now suppose $j < i$ and hence $(\alpha_j) = sL_j \supset sL_i = (\alpha_i)$. Then $|\alpha_i \alpha_j^{-1}|_2 < 1$ and

$$sL_j^{(\alpha_i)} = s(\alpha_i \alpha_j^{-1} L_j) = (\alpha_i \alpha_j^{-1})^2 sL_j \subset (\alpha_i \alpha_j^{-1}) sL_j = sL_i.$$

It follows that L_i is the leading component of a Jordan splitting for $L^{(\alpha_i)}$; that is, there is a Jordan splitting of the form $L^{(\alpha_i)} = L_i \perp \cdots$. Similarly, writing $L = L'_1 \perp \cdots \perp L'_t$ gives a Jordan splitting $L^{(\alpha_i)} = L'_i \perp \cdots$. Therefore $nL_i = nL^{(\alpha_i)} = nL'_i$, and the theorem is proved □

The following theorem of Eichler ([**Ei2**], Satz 9.6) will be important not only in proving the subsequent theorem, but in giving a new perspective on sums of squares in Chapter 9.

Theorem 8.8 (Eichler's Maximal Lattice Theorem). *Let F be a field with a complete discrete valuation $|\ \ |$, and let R be the associated valuation ring. Suppose V is a regular quadratic F-space and A is a fractional R-ideal. Then there is only one isometry class of A-maximal R-lattices on V. Moreover, if V is anisotropic, then there is only one A-maximal lattice on V.*

Proof. As in Lemmas 6.34, 6.35, and 6.36, by scaling as needed we can and will assume that $A = R$ without loss of generality.

First suppose V is anisotropic. Define

$$X = \{\, x \in V \mid q(x) \in R \,\}.$$

We claim that X is closed under addition. To this end, let $x, y \in X$. If we had $|q(x+y)| > 1$ then $|2B(x,y)| = |q(x+y) - q(x) - q(y)| > 1$. But $|q(x)q(y)| \leq 1$, and so $\left|\dfrac{q(x)q(y)}{(B(x,y))^2}\right| < |4|$. This gives

$$\det\begin{pmatrix} q(x) & B(x,y) \\ B(x,y) & q(y) \end{pmatrix} = -(B(x,y))^2\left(1 - \frac{q(x)q(y)}{(B(x,y))^2}\right),$$

which is in $-F^{*2}$ by the Local Square Theorem. Had x and y been linearly dependent, the displayed determinant would have been 0 instead; so $Fx+Fy$ is actually a hyperbolic plane, contradicting the fact that V is anisotropic. Therefore X is closed under "+" as claimed, and hence is an R-module.

If L is an R-maximal lattice on V, then clearly $L \subseteq X$. If there were a vector $w \in X - L$ then $L + Rw$ would also be an R-lattice (being finitely generated) and would satisfy $n(L + Rw) \subseteq R$, contradicting the R-maximality of L. So there can be no such w, hence $L = X$. This finishes the proof in the anisotropic case, and in fact shows that there is a *unique* R-maximal lattice on V in that case.

Now suppose V is isotropic, and suppose L and M are R-maximal lattices on V. If V is hyperbolic, then by iterations of Lemma 6.36 there are hyperbolic planes $H_1, \ldots, H_t$ such that

$$L = \mathop{\perp}_{i=1}^{t} (H_i \cap L) \quad \text{and} \quad M = \mathop{\perp}_{i=1}^{t} (H_i \cap M)$$

and by Lemma 6.35 we have $H_i \cap L \cong H_i \cap M$ for all i, giving the result.

If V is not hyperbolic, then there is a hyperbolic space H such that $L = (H \cap L) \perp L_1$ and $M = (H \cap M) \perp M_1$, with L_1 and M_1 anisotropic. But then $H \cap L \cong H \cap M$ by the hyperbolic case already considered, and $L_1 = M_1$ by the first part of the proof. □

We now focus our attention on unimodular lattices over $\mathbb{Z}_2$. In the proof of the next theorem we will make frequent use of the following observation (actually true for lattices over an arbitrary principal ideal domain in place of $\mathbb{Z}_2$): If in an integral $\mathbb{Z}_2$-lattice L the vector v satisfies $q(v) = \varepsilon \in \mathbb{Z}_2^*$, then v is a primitive vector in L and splits L; that is, $L = \mathbb{Z}_2 v \perp (\quad)$. The fancy reason for this is that a unimodular sublattice of an integral lattice splits the lattice; but more simply it is an immediate consequence of the Gram–Schmidt process.

Theorem 8.9. *Let L be a unimodular $\mathbb{Z}_2$-lattice. If $nL = \mathbb{Z}_2$ (equivalently, L represents an element of $\mathbb{Z}_2^*$) then L has an orthogonal basis; indeed, L is an orthogonal sum of sublattices of one or more of the types $\langle 1\rangle, \langle 3\rangle, \langle 5\rangle, \langle 7\rangle$. If $nL = 2\mathbb{Z}_2$, then L is an orthogonal sum of components of the form $\left\langle \begin{pmatrix} 0 & 1 \\ 1 & 0 \end{pmatrix} \right\rangle$ or $\left\langle \begin{pmatrix} 2 & 1 \\ 1 & 2 \end{pmatrix} \right\rangle$.*

Proof. We already know that L is an orthogonal sum of components of rank ≤ 2. Any components of rank 1 must be as claimed, because $\{1, 3, 5, 7\}$ is a complete set of representatives of $\mathbb{Z}_2^*/\mathbb{Z}_2^{*2}$. And a binary component not representing a unit (equivalently, with norm ideal $2\mathbb{Z}_2$) must have the form $\left\langle \begin{pmatrix} 2\alpha & 1 \\ 1 & 2\beta \end{pmatrix} \right\rangle$ for some $\alpha, \beta \in \mathbb{Z}_2$.

If $nL = \mathbb{Z}_2$ then $q(v_1) = \varepsilon \in \mathbb{Z}_2^*$ for some $v_1 \in L$, and hence there is a splitting $L \cong \mathbb{Z}_2 v_1 \perp (\quad)$. Suppose $(\mathbb{Z}_2 v_1)^\perp$ has a component of the form $\left\langle \begin{pmatrix} 2\alpha & 1 \\ 1 & 2\beta \end{pmatrix} \right\rangle$; then consider L's sublattice

$$J = \mathbb{Z}_2 v_1 \perp (\mathbb{Z}_2 v_2 + \mathbb{Z}_2 v_3) \cong \langle \varepsilon \rangle \perp \begin{pmatrix} 2\alpha & 1 \\ 1 & 2\beta \end{pmatrix}.$$

This lattice J in turn has a sublattice $\mathbb{Z}_2(v_1+v_2)+\mathbb{Z}_2 v_3 \cong \begin{pmatrix} \varepsilon + 2\alpha & 1 \\ 1 & 2\beta \end{pmatrix}$, which is unimodular and hence splits J; say

$$J \cong \langle \varepsilon' \rangle \perp \left\langle \begin{pmatrix} \varepsilon + 2\alpha & 1 \\ 1 & 2\beta \end{pmatrix} \right\rangle$$

for some $\varepsilon' \in \mathbb{Z}_2^*$. And then, since $\varepsilon + 2\alpha \in \mathbb{Z}_2^*$, it follows that J has an orthogonal basis. Repetition of this argument as often as necessary eventually yields an orthogonal basis for L.

Now suppose $nL = 2\mathbb{Z}_2$; that is, L represents no units. We must show that L's binary components are as claimed. Thus it suffices to show that if

M is a binary unimodular $\mathbb{Z}_2$-lattice with $nM = 2\mathbb{Z}_2$, then $M \cong \begin{pmatrix} 0 & 1 \\ 1 & 0 \end{pmatrix}$ or $M \cong \begin{pmatrix} 2 & 1 \\ 1 & 2 \end{pmatrix}$.

First suppose M is isotropic. Then $M \cong \begin{pmatrix} 0 & 1 \\ 1 & 2\beta \end{pmatrix}$, with $\beta \in \mathbb{Z}_2$, with respect to some basis $\{v_1, v_2\}$, and therefore $M \cong \begin{pmatrix} 0 & 1 \\ 1 & 0 \end{pmatrix}$ with respect to $\{v_1, -\beta v_1 + v_2\}$.

Now suppose M is anisotropic, say $M \cong \begin{pmatrix} 2\alpha & 1 \\ 1 & 2\beta \end{pmatrix}$, with $\alpha, \beta \in \mathbb{Z}_2$. Then $dM = -1 + 4\alpha\beta$, so the combination of the Local Square Theorem and the fact that M is anisotropic tells us that $\alpha, \beta \in \mathbb{Z}_2^*$. Applying a microscope to the discriminant then yields more completely that

$$dM = -1 + 4\alpha\beta = \underbrace{1 + 2 + 4 + 8 + \cdots}_{-1} + 4\underbrace{(1 + 2\mu)}_{\alpha}\underbrace{(1 + 2\tau)}_{\beta} = 3 + 8\gamma$$

for some $\mu, \tau, \gamma \in \mathbb{Z}_2$, and hence $dM = 3$ by the Local Square Theorem. Therefore, after relabeling we can write $M \cong \begin{pmatrix} 2\alpha & \varepsilon \\ \varepsilon & 2\beta \end{pmatrix}$, with $\alpha, \beta, \varepsilon \in \mathbb{Z}_2^*$ and $\det \begin{pmatrix} 2\alpha & \varepsilon \\ \varepsilon & 2\beta \end{pmatrix} = 3$. Suppose $\varepsilon = 1 + 2\theta$. If $\theta \neq 0$ then upon replacing v_2 by $v_2 - \dfrac{\theta}{\alpha} v_1$ we see that without loss of generality we can assume that $\varepsilon = 1$ in this matrix and hence $\beta = \alpha^{-1}$. Thus

$$M \cong \begin{pmatrix} 2\alpha & 1 \\ 1 & 2\alpha^{-1} \end{pmatrix} \text{ in } \{v_1, v_2\}.$$

Since $1 - \alpha \in 2\mathbb{Z}_2$, the element $-3 + 4\alpha = 1 - 4(1 - \alpha)$ is a square in $\mathbb{Z}_2^*$, and $-1 + \sqrt{-3 + 4\alpha} \in 2\mathbb{Z}_2$. Set

$$\lambda = \frac{-1 + \sqrt{-3 + 4\alpha}}{2\alpha}.$$

Then $\lambda \in \mathbb{Z}_2$, and $q(\lambda v_1 + v_2) = 2$. Therefore

$$M \cong \begin{pmatrix} 2 & 1 \\ 1 & 2\beta \end{pmatrix}$$

for some $\beta \in \mathbb{Z}_2^*$. The underlying space $\mathbb{Q}_2 M$ has discriminant 3 and represents 2, and therefore

$$\mathbb{Q}_2 M \cong \begin{pmatrix} 2 & 1 \\ 1 & 2\beta \end{pmatrix} \cong \begin{pmatrix} 2 & 1 \\ 1 & 2 \end{pmatrix}.$$

Thus there is a lattice M' on $\mathbb{Q}_2M$ satisfying $M' \cong \begin{pmatrix} 2 & 1 \\ 1 & 2 \end{pmatrix}$. But then M and M' are $2\mathbb{Z}_2$-maximal lattices on $\mathbb{Q}_2M$, so by Theorem 8.8 we have

$$M \cong M' \cong \begin{pmatrix} 2 & 1 \\ 1 & 2 \end{pmatrix}$$

as claimed. □

Corollary 8.10. *Let L be a unimodular $\mathbb{Z}_2$-lattice of rank $2t \geq 4$ having norm ideal $nL = 2\mathbb{Z}_2$. Then there is a splitting $L = L_1 \perp \cdots \perp L_t$ in which $L_i \cong \begin{pmatrix} 0 & 1 \\ 1 & 0 \end{pmatrix}$ for $1 \leq i \leq t-1$ and either $L_t \cong \begin{pmatrix} 0 & 1 \\ 1 & 0 \end{pmatrix}$ or $L_t \cong \begin{pmatrix} 2 & 1 \\ 1 & 2 \end{pmatrix}$.*

Proof. Consider $\mathbb{Z}_2$-lattices

$$K \cong \begin{pmatrix} 0 & 1 \\ 1 & 0 \end{pmatrix} \perp \begin{pmatrix} 0 & 1 \\ 1 & 0 \end{pmatrix} \quad \text{and} \quad M \cong \begin{pmatrix} 2 & 1 \\ 1 & 2 \end{pmatrix} \perp \begin{pmatrix} 2 & 1 \\ 1 & 2 \end{pmatrix}.$$

From the theorem, it suffices to prove that $K \cong M$, and for this we give two arguments.

(i) A check of discriminants and Hasse symbols shows that there is an isometry $\mathbb{Q}_2K \cong \mathbb{Q}_2M$. Hence M is isotropic, so from the theorem there is a splitting of the form $M \cong \begin{pmatrix} 0 & 1 \\ 1 & 0 \end{pmatrix} \perp M'$. But then $dM' = -1$, which forces $M' \cong \begin{pmatrix} 0 & 1 \\ 1 & 0 \end{pmatrix}$.

(ii) First observe that the Local Square Theorem (coupled with the equation $-7 = 1 - 8$) tells us that $-7 = \varepsilon^2$ for some $\varepsilon \in \mathbb{Z}_2^*$. Now suppose $M_1 \cong \begin{pmatrix} 2 & 1 \\ 1 & 2 \end{pmatrix}$ in the basis $\{w_1, w_2\}$. Then $q(3w_1 - w_2) = -2\varepsilon^2$, and hence $-2 \to M_1$. From this it follows immediately that M is isotropic, and now continue as in (i). □

In the 1950s Timothy O'Meara solved the classification problem over the ring of integers of an arbitrary local field, and the nondyadic case is covered in our Theorem 8.5. We conclude this chapter with a brief summary of O'Meara's dyadic classification results in the case of interest to us here, namely over $\mathbb{Z}_2$. For full coverage of all the details, see Chapter 9 of [**O'M1**].

We saw in Theorem 8.7 that given a Jordan splitting $L = L_1 \perp \cdots \perp L_t$, the ranks, scales, and norms of the L_i are invariants for L, in the sense that they will be the same in any other Jordan splitting. Now define

$$L_{(i)} = L_1 \perp \cdots \perp L_i,$$

for $1 \leq i \leq t$. Then the sequence $L_{(1)} \subset \cdots \subset L_{(t)}$ is the **Jordan chain** associated with the given splitting. We can now state the classification theorem.

Theorem 8.11 (O'Meara's 2-adic Classification). *Let L and M be $\mathbb{Z}_2$-lattices on a regular $\mathbb{Q}_2$-space V, with Jordan splittings*

$$L = \perp_{1 \leq i \leq t} L_i \quad \textit{and} \quad M = \perp_{1 \leq i \leq t} M_i$$

and associated Jordan chains $\{L_{(i)}\}_{1 \leq i \leq t}$ and $\{M_{(i)}\}_{1 \leq i \leq t}$. Assume that

$$\operatorname{rank} L_i = \operatorname{rank} M_i, \quad sL_i = sM_i, \quad \textit{and} \quad nL_i = nM_i$$

for all i. Then

(I) *If $t = 1$, then $L \cong M$.*

(II) *Now suppose $t \geq 2$; and, for $1 \leq i \leq t$, set $s_i = sL_i = sM_i$ and $n_i = nL_i = nM_i$. Then $L \cong M$ if and only the following conditions hold:*

(i) *There is no index i for which*

$$\frac{n_i n_{i+1}}{s_i^2} \subseteq (4) \quad \textit{while} \quad \frac{dL_{(i)}}{dM_{(i)}} \equiv 3 \pmod 4.$$

(ii) *Whenever $n_{i+1} \subseteq 4n_i$, there is a representation*

$$\mathbb{Q}_2 L_{(i)} \longrightarrow \mathbb{Q}_2 M_{(i)} \perp \langle 2^{\nu_2(n_i)} \rangle.$$

We remark that our proof of Corollary 8.10 proved assertion (I) of Theorem 8.11 in the event that $nL = 2sL$.

Exercises

(1) With $\mathfrak{o}$ the ring of integers of a local field F, let L be an $\mathfrak{o}$-lattice on a regular quadratic F-space V.
(i) Given an anisotropic vector $y \in V - \{0\}$, how can we determine whether the symmetry τ_y is in the orthogonal group $O(L)$?
(ii) If L has an orthogonal basis $\{v_1, \ldots, v_n\}$, show that $\tau_{v_i} \in O(L)$ for all i.
(iii) Now suppose L is a $\mathbb{Z}_2$-lattice and $L = L_1 \perp \cdots \perp L_t$ is a Jordan splitting with the property that each L_i for which $nL_i \subset sL_i$ satisfies the discriminant condition $d(\mathbb{Q}_2 L_i) = 1$. Show that there is a basis $\{v_1, \ldots, v_n\}$ associated with this splitting with the property that $\tau_{v_i} \in O(L)$ for all i. Here the phrase "associated with this splitting" means that the first few v_i form a basis for L_1, the next few for L_2, and so on.
(iv) Give an example of a $\mathbb{Z}_2$-lattice having *no* basis $\{v_1, \ldots, v_n\}$ with the property that $\tau_{v_i} \in O(L)$ for all i.

(2) Suppose L is a $\mathbb{Z}$-lattice on a regular quadratic $\mathbb{Q}$-space V; say $L \cong A \in GL_n(\mathbb{Q})$. What is the connection between the Smith normal form $\Sigma(A)$ and the Jordan splittings of the localizations L_p?

(3) From the Invariant Factor Theorem we saw in Chapter 6 that if L and M are $\mathbb{Z}$-lattices on the same quadratic $\mathbb{Q}$-space, then $L_p = M_p$ for almost all primes p, from which it obviously follows that $L_p \cong M_p$ for almost all primes p. But suppose we have a memory lapse and forget the Invariant Factor Theorem. Show that we could still deduce that $L_p \cong M_p$ for almost all primes p.

(4) Let p be a prime number, and let L be a $\mathbb{Z}_p$-lattice with Jordan splitting $L = L_1 \perp \cdots \perp L_t$, where $sL_1 \supset \cdots \supset sL_t$. Put $sL_i = (\sigma_i)$ and $nL_i = (\nu_i)$, with $\sigma_i, \nu_i \in \mathbb{Q}_p^*$. Describe a Jordan splitting for the dual lattice $L^\sharp$. Include the ranks, scales, and norms of its modular components in your description.

(5) The **Möbius function** $\mu : \mathbb{N} \to \{0, 1, -1\}$ is defined as follows:

$$\mu(n) = \begin{cases} 1 & \text{if } n = 1, \\ (-1)^r & \text{if } n = p_1 \cdots p_r \text{ with the } p_i \text{ distinct primes,} \\ 0 & \text{otherwise.} \end{cases}$$

It is known that the number of prime polynomials in $\mathbb{F}_q[x]$ of degree k is $N_q(k) = \frac{1}{k}\sum_{d|k} \mu(d) q^{\frac{k}{d}}$, where the sum is understood to be over the *positive* divisors of k. (For a proof, see [**IR**] or [**LN**].) Now suppose we are given an $\mathbb{F}_q[x]$-lattice $L \cong (a_{ij}) \in M_n(\mathbb{F}_q[x])$, and suppose further that $\max_{i,j} \partial a_{ij} = 2$.
(i) Put an upper bound on the number of prime polynomials p for which the localization L_p has *no* unimodular component in a Jordan splitting.
(ii) If at least one a_{ij} has nonzero constant term, how does this affect the answer to the question in part (i)?

(6) Consider a $\mathbb{Z}$-lattice

$$L \cong \begin{pmatrix} 41 & 23 & 0 & 0 \\ 23 & 14 & 0 & 0 \\ 0 & 0 & 0 & 12 \\ 0 & 0 & 12 & 0 \end{pmatrix}$$

in the basis $\{v_1, v_2, v_3, v_4\}$. For each prime number p, determine a Jordan splitting for the $\mathbb{Z}_p$-lattice L_p.

(7) Let p be an odd prime. Is there a quadratic $\mathbb{Q}_p$-space V (of some positive dimension) on which there are $\mathbb{Z}_p$-lattices L and M satisfying the following three properties simultaneously?

$$sL = sM = \mathbb{Z}_p, \qquad \text{ord}_p(dL) = \text{ord}_p(dM) = 1, \qquad L \not\cong M.$$

(8) (i) Suppose L is a $\mathbb{Z}_2$-lattice satisfying

$$L \cong \langle 3 \rangle \perp \begin{pmatrix} 6 & 5 \\ 5 & 8 \end{pmatrix} \perp \begin{pmatrix} 4 & 3 \\ 3 & 2 \end{pmatrix}$$

in the basis $\{v_1, \ldots, v_5\}$. Construct an orthogonal basis for L.
(ii) Do the same for the $\mathbb{Z}_2$-lattice

$$M \cong \langle 3 \rangle \perp \begin{pmatrix} 6 & 10 \\ 10 & 8 \end{pmatrix} \perp \begin{pmatrix} 4 & 3 \\ 3 & 2 \end{pmatrix}$$

in the basis $\{w_1, \ldots, w_5\}$.

(9) Over the ring $\mathbb{Z}_2$ of 2-adic integers, give an example of a lattice L having two Jordan splittings $L = L_1 \perp L_2 = L'_1 \perp L'_2$ (with $sL_i = sL'_i$ for $i = 1, 2$) such that $L_i \not\cong L'_i$.

(10) (i) Let p be an odd prime. Show that if L_1, L_2, and L_3 are $\mathbb{Z}_p$-lattices on regular $\mathbb{Q}_p$-spaces and $L_1 \perp L_2 \cong L_1 \perp L_3$, then $L_2 \cong L_3$. (So Witt's Cancellation Theorem extends to the context of $\mathbb{Z}_p$-lattices.)
(ii) Show that over the ring $\mathbb{Z}_2$ of 2-adic integers there is a lattice isometry

$$\begin{pmatrix} 0 & 1 \\ 1 & 0 \end{pmatrix} \perp \langle 1 \rangle \cong \begin{pmatrix} 1 & 1 \\ 1 & 0 \end{pmatrix} \perp \langle 1 \rangle$$

although $\begin{pmatrix} 0 & 1 \\ 1 & 0 \end{pmatrix} \not\cong \begin{pmatrix} 1 & 1 \\ 1 & 0 \end{pmatrix}$. (So here there is no cancellation à la Witt.)
(iii) Under what condition on a principal ideal domain R will the two results (the stated isometry and the cancellation failure) of part (ii) hold for R-lattices?

Chapter 9

The Local-Global Approach to Lattices

Let R be a principal ideal domain with quotient field $F \supset R$, and let p be a prime element of R. Associated with p is the **p-adic valuation** on F, denoted $|\ |_p$, modeled after the p-adic valuation on $\mathbb{Q}$: Fix a real number C satisfying $0 < C < 1$. Then if $0 \neq a \in F$ and $a = p^\nu \frac{r}{s}$ with $r, s \in R$ relatively prime to p, define $|a|_p = C^\nu$. This valuation determines the **p-adic spot**—call it $\mathfrak{p}$ for now—on F. This spot is clearly nonarchimedean and in fact discrete, and so most of the development in Chapter 3 applies, giving an associated valuation ring $\mathfrak{o}(\mathfrak{p})$ in F with prime ideal $\mathfrak{m}(\mathfrak{p})$, the residue class field $R/(p) \approx \mathfrak{o}(\mathfrak{p})/\mathfrak{m}(\mathfrak{p})$, a completion F_p (to which our beloved Local Square Theorem applies), and the local ring R_p, which is the topological closure of R in F_p. If $A = (\alpha)$ is a fractional R-ideal, by its **localization** A_p at p we mean the fractional R_p-ideal (α) in F_p; thus if $\operatorname{ord}_p \alpha = \nu$ then $A_p = (p^\nu)$. It is important to keep in mind that the material in the latter part of Chapter 3 applies to *local* fields: fields with a complete discrete spot *whose associated residue class field is finite.* And it was finiteness of the residue class field that led to finiteness of (and formulas for) the group indices $F_\mathfrak{p}^*/F_\mathfrak{p}^{*2}$ and $\mathfrak{u}(\mathfrak{p})/\mathfrak{u}(\mathfrak{p})^{*2}$. The finiteness of these numbers was crucial in the classification of quadratic spaces over $\mathfrak{p}$ in Chapter 4 and in the classification of lattices over the rings of integers of nondyadic local fields and over $\mathbb{Z}_2$ in Chapter 8. Nevertheless, we will see that much of the local machinery can be useful even when the residue class field is not finite.

Example 9.1. Let K be a field, and let $R = K[x]$ and $F = K(x)$. If p is a prime polynomial in R, then the associated p-adic spot on F leads to a completion $F_\mathfrak{p}$, and this is a local field if and only if K is finite, since the residue class field at $\mathfrak{p}$ is isomorphic to $K[x]/(p(x))$.

Notation 9.2. From now on we will use the symbol S in three ways: for the set of all nonzero prime ideals of R, for the associated set of discrete spots on F, and for a complete set of nonassociated prime elements of R (that is, one prime element from each prime ideal in S). In particular, if $R = \mathbb{Z}$, then S will denote the set of prime numbers; and if R is a polynomial ring $k[x]$, then S will denote the set of all prime (i.e., monic and irreducible) polynomials in R. We continue to use gothic letters $(\mathfrak{p}, \mathfrak{q}, \dots)$ for prime ideals or spots in S, and lower case italic $(p, q, \dots)$ for prime elements in S.

9.1. Localization

Let L be an R-lattice on the n-dimensional vector space V over F; say $L = \bigoplus_{i=1}^n Rv_i$. For $p \in S$, we define the **localization** L_p of L at p to be the R_p-lattice on V_p generated by L; thus $L = \bigoplus_{i=1}^n R_p v_i$. So $V \cap L_p = \bigoplus_{i=1}^n (R_p \cap F)v_i$, and therefore

$$\bigcap_{p\in S}(V \cap L_p) = \bigoplus_{i=1}^{n}\Big(F \cap \bigcap_{p\in S} R_p\Big)v_i = \bigoplus_{i=1}^{n} Rv_i = L.$$

It follows from this that if M is another R-lattice on V, then

$$L = M \quad \text{if and only if} \quad L_p = M_p \text{ for all } p \in S.$$

In fact, we claim that from invariant factor theory we have

$$L \subseteq M \quad \text{if and only if} \quad L_p \subseteq M_p \text{ for all } p \in S.$$

To see this, observe that if $M = \bigoplus_i Rw_i$ and $L = \bigoplus_i (s_i)w_i$, with $(s_1) \supseteq \cdots \supseteq (s_n)$, then $L_p = \bigoplus_i (s_i)_p w_i$, and $(s_i) \subseteq R$ if and only if $(s_i)_p \subseteq R_p$ for all $p \in S$. Notice also that if $\operatorname{ord}_p s_i = \nu_i(p)$, then the invariant factors of L_p in M_p are $(p^{\nu_1(p)}) \supseteq \cdots \supseteq (p^{\nu_n(p)})$. Since for $1 \le i \le n$ we have $\operatorname{ord}_p s_i = 0$ for almost all p, it follows that

$$L_p = M_p \text{ for almost all } p \in S.$$

Lemma 9.3. *Let L be an R-lattice on the quadratic F-space V, and let $p \in S$. Then $(nL)_p = nL_p$ and $(sL)_p = sL_p$.*

Proof. We prove the first claim; proof of the second is similar and is left to the reader.

Inclusion "$\subseteq$" is clear, since $L \subseteq L_p$. For the converse, suppose $nL_p = (\lambda)$. Then $\lambda = \sum_i \lambda_i q(x_i)$, for some $x_i \in L_p$ and $\lambda_i \in R_p$. Fix a basis $\{v_i\}$ for L, and write $x_j = \sum_i \beta_{ij} v_i$, with $\beta_{ij} \in R_p$. Approximate the β_{ij} and λ_i p-adically with elements of R. If our approximations are good enough, we obtain $\lambda' \in nL$ such that $|\lambda' - \lambda|_p < |\lambda|_p$ and hence $|\lambda'|_p = |\lambda|_p$, proving the claim. $\square$

The Hasse–Minkowski Theorem demonstrated that the structure of a quadratic F-space V, with $F = \mathbb{Q}$ or $\mathbb{F}_q(x)$, is completely determined by the local structures V_p. We will see in the next section that for R-lattices, with $R = \mathbb{Z}$ or $\mathbb{F}_q[x]$, respectively, on quadratic F-spaces the analogue of Hasse–Minkowski is not generally true. That is, R-lattices on V having what appears to be essentially the same local structure—meaning that $L_p \cong M_p$ at all $p \in S$—need not be isometric. [The jargon: the local does not determine the global.] So something mysterious can happen to the quadratic form on a lattice L when we make a change in a localization L_p, even when the local change keeps the isometry class of L_p intact and leaves all other localizations of L completely unchanged. Analysis of the effect of local changes on the isometry class of a lattice is a subject of ongoing research.

We have already seen that two lattices L and M on the same space are equal "almost everywhere", meaning that $L_p = M_p$ for all but finitely many $p \in S$, with the set $\{p \in S \mid L_p \neq M_p\}$ determined by the invariant factors of one lattice in the other. The following theorem is a valuable tool in understanding the process of changing one lattice into another in a step-like fashion. Here we return to the general situation in which R is a principal ideal domain and F is its quotient field.

Theorem 9.4. *Let V be an n-dimensional vector space over F, and let L be an R-lattice on V. Suppose T is a finite subset of S, and suppose that for each $p \in T$ an R_p-lattice $M_{(p)}$ is given on V_p. Then there is an R-lattice L' on V such that*

$$L'_p = \begin{cases} M_{(p)} & \text{if } p \in T, \\ L_p & \text{if } p \in S - T. \end{cases}$$

Proof. It is sufficient to prove the result when T consists of a single prime, say $T = \{q\}$; in other words, it is sufficient to show how to change one localization of L at a time. (Strictly speaking, the argument is by induction on $|T|$.)

(I) We claim that $M_{(q)}$ has a basis that is actually a basis for V.

Let $\{z_1, \ldots, z_n\}$ be a basis for $M_{(q)}$, and let $\{x_1, \ldots, x_n\}$ be a basis for V contained in $M_{(q)}$. (For this, start with an arbitrary basis of V, write each of its vectors as a linear combination of $z_1, \ldots, z_n$, and then clear denominators by multiplying by q^ν with ν sufficiently high.) Say $z_j = \sum_i \alpha_{ij} x_i$ with $\alpha_{ij} \in F_q$ for all i, j. Since F is dense in F_q, there exist $a_{ij} \in F$ such that $|a_{ij} - \alpha_{ij}|_q < 1$ and in fact with $a_{ij} = \alpha_{ij} + q^\lambda \gamma_{ij}$, where $\gamma_{ij} \in R_q$ and λ is large enough so that $|\det(a_{ij})|_q = |\det(\alpha_{ij})|_q$. (For example, taking

$\lambda > \mathrm{ord}_p(\det(\alpha_{ij}))$ is sufficient.) Now define $y_j = \sum_i a_{ij}x_i$ for $1 \le j \le n$. Then

$$y_j - z_j = \sum_i (a_{ij} - \alpha_{ij})x_i \in M_{(q)}$$

and hence $y_j \in M_{(q)}$ for $1 \le j \le n$. Say $y_j = \sum_i \beta_{ij} z_i$ with $\beta_{ij} \in R_q$. Therefore we have

$$y_j = \sum_i \beta_{ij}\left(\sum_k \alpha_{ki}x_k\right) = \sum_k \left(\sum_i \alpha_{ki}\beta_{ij}\right) x_k$$

and hence the matrix equation $(a_{kj}) = (\alpha_{ki})(\beta_{ij})$, from which we see that $|\det(\beta_{ij})|_q = 1$. Therefore $\{y_1, \ldots, y_n\}$ is a basis for $M_{(q)}$ contained in V, finishing the proof of (I).

(II) Now consider the R-lattice $M = Ry_1 \oplus \cdots \oplus Ry_n$. Then $M_q = M_{(q)}$, and by the Invariant Factor Theorem there is a basis $\{v_i\}$ for V and there are fractional R-ideals $(s_1) \supseteq \cdots \supseteq (s_n)$ such that

$$\begin{aligned} L &= Rv_1 \oplus \cdots \oplus Rv_n, \\ M &= (s_1)v_1 \oplus \cdots \oplus (s_n)v_n, \end{aligned}$$

where $s_i = q^{\nu_i}u_i$ with $|u_i|_q = 1$. Then the lattice

$$L' = R(q^{\nu_1}v_1) \oplus \cdots \oplus R(q^{\nu_n}v_n)$$

has the desired properties. □

9.2. The Genus

With V, L, and M as in the preceding section, suppose further that V is actually a *quadratic* F-space. If $L \cong M$ then clearly $L_p \cong M_p$ for all $p \in S$, since $GL_n(R) \subseteq GL_n(R_p)$ for all p. That is, local isometry $L_p \cong M_p$ for all $p \in S$ is a *necessary* condition for global isometry $L \cong M$. With our happy memories of the Hasse–Minkowski Theorem, it is only natural to expect that when $R = \mathbb{Z}$ or $\mathbb{F}_q[x]$ a complete family of local lattice isometries would also be a *sufficient* condition for global isometry; that is, isometries $L_p \cong M_p$ for all $p \in S$ would force an isometry $L \cong M$. But the following examples (the first is a variation of an example due to David Leep [**Lp**] in 2005) should immediately squelch these expectations.

Example 9.5. Suppose d is an integer, $d > 1$, and suppose further that each prime $p \mid d$ satisfies $p \equiv 1$ or $7 \pmod 8$. Consider $\mathbb{Z}$-lattices

$$L \cong \langle 1, 2d\rangle \quad \text{and} \quad M \cong \langle 2, d\rangle$$

on respective quadratic $\mathbb{Q}$-spaces V and W. Clearly $V_\infty \cong W_\infty$. We will show that $L_p \cong M_p$ (and hence $V_p \cong W_p$) for all primes $p \neq \infty$; so by Hasse–Minkowski if follows that V and W are isometric. Thus there is no loss of generality in assuming that L and M are on the same $\mathbb{Q}$-space. But trivially $1 \to L$ and $1 \not\to M$, so $L \not\cong M$. Therefore the analogue of the Hasse–Minkowski Theorem over $\mathbb{Z}$ is false; in other words, *local isometry of two $\mathbb{Z}$-lattices at all spots does not imply global isometry.*

Now let us proceed with the argument. We want to show that

$$L_p \cong \langle 1, 2d \rangle \underset{\mathbb{Z}_p}{\cong} \langle 2, d \rangle \cong M_p$$

for all primes $p \neq \infty$. If $p \nmid 2d$ then, since p is odd and $2d \in \mathbb{Z}_p^*$, the result is clear from the classification of unimodular $\mathbb{Z}_p$-lattices.

Now suppose $p \mid d$. So $p \equiv 1, 7 \pmod 8$. Then $\left(\frac{2}{p}\right) = (-1)^{\frac{p^2-1}{8}} = 1$ and therefore (from Chapter 3) $2 \in \mathbb{Z}_p^{*2}$, from which $L_p \cong M_p$ is an immediate consequence.

Finally, suppose $p = 2$. If $d \equiv 1 \pmod 8$ then $d \in \mathbb{Z}_p^{*2}$ by the Local Square Theorem, giving $L_2 \cong M_2$; so now suppose $d \equiv 7 \pmod 8$. Then $-d \equiv -7 \equiv 1 \pmod 8$, and therefore $-d \in \mathbb{Z}_2^{*2}$; that is, $d = -\varepsilon^2$ for some $\varepsilon \in \mathbb{Z}_2^{*2}$. Therefore

$$L_2 \cong \langle 1, -2 \rangle \quad \text{and} \quad M_2 \cong \langle 2, -1 \rangle$$

and, since it is now clear that $1 \to M_2$, we have $M_2 \cong \langle 1, -2 \rangle \cong L_2$, as desired.

We note that there are representations $2 \longrightarrow M$ and $2 \longrightarrow L_p$ for all primes p; but $2 \not\longrightarrow L$.

And now for a function field example with the same basic flavor as the preceding example over $\mathbb{Z}$. Here and later in this chapter we will make reference to the fact that both the Hilbert Reciprocity Law and the Hasse–Minkowski Theorem hold for quadratic $\mathbb{F}_q(x)$-spaces as well as for $\mathbb{Q}$-spaces. We ask the reader to accept this for now; we will have more to say on this topic in Chapter 10.

Example 9.6. Let $k = \mathbb{F}_q$, and let t be an odd integer, $t \geq 3$. Consider a $k[x]$-lattice

$$L \cong \begin{pmatrix} x & 1 \\ 1 & x^{t-1} \end{pmatrix} \text{ in a basis } \{v_1, v_2\} \text{ on a quadratic } k(x)\text{-space } V$$

and a lattice

$$M \cong \langle 1, x^t - 1 \rangle \text{ on a } k(x)\text{-space } W.$$

Since dL and x are relatively prime, the localized lattice L_x is unimodular, and therefore $L_x \cong \langle 1, x^t - 1 \rangle$; and since $|x^{t-1}|_p = 1$ for all $p \in S - \{x\}$, and x^{t-1} is obviously a square, Gram–Schmidt tells us that actually

$$L_p \cong \langle 1, x^t - 1 \rangle \cong M_p \text{ for all } p \in S.$$

Then, since $V_p \cong W_p$ for all $p \in S$, by Hilbert's Reciprocity Law we must have $S_\infty V_\infty = S_\infty W_\infty$, and therefore (since $dV = dW$) $V_\infty \cong W_\infty$. So $V \cong W$ by the function field version of the Hasse–Minkowski Theorem. Thus there is a lattice isometric to M on V, so without loss of generality we can suppose that M is actually on V.

We claim that $L \not\cong M$, and—as in the preceding example—it is enough to show that $1 \not\longrightarrow L$. In fact we show that L represents no elements of k^*.

To this end, suppose $0 \neq v \in L$; say $v = fv_1 + gv_2$ with $f, g \in k[x]$. Then $Q(v) = xf^2 + x^{t-1}g^2 + 2fg$. If $\partial f \neq \partial g$, then

$$\partial(xf^2 + x^{t-1}g^2) \geq 2\max\{\partial f, \partial g\} + 1;$$

while if $\partial f = \partial g$, then

$$\partial(xf^2 + x^{t-1}g^2) = 2\partial g + t - 1 > \partial fg.$$

In both cases $\partial Q(v) \geq 1$, as claimed.

Thus for $\mathbb{Z}$-lattices and $\mathbb{F}_q[x]$-lattices, local isometry or local representation at all spots does not guarantee the corresponding global result. But it is an important precondition that can be checked by the methods of the preceding chapter, and with this in mind we make the following definition.

Definition 9.7. Let $R = \mathbb{Z}$ or $\mathbb{F}_q[x]$, let $F = \mathbb{Q}$ or $\mathbb{F}_q(x)$, respectively, and let S be the corresponding set of nonassociated primes. With L an R-lattice on a quadratic F-space V, the **genus** of L, denoted $\operatorname{gen} L$, is the set of all R-lattices M on V satisfying $M_p \cong L_p$ for all $p \in S$.

Given two R-lattices L and M of the same rank—or given Gram matrices for them—we claim that it is a finite task to determine whether they are in the same genus; more precisely, to determine whether one of the lattices is isometric to a lattice in the same genus as the other. By scaling if necessary, there is no loss of generality in assuming that the lattices are integral; that is, $sL, sM \subseteq R$. Start by checking that $dL = dM$. This being the case, we immediately know that at all nondyadic primes p not dividing the common discriminant dL we have the Hasse symbol equation $S_p(FL)_p = S_p(FM)_p = 1$, and therefore $(FL)_p \cong (FM)_p$. Thus the underlying F_p-spaces are isometric; indeed, $L_p \cong M_p$ at all such p, since L_p and M_p are unimodular and have the same rank and discriminant. Then check that $L_p \cong M_p$ when $p \mid dL$ (and, if $R = \mathbb{Z}$, when $p = 2$), using the results of Chapter 8. Once this has been done, we will know that $S_\infty(FL) = S_\infty(FM)$, by Hilbert's Reciprocity Law. If $R = \mathbb{F}_q[x]$ this gives $(FL)_\infty \cong (FM)_\infty$, and hence we will have settled the matter affirmatively, by the Hasse–Minkowski Theorem. If $R = \mathbb{Z}$ the equality $S_\infty(\mathbb{Q}L) = S_\infty(\mathbb{Q}M)$ does not guarantee that $(\mathbb{Q}L)_\infty \cong (\mathbb{Q}M)_\infty$, so we need to go further and check that the $\mathbb{R}$-spaces $(\mathbb{Q}L)_\infty$ and $(\mathbb{Q}M)_\infty$ have equal Witt indices to finish our confirmation process.

We will write $\operatorname{cls} L$ for the **class** of an R-lattice L: the collection of all R-lattices on the space $V = FL$ that are isometric to L. With $R = \mathbb{Z}$ or $\mathbb{F}_q[x]$, clearly $\operatorname{cls} L \subseteq \operatorname{gen} L$, since an isometry of L onto a lattice M extends uniquely to isometries of L_p onto M_p for all $p \in S$. (Remember: Every basis for L over R is also a basis for L_p over R_p.) The **class number** of L, denoted $h(L)$, is the number of classes in $\operatorname{gen} L$. In particular, $h(L) = 1$ if and only if every lattice on V locally isometric to V at all $p \in S$ is actually isometric to L; in other words, the integral analogue of Hasse–Minkowski holds for such a lattice L. In fact we have already seen (in the corollary to Hermite's Theorem) that if L is an anisotropic unimodular $\mathbb{Z}$-lattice of rank ≤ 5 then $h(L) = 1$; and by Harder's Theorem, $h(L) = 1$ if L is *any* unimodular $\mathbb{F}_q[x]$-lattice. It can be shown that *all R-lattices have finite class number.* We will prove this only in the anisotropic case. Observe that if $M \in \operatorname{gen} L$ then $|dL|_p = |dM|_p$ for all $p \in S$, and hence $dL = \varepsilon dM$ for some $\varepsilon \in R^*$. But M and L are on the same space, and therefore $\varepsilon \in F^{*2} \cap R^* = R^{*2}$, giving $dM = dL$. Moreover,

$$(sL)_p = sL_p = sM_p = (sM)_p \text{ for all } p \in S$$

and hence $sL = sM$. But we saw in Chapter 7 that there are only finitely many classes of anisotropic R-lattices with given rank, discriminant, and scale. Thus $h(L) < \infty$, as claimed. However, a large class number $h(L)$—and we will see examples later—indicates a major failure of the integral analogue of Hasse–Minkowski within the genus of L. In other words, a large class number marks a major gap between local (over R_p for all $p \in S$) isometry and global (over R) isometry.

9.3. Maximal Lattices and the Cassels–Pfister Theorem

The main goal of this section is to prove the Cassels–Pfister theorem, which has the following history. Artin proved in a 1927 paper that if k is a field of characteristic not 2, and a polynomial $f(x) \in k[x]$ is a sum of squares of rational functions, then $f(x)$ is actually a sum of squares of polynomials. In the 1964 paper [**C1**], Cassels showed that the same number of squares suffices, and then a year later Pfister [**Pf**] extended Cassels's result to the form in which it will appear here. The theorem was the starting point for Pfister's remarkable work on quadratic forms over arbitrary fields that initiated the explosive development of the algebraic theory of quadratic forms mentioned in the historical sketch in Chapter 1. Our proof (which first appeared in [**G2**]) will be quite different from that of Cassels and Pfister, in that it will be based on properties of lattices on quadratic spaces—a theory with strong number-theoretic roots—instead of on the more "algebraically geometric" argument of the original proof. For the original argument by Cassels and Pfister, see [**Lam**] (pp. 300–302) or the papers cited here.

We will need some properties of maximal lattices. Let R be any principal ideal domain properly contained in its quotient field F (as usual, of characteristic not 2), with S a representative set of nonassociated primes in R. Recall that if A is a fractional R-ideal, then an R-lattice L on a quadratic F-space V is **A-maximal** if L is maximal among those R-lattices on V whose norm ideal is contained in A.

Lemma 9.8. *Let L be an R-lattice on V. Then L is A-maximal if and only if L_p is A_p-maximal at all $p \in S$.*

Proof. Assume L is A-maximal. Then $nL_p = (nL)_p \subseteq A_p$ for all $p \in S$. If for some $q \in S$ there is an R_q-lattice $J_{(q)}$ such that $L_q \subset J_{(q)}$, and $J_{(q)} \subseteq A_q$, then define L' on V by

$$L'_p = \begin{cases} L_p & \text{if } p \in S - \{q\}, \\ J_{(q)} & \text{if } p = q. \end{cases}$$

But then $L' \supset L$ and $nL' \subseteq A$, contradicting the A-maximality of L; hence there is no such prime q.

Conversely, if L_p is A_p-maximal at all $p \in S$, then $(nL)_p = nL_p \subseteq A_p$ for all $p \in S$ and hence $nL \subseteq A$. If we had $L \subset M$ and $nM \subseteq A$, then $L_q \subset M_q$ for some $p \in S$; and since $nM_q \subseteq A_q$ this would contradict the A_q-maximality of L_q. □

The following theorem was the starting point for much of the development of the algebraic theory of quadratic forms in the 1960s.

Theorem 9.9 (Cassels–Pfister Theorem). *Let k be a field of characteristic not 2, and let $f(x) \in k[x]$. Suppose there are scalars $\alpha_1, \dots, \alpha_n \in k$ and rational functions $r_1(x), \dots, r_n(x) \in k(x)$ such that*

$$f(x) = \alpha_1\big(r_1(x)\big)^2 + \cdots + \alpha_n\big(r_n(x)\big)^2.$$

Then there are polynomials $p_1(x), \dots, p_n(x) \in k[x]$ such that

$$f(x) = \alpha_1\big(p_1(x)\big)^2 + \cdots + \alpha_n\big(p_n(x)\big)^2.$$

Proof. Let S denote the set of prime polynomials in the principal ideal domain $k[x]$. Consider a quadratic $k(x)$-space $V \cong \langle \alpha_1, \dots, \alpha_n \rangle$ with respect to a basis $\{v_1, \dots, v_n\}$. By the hypothesis there is a vector $v \in V$ such that $q(v) = f(x)$, so there is a $k[x]$-maximal lattice L on V such that $v \in L$. The lattice $M = k[x]v_1 \perp \cdots \perp k[x]v_n$ is a unimodular lattice on V and, since $2 \in k[x]^*$, it follows that M is $k[x]$-maximal as well. Hence M_p is $k[x]_p$-maximal at all $p \in S$. By Eichler's Theorem (8.8), we know that $L_p \cong M_p$ at all $p \in S$, and therefore L is unimodular. So by Harder's Theorem (7.13) and the subsequent remarks (see §7.3) it follows that $L \cong \langle \alpha_1, \dots, \alpha_n \rangle$ over $k[x]$, which proves the theorem. □

Corollary 9.10. *Suppose k is a field of characteristic not 2, with $u(k) < \infty$. Then every element of $k[x]$ is a sum of $u(k)+1$ polynomial squares.*

Proof. The $\big(u(k)+1\big)$-dimensional quadratic $k(x)$-space $V \cong \langle 1, 1, \ldots, 1\rangle$ is isotropic. (In fact $\langle 1, 1, \ldots 1\rangle$ is isotropic over k.) Therefore V is universal; in particular, V represents every element of $k[x]$. The conclusion now follows from the Cassels–Pfister Theorem. □

Example 9.11. Every element of $\mathbb{F}_q[x]$ is a sum of three polynomial squares.

Example 9.12. Every element of $\mathbb{C}[x]$ is a sum of two polynomial squares.

Recall from the exercises in §2.4 that for a field F the **level** or **stufe** $s(F)$ is the smallest integer n such that -1 is a sum of n squares in F. (If -1 is not a sum of squares in F, then F is said to be **formally real** and we write $s(F) = \infty$.) In 1965 Pfister showed that a positive integer n is a level of some field if and only if n is a power of 2, and the Cassels–Pfister Theorem was a central cog in the machinery of his argument. That argument involved delicate interactions between representations by quadratic spaces over a given field and representations by spaces over assorted function-field extensions of that field.

It is natural to ask whether the Cassels–Pfister Theorem holds for polynomials and rational functions in more than one variable; in particular, if $f(x,y) \in k[x,y]$ is a sum of n squares in $k(x,y)$, must it be a sum of n squares in $k[x,y]$? The correct answer is "maybe" (see [**G3**] for discussion and references). Slightly more explicitly, the answer is "yes" if certain further conditions are imposed on k or on n; but the answer in general is "no" as the following celebrated example will demonstrate. Incidentally, in [**G3**] it is shown that if conditions on k are such that Harder's Theorem applies to unimodular $k[x,y]$-lattices (in other words, that every unimodular $k[x,y]$-lattice is extended from a quadratic k-space), then Cassels–Pfister would apply as well. But the difficulty is that generally Harder's Theorem does *not* work over $k[x,y]$, and some of the blame for this failure can be attributed to $k[x,y]$ not being a Euclidean ring or even a principal ideal domain.

Example 9.13. Consider the polynomial

$$f(x,y) = 1 + x^2(x^2-3)y^2 + x^2y^4 \in \mathbb{R}[x,y].$$

(This is the **Motzkin polynomial**, first considered by Theodore Motzkin in 1967.) This is a sum of four squares in $\mathbb{R}(x,y)$; in grisly detail:

$$f(x,y) = \frac{(1+x^2-2x^2y^2)^2 + \big(x(1-x^2)y\big)^2 + \big(x(1-x^2)y^2\big)^2 + \big(x^2(1-x^2)y^2\big)^2}{(1+x^2)^2}.$$

We claim that $f(x,y)$ cannot be expressed as a sum of *any* number of squares in $\mathbb{R}[x,y]$. For suppose $f(x,y) = \sum_{i=1}^{n} p_i(x,y)^2$ with $p_i(x,y) \in \mathbb{R}[x,y]$. Since

$\deg f = 6$ we must have $\deg p_i \le 3$ for $1 \le i \le n$. View $f(x,y)$ as a function on $\mathbb{R}^2$. Then the observation that $f(x,0) = f(0,y) = 1$ for all $x, y \in \mathbb{R}$ tells us that $|p_i(x,y)| \le 1$ whenever (x,y) is on the x- or y-axis. It follows that every term of p_i other than the constant term must be divisible by xy; thus

$$p_i(x,y) = a_i + xyl_i(x,y)$$

for some $a_i \in \mathbb{R}$ and some linear polynomial l_i. Upon carrying out the indicated squarings in $f(x,y) = \sum_{i=1}^{n} p_i(x,y)^2$, we now see that the coefficient of x^2y^2 in $f(x,y)$ must be $\sum_{i-1}^{n} (l_i(0,0))^2$, a nonnegative real number. But a glance back at $p(x,y)$ tells us that that coefficient is -3. So our assumption that $f(x,y)$ is a sum of polynomial squares must have been wrong.

9.4. Sums of Integer Squares (Second Version)

In this section we will use our work on lattices to get a new perspective on the topic of representing positive integers as sums of squares, and in particular we obtain new proofs of the "two-square," "three-square," and "four-square" theorems of Fermat, Gauss, and Lagrange, respectively. Theorem 9.14 will provide the key. Fundamentally the idea is the same one we used in proving the Cassels–Pfister Theorem in the preceding section. But here life gets more complicated because 2 is no longer a unit in our coefficient ring, and consequently a modular lattice need not be maximal.

Theorem 9.14. *If $1 \le n \le 4$, then a positive integer is a sum of n integer squares if and only if it is a sum of n rational squares.*

Proof. We can assume $2 \le n \le 4$, and it is enough to prove direction "$\Longleftarrow$."

Consider a $\mathbb{Z}$-lattice $L \cong \langle 1, \dots, 1\rangle$ on a quadratic $\mathbb{Q}$-space V. We claim that it suffices to show that L is $\mathbb{Z}$-maximal. Suppose for now that this has been achieved, and let m be a positive integer represented by V. Then m is represented by some $\mathbb{Z}$-maximal lattice M on V. But by Theorem 8.8 and Lemma 9.8, all $\mathbb{Z}$-maximal lattices on V are in the same genus; namely, the genus of L. Therefore M is unimodular, and hence $M \cong \langle 1, \dots, 1\rangle$ by Hermite's Theorem. The result then follows. So it remains to show that L_p is $\mathbb{Z}_p$-maximal for all primes p.

If $p \ne 2$ then $vL_p = sL_p = nL_p = \mathbb{Z}_p$. If L_p were not $\mathbb{Z}_p$-maximal, then there would be a lattice $J_p \supset L_p$ such that $nJ_p \subseteq \mathbb{Z}_p$. But then $vJ_p \supset vL_p = \mathbb{Z}_p$; and since $vJ_p \subseteq (sJ_p)^n$ we would have $nJ_p = sJ_p \supset \mathbb{Z}_p$, contradicting the fact that $nJ_p \subseteq \mathbb{Z}_p$. Therefore L_p is $\mathbb{Z}_p$-maximal for all $p \ne 2$. To finish the proof, we need to show that L_2 is $\mathbb{Z}_2$-maximal.

(i) Assume $n = 2$. If L_2 is not $\mathbb{Z}_2$-maximal, there is a $\mathbb{Z}_2$-maximal lattice M_2 properly containing L_2. So $sM_2 \supset nM_2 = \mathbb{Z}_2$, which gives $sM_2 = \left(\frac{1}{2}\right)$.

Therefore

$$M_2 \cong \begin{pmatrix} 0 & \frac{1}{2} \\ \frac{1}{2} & 0 \end{pmatrix} \quad \text{or} \quad M_2 \cong \begin{pmatrix} 1 & \frac{1}{2} \\ \frac{1}{2} & 1 \end{pmatrix},$$

by Theorem 8.9. (Note that M_2 cannot have an orthogonal basis, because that would force the equality $sM_2 = nM_2$.) But both of these contradict the fact that $dV = 1$.

(ii) Suppose $n = 3$. If L_2 is not $\mathbb{Z}_2$-maximal then, reasoning as in the case $n = 2$, we would have

$$V_2 \cong \begin{pmatrix} 0 & \frac{1}{2} \\ \frac{1}{2} & 0 \end{pmatrix} \perp \langle -1 \rangle \quad \text{or} \quad V_2 \cong \begin{pmatrix} 1 & \frac{1}{2} \\ \frac{1}{2} & 1 \end{pmatrix} \perp \langle 3 \rangle;$$

however, the $\mathbb{Q}_2$-space displayed on the left is isotropic, hence isometric to $\langle 1, -1, -1 \rangle$, which has Hasse symbol -1; and the space on the right is isometric to $\langle 1, 3, 3 \rangle$, which also has Hasse symbol -1. Both possibilities contradict the fact that $S_2 V = 1$. Therefore L_2 must be $\mathbb{Z}_2$-maximal, as claimed.

(iii) The case $n = 4$. If L_2 were not $\mathbb{Z}_2$-maximal, then V_2 would be isometric to one of the following:

$$\begin{pmatrix} 0 & \frac{1}{2} \\ \frac{1}{2} & 0 \end{pmatrix} \perp \begin{pmatrix} 0 & \frac{1}{2} \\ \frac{1}{2} & 0 \end{pmatrix}, \quad \begin{pmatrix} 0 & \frac{1}{2} \\ \frac{1}{2} & 0 \end{pmatrix} \perp \langle a, b \rangle, \quad \begin{pmatrix} 1 & \frac{1}{2} \\ \frac{1}{2} & 1 \end{pmatrix} \perp \langle a, b \rangle$$

with $a, b \in \mathbb{Z}_2^*$. (Here we have used Corollary 8.10.) The first of these is hyperbolic and has Hasse symbol -1, so is not a possibility. A discriminant argument shows that the second space is also hyperbolic, hence disallowed. In the final case we would have $V_2 \cong \langle 1, 3, \lambda, 3\lambda \rangle$ for some $\lambda \in \mathbb{Z}_2^*$, which also yields $S_2 V = -1$. (We need only check this for $\lambda \in \{1, 3, 5, 7\}$, a representative set for $\mathbb{Z}_2^*/\mathbb{Z}_2^{*2}$.) This is again a contradiction. So L_2 is indeed $\mathbb{Z}_2$-maximal, and we are done. □

Corollary 9.15 (Lagrange's Four-Square Theorem). *Every positive integer is a sum of four integer squares.*

Proof. Let $m \in \mathbb{N}$, and let V be the $\mathbb{Q}$-space $\langle 1, 1, 1, 1 \rangle$. Then $V \perp \langle -m \rangle$ is isotropic by the Hasse–Minkowski Theorem, and hence $m \to V$. Now apply the theorem. □

Corollary 9.16 (Fermat's Two-Square Theorem). *Let $m \in \mathbb{N}$. Then m is a sum of two integer squares if and only if for every prime p satisfying $p \equiv 3 \pmod 4$ the value $\operatorname{ord}_p m$ is even.*

Proof. By the theorem and Hasse–Minkowski, there is a representation $m \underset{\mathbb{Z}}{\to} \langle 1, 1 \rangle$ if and only if there are representations $m \underset{\mathbb{Q}_p}{\to} \langle 1, 1 \rangle$ for all p. Moreover, without loss of generality we can assume that m is squarefree.

(Part I) First suppose $p \neq 2$.

(i) If $p \nmid m$ then m is a p-adic unit, and hence $m \underset{\mathbb{Q}_p}{\rightarrow} \langle 1,1\rangle$. (In fact $\langle 1,1\rangle \cong \langle m, m^{-1}\rangle$.)

(ii) Now suppose $\operatorname{ord}_p m = 1$. If $p \equiv 1 \pmod 4$ then $-1 \in \mathbb{Q}_p^{*2}$, so $\langle 1,1\rangle$ is hyperbolic over $\mathbb{Q}_p$, and therefore $m \underset{\mathbb{Q}_p}{\rightarrow} \langle 1,1\rangle$. So we know that any further conditions on m that are necessary and sufficient to guarantee that m is a sum of two squares will involve at most the odd primes $p \equiv 3 \pmod 4$ and the prime 2.

If $p \equiv 3 \pmod 4$ then $-1 \notin \mathbb{Q}_p^{*2}$ and hence $\langle 1,1\rangle$ is anisotropic over $\mathbb{Q}_p$. Since $\operatorname{ord}_p m = 1$, we have $m = p\varepsilon$ for some p-adic unit ε. Then

$$\begin{aligned} m \underset{\mathbb{Q}_p}{\rightarrow} \langle 1,1\rangle &\iff \langle 1,1,-p\varepsilon\rangle \text{ is isotropic over } \mathbb{Q}_p \\ &\iff \langle 1,1,-p\varepsilon\rangle \underset{\mathbb{Q}_p}{\cong} \langle 1,-1,p\varepsilon\rangle \\ &\iff 1 = (-1,p)_p = \left(\frac{-1}{p}\right), \end{aligned}$$

contradicting the fact that $(-1,p)_p = -1$ when $p \equiv 3 \pmod 4$. Therefore there can be no such prime.

Summary up to here: In order for m to be a sum of two squares (rational or integer), the necessary condition on the odd primes p is that those such that $\operatorname{ord}_p m$ is odd satisfy the congruence $p \equiv 1 \pmod 4$. Therefore, when considering what further conditions involving the prime 2 need to be imposed on m, without loss of generality we can and will assume that m is squarefree with no prime factors p satisfying $p \equiv 3 \pmod 4$.

(Part II) With m as just described, we have

$$\begin{aligned} m \underset{\mathbb{Q}_2}{\rightarrow} \langle 1,1\rangle &\iff \langle 1,1,-m\rangle \underset{\mathbb{Q}_2}{\cong} \langle 1,-1,m\rangle \\ &\iff (-1,m)_2 = 1. \end{aligned}$$

If $2 \nmid m$ then $m \equiv 1 \pmod 4$, and hence (in the notation of Theorem 4.12) we have the Hilbert symbol calculation

$$(-1,m)_2 = (-1)^{\varepsilon(-1)\varepsilon(m)} = (-1)^{\varepsilon(m)} = 1.$$

On the other hand, if $2 \mid m$, write $m = 2k$, with $k \equiv 1 \pmod 4$. Then

$$(-1,m)_2 = (-1)^{\varepsilon(-1)\varepsilon(k)+\omega(-1)} = (-1)^{\varepsilon(k)} = 1.$$

Thus with m as conditioned in Part (I) of this proof, we are guaranteed a representation $m \underset{\mathbb{Q}_2}{\rightarrow} \langle 1,1\rangle$ and hence $m \underset{\mathbb{Z}}{\rightarrow} \langle 1,1\rangle$. □

Corollary 9.17 (Gauss's Three-Square Theorem). *A positive integer n is a sum of three integer squares if and only if n is not of the form $4^\nu k$ with $k \equiv 7 \pmod 8$.*

Proof. The proof has the same basic outline as the proof given in Chapter 5 (Theorem 5.21), except now use Theorem 9.14 instead of Theorem 5.20 to reduce the problem to a representation problem over $\mathbb{Q}$. □

Exercises

(1) Let L be an R-lattice on the n-dimensional quadratic F-space V, and let $p_0 \in S$. Suppose M is an R_{p_0}-lattice on V_{p_0}, and suppose the lattice L' on V is defined by

$$L'_p = \begin{cases} M \text{ if } p = p_0, \\ L_p \text{ if } p \neq p_0. \end{cases}$$

How are the discriminants dL and dL' related?

(2) Show that two $\mathbb{Z}$-lattices in the same genus must have the same discriminant.

(3) Consider $\mathbb{Z}$-lattices

$$L \cong \begin{pmatrix} 60 & 54 & 30 \\ 54 & 67 & 19 \\ 30 & 19 & 19 \end{pmatrix} \quad \text{and} \quad M \cong \begin{pmatrix} 48 & 0 & 24 \\ 0 & -1 & 2 \\ 24 & 2 & -4 \end{pmatrix}.$$

(i) Show that L and M are not in the same genus. [Note: This can be done with or without the help of part (ii).]
(ii) For exactly which primes p is it true that $L_p \cong M_p$?

(4) Suppose L and M are integral $\mathbb{Z}$-lattices in the same genus on a regular $\mathbb{Q}$-space. Show that if L represents at least one odd integer, then M also represents at least one odd integer.

(5) Consider a $\mathbb{Z}$-lattice $L \cong \langle 1, 1 \rangle$ on the quadratic $\mathbb{Q}$-space V. We want to define a new integral $\mathbb{Z}$-lattice L' on V with the property that $dL' = 9$ and $L'_p = L_p$ for all primes $p \neq 3$. How many possible genera are there for such a lattice L'?

(6) Let k be a field of characteristic not 2, and let $f(x) \in k[x]$. Suppose there are rational functions $r_1, \dots, r_{n+1} \in k(x)$ such that

$$f(x) = \left(\sum_{i=1}^{n} r_i^2 \right) + x r_{n+1}^2.$$

Show that there are polynomials $p_1, \dots, p_{n+1} \in k[x]$ such that

$$f(x) = \left(\sum_{i=1}^{n} p_i^2 \right) + x p_{n+1}^2.$$

(7) (i) Prove that if $q \equiv 1 \pmod 4$, then every polynomial in $\mathbb{F}_q[x]$ is a sum of two squares in $\mathbb{F}_q[x]$.
(ii) For some q, obtain an explicit expression of x as a sum of two squares in $\mathbb{F}_q[x]$.

(8) Show that the $\mathbb{Z}$-lattice $\mathbb{Z}^8$ (with the standard dot product as the inner product) is not a $\mathbb{Z}$-maximal lattice.

9.5. Indefinite Unimodular $\mathbb{Z}$-Lattices

An integral $\mathbb{Z}$-lattice L is said to be **even** or of **type II** if $nL \subseteq 2\mathbb{Z}$; and L is **odd** or of **type I** otherwise. Accordingly, one speaks of the **parity** or **type** of the lattice.

Theorem 9.18. *Let V be an n-dimensional quadratic $\mathbb{Q}$-space. Then*

(i) *There is a unimodular $\mathbb{Z}$-lattice on V if and only if*

$$V \cong \langle \underbrace{1, \ldots, 1}_{r}, \underbrace{-1, \ldots, -1}_{s} \rangle$$

for some $r, s \geq 0$. In particular, an indefinite unimodular $\mathbb{Z}$-lattice is isotropic.

(ii) *If $V \cong \langle 1, \ldots, 1 \rangle$, then there is an even unimodular $\mathbb{Z}$-lattice on V if and only if $n \equiv 0 \pmod 8$.*

Proof. (i) The second statement in part (i) is an immediate consequence of the first statement, so it is enough to prove the first.

Direction "$\Leftarrow$" is clear. Conversely, suppose L is unimodular on V. So $dL = \pm 1$ and $V_\infty \cong \langle \underbrace{1, \ldots, 1}_{r}, \underbrace{-1, \ldots, -1}_{s} \rangle$ with $(-1)^s = dL$. By the Hasse–Minkowski theorem it is enough to prove that all localizations V_p have this same Gram matrix. Since L is unimodular this is clear for all $p \neq 2$, and then the fact that also V_2 has this form follows from Hilbert's Reciprocity Law and the fact that $dV = (-1)^s$.

(ii) Assume $V \cong \langle 1, \ldots, 1 \rangle$. If $n = 8$ then the local classification theory tells us that

$$V_2 \cong \begin{pmatrix} 0 & 1 \\ 1 & 0 \end{pmatrix} \perp \cdots \perp \begin{pmatrix} 0 & 1 \\ 1 & 0 \end{pmatrix},$$

so there is a $\mathbb{Z}_2$-lattice J_2 on V_2 satisfying $J_2 \cong \begin{pmatrix} 0 & 1 \\ 1 & 0 \end{pmatrix} \perp \cdots \perp \begin{pmatrix} 0 & 1 \\ 1 & 0 \end{pmatrix}$. Now take a lattice $M \cong \langle 1, \ldots, 1 \rangle$ on V. Then by Theorem 9.4 there is a $\mathbb{Z}$-lattice L on V satisfying $L_p = M_p$ for all $p \neq 2$ and $L_2 = J_2$. This is an even unimodular lattice, as desired. For arbitrary $n \equiv 0 \pmod 8$ take an orthogonal sum of lattices of the kind just constructed.

Conversely, assume an even unimodular lattice L exists on V. Then from the local theory we know that

$$V_2 \cong \begin{pmatrix} 0 & 1 \\ 1 & 0 \end{pmatrix} \perp \cdots \perp \begin{pmatrix} 0 & 1 \\ 1 & 0 \end{pmatrix} \quad \text{or} \quad V_2 \cong \begin{pmatrix} 0 & 1 \\ 1 & 0 \end{pmatrix} \perp \cdots \perp \begin{pmatrix} 0 & 1 \\ 1 & 0 \end{pmatrix} \perp \begin{pmatrix} 2 & 1 \\ 1 & 2 \end{pmatrix}$$

and therefore $dV_2 \in \{\pm 1, \pm 3\}$. But in fact the given space V satisfies $dV = 1$, hence we must have $V_2 \cong \begin{pmatrix} 0 & 1 \\ 1 & 0 \end{pmatrix} \perp \cdots \perp \begin{pmatrix} 0 & 1 \\ 1 & 0 \end{pmatrix}$, with an even number of components. Therefore $n \equiv 0 \pmod 4$. Now if $n \equiv 4 \pmod 8$ we would have Hasse symbol $S_2 V = -1$, a contradiction, and hence $n \equiv 0 \pmod 8$ as claimed. □

In what follows, by the **signature** of a $\mathbb{Z}$-lattice we mean the signature of the underlying quadratic space.

Theorem 9.19. *If L is an indefinite odd unimodular $\mathbb{Z}$-lattice, then*

$$L \cong \langle \underbrace{1, \ldots, 1}_{r}, \underbrace{-1, \ldots, -1}_{s} \rangle$$

for some $r, s \geq 0$. Therefore two odd unimodular $\mathbb{Z}$-lattices are isometric if and only if they have the same rank and signature.

Proof. We argue by induction on $n = \text{rank } L$, the case $n = 1$ being trivial.

From the preceding theorem we know L is isotropic; let v be a primitive isotropic vector in L. Then there is a vector $w \in L$ such that $B(v, w) = 1$. In brief, here are two ways to see this: (i) Use v as the first vector in a basis for L, and then take w to be the first vector in the dual basis; (ii) argue as in the proof of Jacobi's triple-diagonalization theorem.

If $q(w)$ is odd, say $q(w) = 2m + 1$, define $v_1 = w - mv$ and $v_2 = w - (m+1)v$. Then $M = \mathbb{Z}v_1 + \mathbb{Z}v_2 \cong \langle 1, -1 \rangle$ is a unimodular sublattice of L and therefore an orthogonal component of L. Now apply the induction hypothesis to the orthogonal complement of M.

If $q(w)$ is even, choose a vector $x \in L$ with $q(x)$ odd and replace w by $w' = x + (1 - B(v, x))\, w$. Then $q(w')$ is odd and $B(v, w') = 1$, so the argument of the preceding paragraph applies. □

In what follows we use H for a unimodular $\mathbb{Z}$-lattice spanned by a hyperbolic pair. Thus $H \cong \begin{pmatrix} 0 & 1 \\ 1 & 0 \end{pmatrix}$. We will follow the argument in [**Se**] and [**MH**].

Lemma 9.20. *If L is an indefinite even unimodular $\mathbb{Z}$-lattice, then there is a splitting of the form $L \cong H \perp (\quad)$.*

Proof. We have already shown L to be isotropic; therefore L has a component of the form $M \cong \begin{pmatrix} 0 & 1 \\ 1 & 2a \end{pmatrix}$, say in the basis $\{x, y\}$. Then $M \cong \begin{pmatrix} 0 & 1 \\ 1 & 0 \end{pmatrix}$ in the basis $\{x, y - ax\}$. □

Lemma 9.21. *If L_1 and L_2 are even unimodular $\mathbb{Z}$-lattices and*

$$L_1 \perp \langle 1, -1 \rangle \cong L_2 \perp \langle 1, -1 \rangle,$$

then $L_1 \perp H \cong L_2 \perp H$.

Proof. For any odd unimodular $\mathbb{Z}$-lattice X, we define

$$X_0 = \{\, x \in X \mid q(x) \equiv 0 \pmod{2} \}.$$

Then X_0 is a sublattice of X of index 2, and we have $X_0 \subset X \subseteq X_0^\sharp$. By the Invariant Factor Theorem, X has a basis $\{x_1, \dots, x_n\}$ such that $\{x_1, \dots, x_{n-1}, 2x_n\}$ is a basis for X_0. Then the dual basis $\{x_1^\sharp, \dots, x_n^\sharp\}$ is also a basis for X, while $\{x_1^\sharp, \dots, x_{n-1}^\sharp, \frac{1}{2}x_n^\sharp\}$ is a basis for $X_0^\sharp$. It follows that $(X_0^\sharp : X) = (X : X_0) = 2$, and hence $(X_0^\sharp : X_0) = 4$.

In place of the generic odd unimodular X take a lattice $J \cong \langle 1, -1 \rangle$ in a basis $\{v_1, v_2\}$, and let $M = \mathbb{Z}(v_1 + v_2) \oplus \mathbb{Z}(v_1 - v_2)$. We claim that $J_0 = M$. Inclusion "$\supseteq$" is clear. For the converse, notice that the surjective homomorphism $J \to M$ given by $v_1 \longmapsto v_1 + v_2$ and $v_2 \longmapsto v_1 - v_2$ has determinant 2, and therefore (from Section 6.3) $(J : M) = 2$, proving the claim. It follows that $J_0^\sharp$ has basis $\{\frac{1}{2}(v_1 - v_2), \frac{1}{2}(v_1 + v_2)\}$, so the invariant factors of J_0 in $J_0^\sharp$ are $\{(2), (2)\}$, and hence $J_0^\sharp / J_0$ is the Klein 4-group. So there are three lattices strictly between J_0 and $J_0^\sharp$, one of which is J, and the other two—with bases $\{v_1 - v_2, \frac{1}{2}(v_1 + v_2)\}$ and $\{\frac{1}{2}(v_1 - v_2), v_1 + v_2\}$—are isometric to H.

Now consider one of the two lattices—call it M_1—in the hypothesis; say $M_1 \cong L_1 \perp J$. Since L_1 is even, we have $(M_1)_0 \cong L_1 \perp J_0$ and $(M_1)_0^\sharp \cong L_1 \perp J_0^\sharp$, and therefore there is an isomorphism $(M_1)_0^\sharp/(M_1)_0 \approx J_0^\sharp/J_0$. One of the three properly intermediate (between $(M_1)_0^\sharp$ and $(M_1)_0$) lattices is odd, namely M_1; the other two are even and are isometric to $L_1 \perp H$. All this carries over to $M_2 \cong L_2 \perp J$ via the isometry $M_1 \cong M_2$, and therefore $L_1 \perp H \cong L_2 \perp H$, as claimed. □

Theorem 9.22. *Two indefinite unimodular $\mathbb{Z}$-lattices are isometric if and only if they have the same rank, type, and signature.*

Proof. Only the sufficiency requires proof, and the case of odd lattices was considered earlier. So we may now assume L_1 and L_2 are even lattices having the same rank and signature, and we must show they are isometric.

By Lemma 9.20 there are splittings of the form $L_1 \cong H \perp M_1$ and $L_2 \cong H \perp M_2$, and clearly the components M_1 and M_2 have the same rank and signature. Then there is an isometry $\langle 1, -1\rangle \perp M_1 \cong \langle 1, -1\rangle \perp M_2$ by the classification of odd lattices, and therefore $L_1 \cong L_2$ by Lemma 9.21. $\square$

Recall from Example 7.8 that when $n \equiv 0 \pmod 8$, the $\mathbb{Z}$-lattice

$$K_n \cong \begin{pmatrix} \frac{n}{4} & 1 & & & & \\ 1 & 4 & 2 & & & \\ & 2 & 2 & 1 & & \\ & & 1 & 2 & \ddots & \\ & & & \ddots & \ddots & 1 \\ & & & & 1 & 2 \end{pmatrix}$$

is positive definite, even, unimodular, and indecomposable.

Corollary 9.23. *Let L be an indefinite even unimodular $\mathbb{Z}$-lattice of signature s. If $s = 0$ then $L \cong H \perp \cdots \perp H$; and if $s \neq 0$ then*

$$L \cong \begin{cases} H \perp \cdots \perp H \perp \underbrace{K_8 \perp \cdots \perp K_8}_{\frac{s}{8}} & \text{if } s > 0, \\ H \perp \cdots \perp H \perp \underbrace{(K_8 \perp \cdots \perp K_8)}_{\frac{|s|}{8}}{}^{-1} & \text{if } s < 0. \end{cases}$$

9.6. The Eichler–Kneser Theorem; the Lattice $\mathbb{Z}^n$

Now we turn to definite $\mathbb{Z}$-lattices, and there is no loss of generality in supposing that the lattices are *positive* definite; in other words, we can view our lattices and their underlying quadratic $\mathbb{Q}$-spaces as being contained in an ordinary Euclidean space $\mathbb{R}^n \cong \langle 1, \ldots, 1\rangle$. We begin with one of the fundamental theorems of the subject. It first appeared in [**Ei1**] and [**K1**].

Theorem 9.24 (Eichler–Kneser Theorem). *A definite $\mathbb{Z}$-lattice L has a unique (except for the order of the summands) expression of the form*

$$L = L_1 \perp \cdots \perp L_t$$

in which the sublattices L_i are indecomposable.

Proof. If L is indecomposable then the result is trivial, so we can assume L has at least one nontrivial splitting into indecomposables, and we must show it is unique. By scaling, without loss of generality we can assume that L is *positive* definite, and in fact that $q(v) \in \mathbb{N}$ whenever $0 \neq v \in L$.

Call a nonzero vector $v \in L$ **reducible** if v is a sum of two shorter vectors; that is, there is an expression $v = x + y$ with $x, y \in L$ and $q(x), q(y) < q(v)$; otherwise v is **irreducible**. Clearly L is generated by its set X of irreducible vectors. Define a relation "$\sim$" on X by declaring $x \sim y$ if there exists a sequence $x = x_0, x_1, \ldots, x_r = y$ in X such that $B(x_i, x_{i+1}) \neq 0$ for $0 \leq i \leq r-1$. This is an equivalence relation on X. Let $\{C_j\}_{j \in J}$ be the collection of $\sim$-equivalence classes of irreducible vectors, and let K_j be the sublattice of L generated by C_j. Since the K_j are mutually orthogonal, there are only finitely many of these sublattices, and $L = \perp_j K_j$ because X generates L. On the other hand we have $L = \perp_i L_i$; and each member of X must be in some L_i. Moreover, every member of a given C_j must be in the same L_i, so each K_j is a sublattice of some L_i. Since $L = \bigoplus_i L_i = \bigoplus_j K_j$ it follows that each L_i is the orthogonal sum of all the sublattices K_j it contains. Hence, since the L_i are indecomposable, each L_i is a K_j. □

Remark 9.25. One cannot expect uniqueness of splitting into indecomposables for $\mathbb{Z}$-lattices in the indefinite case. For example, we leave it to the reader to check the $\mathbb{Z}$-lattice isometry $\langle 2, -1 \rangle \cong \langle 1, -2 \rangle$.

Example 9.26. We will now use the Eichler–Kneser Theorem to construct definite unimodular lattices with large class number. First recall from Theorem 9.18 that a positive definite unimodular $\mathbb{Z}$-lattice L can exist only on an underlying quadratic $\mathbb{Q}$-space $V = \mathbb{Q}L$ of the form $\langle 1, \ldots, 1 \rangle$. Also notice that if $\dim V \equiv 0 \pmod 8$ then

$$V_2 \cong \begin{pmatrix} 0 & 1 \\ 1 & 0 \end{pmatrix} \perp \cdots \perp \begin{pmatrix} 0 & 1 \\ 1 & 0 \end{pmatrix},$$

so such a space V supports an even unimodular lattice. Moreover, an even unimodular $\mathbb{Z}$-lattice is $2\mathbb{Z}$-maximal, since any lattice with norm ideal $\subseteq 2\mathbb{Z}$ has integral scale and hence has integral volume as well, and therefore all even unimodular $\mathbb{Z}$-lattices on a given $\mathbb{Q}$-space are in the same genus.

Now consider a quadratic $\mathbb{Q}$-space $V \cong \langle 1, \ldots, 1 \rangle$ of dimension $n = 8k$. By a **partition** of k we mean an expression $k = k_1 + \cdots + k_t$, with $k_i \in \mathbb{N}$ and $k_1 \geq \cdots \geq k_t$. We write $p(k)$ for the number of partitions of the integer k; this is the **partition function**, which has been heavily studied. With each partition $k = k_1 + \cdots + k_t$ we now associate a unimodular lattice of the form

$$K_{8k_1} \perp K_{8k_2} \perp \cdots \perp K_{8k_t}$$

on V. Positive definite even unimodular $\mathbb{Z}$-lattices are all $2\mathbb{Z}$-maximal and hence in the same genus and, by the Eichler–Kneser theorem, lattices of this kind that are associated with different partitions of k are not isometric. Therefore the class number of a positive definite even $\mathbb{Z}$-lattice of rank $n = 8k$ is at least $p(k)$. It is well known that $p(k) \to \infty$ as $k \to \infty$, with the growth rate exponential.

Example 9.27. Obtaining the orthogonal decomposition described in the Eichler–Kneser theorem can be very difficult to carry out in practice, and achieving an effective algorithm for this would be a valuable research contribution. For example, consider the so-called "standard" unimodular lattice in $\mathbb{R}^n$; that is, the lattice $\mathbb{Z}^n \cong \langle 1, \dots, 1 \rangle$ in the orthonormal basis $\{e_1, \dots, e_n\}$. Because the indecomposable components are the sublattices $\mathbb{Z}e_i$, the only irreducible vectors are the vectors $\pm e_i$; that is, the vectors $v \in \mathbb{Z}^n$ satisfying $q(v) = 1$. But how should we locate these vectors if we are given a basis associated with a Gram matrix that looks nothing like the identity matrix? Indeed, given a positive definite unimodular $\mathbb{Z}$-lattice L, no efficient method is known for determining whether L is actually "standard". For example, using the "neighbor lattice" concept—which will be introduced later in this chapter—it can be shown that a positive definite unimodular lattice L with Gram matrix

$$\begin{pmatrix} 5 & 6 & 2 & & & & & \\ 6 & 21 & 12 & 2 & & & & \\ 2 & 12 & 9 & 4 & 1 & & & \\ & 2 & 4 & 6 & 4 & 1 & & \\ & & 1 & 4 & 6 & 4 & 1 & \\ & & & 1 & 4 & 6 & 4 & 1 \\ & & & & 1 & 4 & 6 & 4 \\ & & & & & 1 & 4 & 5 \end{pmatrix}$$

is isometric to $\mathbb{Z}^8$. But the neighbor lattice approach does not lead to an explicit construction of an orthonormal basis for L. And if a higher-dimensional example were given of a positive definite unimodular $\mathbb{Z}$-matrix, how would we proceed then? Would we even be able to recognize whether it was isometric to $\mathbb{Z}^n$?

Let us pursue the issue raised in the preceding example by describing two nonalgorithmic characterizations of $\mathbb{Z}^n$.

Definition 9.28. A vector v in an integral $\mathbb{Z}$-lattice L is a **characteristic** element of L if

$$B(v, x) \equiv q(x) \pmod 2 \quad \text{for all } x \in L.$$

Clearly 0 is a characteristic element of L if and only if L is an even lattice. The following result (from [**G7**]) shows how to find such an element in a unimodular lattice.

Theorem 9.29. *Let L be a unimodular $\mathbb{Z}$-lattice with basis $\{v_1, \dots, v_n\}$ and dual basis $\{v_1^\#, \dots, v_n^\#\}$, and let $v = \sum_{i=1}^n \alpha_i v_i \in L$. Then v is characteristic if and only if*

$$\alpha_i \equiv q(v_i^\#) \pmod 2 \quad \textit{for } 1 \le i \le n.$$

In particular, the vector $\sum_{i=1}^n q(v_i^{\#})v_i$ *is characteristic, and the full set of characteristic vectors is the coset of this vector in* $L/2L$.

Proof. Assume $v = \sum_{i=1}^n \alpha_i v_i$ is characteristic. Then for $1 \le i \le n$ we have

$$\alpha_i = B(v, v_i^{\#}) \equiv q(v_i^{\#}) \pmod 2.$$

Conversely, given integers $\alpha_i \equiv q(v_i^{\#}) \pmod 2$ for $1 \le i \le n$, set $v = \sum_{i=1}^n \alpha_i v_i$, and let $x \in L$, say $x = \sum_{i=1}^n \beta_i v_i^{\#}$. Then

$$q(x) = \sum_{i=1}^n \beta_i^2 v_i^{\#} + 2\sum_{i<j} \beta_i\beta_j B(v_i^{\#}, v_j^{\#}) \equiv \sum_{i=1}^n \beta_i q(v_i^{\#}) \pmod 2.$$

On the other hand,

$$B(v,x) = B\left(\sum_{i=1}^n \alpha_i v_i, \sum_{j=1}^n \beta_j v_j^{\#}\right) = \sum_{i=1}^n \alpha_i\beta_i \equiv \sum_{i=1}^n \beta_i q(v_i^{\#}) \equiv q(x) \pmod 2,$$

and therefore v is characteristic.

The second statement is now clear. □

Corollary 9.30. *If* v *and* v' *are characteristic elements of the unimodular lattice* L, *then* $q(v) \equiv q(v') \pmod 8$.

Corollary 9.31 (van der Blij [Bl]). *If* L *is unimodular and* $v \in L$ *is characteristic, then* $q(v) \equiv \operatorname{sig} L \pmod 8$.

Proof. Consider the orthogonal adjunction $M = L \perp J$, where $J \cong \langle 1, -1\rangle$ in a basis $\{v_1, v_2\}$. Then $\{v_1, -v_2\}$ is the basis for J dual to $\{v_1, v_2\}$, so by the theorem $w = v_1 + v_2$ is characteristic in J, and hence $v + w$ is characteristic in M. But $q(v+w) = q(v)$, and sig $L =$ sig M. Therefore without loss of generality we can assume that L is odd and indefinite, and hence that

$$L \cong \langle \underbrace{1, \dots, 1}_{r}, -1, \dots, -1\rangle$$

in some basis $\{x_1, \dots, x_n\}$. The corresponding dual basis is

$$\{x_1, \dots, x_r, -x_{r+1}, \dots, -x_n\}$$

so $z = \sum_1^n x_i$ is characteristic. But $q(z) =$ sig L, and hence we are done. □

Remark 9.32. Notice that the preceding corollary gives another proof of our earlier result—obtained in Theorem 9.18 with the help of local structure theory—that an even definite unimodular $\mathbb{Z}$-lattice of rank n exists only if $n \equiv 0 \pmod 8$.

Example 9.33. In the standard unimodular lattice $\mathbb{Z}^n$ with orthonormal basis $\{e_1, \dots, e_n\}$ (which is self-dual), the vector $e_1+\cdots+e_n$ is characteristic. In fact the minimal characteristic elements are precisely the vectors

$$e_1 \pm \cdots \pm e_n,$$

and these all satisfy $q(e_1 \pm \cdots \pm e_n) = n$.

The following theorem of Elkies [**El1**], which we will not prove here, shows that the property displayed in the preceding example is unique to $\mathbb{Z}^n$.

Theorem 9.34 (Elkies, 1995). *If L is a positive definite unimodular $\mathbb{Z}$-lattice of rank n, then $L \cong Z^n$ if and only if L has no characteristic vector v satisfying $q(v) < n$.*

In view of Elkies' theorem, given a unimodular lattice L and a Gram matrix for it, it would be interesting to have an algorithm for determining the length of a minimal characteristic vector in L. We can find a characteristic element v using the theorem given earlier in this section, but then the difficulty lies in finding a shortest element in the coset $v + 2L \in L/2L$. We note that subsequent papers by Elkies [**El2**] and Gaulter [**Ga**] suggest that a positive definite unimodular lattice L or rank n whose shortest characteristic vectors have squared-length *near* n comes close to having an orthogonal basis, in the sense that L has a sublattice isometric to $\mathbb{Z}^m$ with $n-m$ small; but the precise nature of this relationship is not yet completely understood.

Other nonalgorithmic characterizations of lattices $L \cong \mathbb{Z}^n$ were given by Fintushel and Stern [**FS**] in 1984, motivated by topological applications, and interestingly these also involve properties of short vectors in the cosets of $L/2L$. We will give one of these.

Definition 9.35. A vector v in a $\mathbb{Z}$-lattice L is **coset-minimal** if $q(v)$ is minimal for vectors in the coset $v+2L \in L/2L$. The symbol $\gamma(v)$ denotes the number of distinct lines (i.e., one-dimensional subspaces) containing coset-minimal vectors in $v + 2L$.

Example 9.36. Suppose $L \cong \mathbb{Z}^n$, with $\{e_1, \dots, e_n\}$ an orthonormal basis. Then a nonzero vector $v \in L$ is coset-minimal if and only if there is a nonempty subset $T \subseteq \{1, \dots, n\}$ such that $v = \sum_{i \in T} \varepsilon_i e_i$ with $\varepsilon_i \in \{1, -1\}$ for all $i \in T$. If $|T| = k$ then there are exactly 2^k coset-minimal vectors in $v + 2L$; and two such vectors v, v' are in the same line if and only if $v' = \pm v$. Therefore $\gamma(v) = 2^{k-1}$; so $\gamma(v)$ is even unless $q(v) = 1$, in which case $v = \pm e_i$ for some i and $\gamma(v) = 1$.

Theorem 9.37 (Fintushel and Stern). *Let L be a positive definite unimodular $\mathbb{Z}$-lattice of rank $n \geq 2$. Then $L \cong \mathbb{Z}^n$ if and only if every coset-minimal vector v with $q(v) > 1$ satisfies $\gamma(v) \geq 2$.*

Proof. Direction "$\Rightarrow$" is covered by the preceding example. Conversely, with the stated conditions on coset-minimal vectors in place, suppose $L \not\cong \mathbb{Z}^n$, and we seek a contradiction. Since $L \not\cong \mathbb{Z}^n$, we know L has an indecomposable orthogonal component M of rank ≥ 2; so, by Gram–Schmidt, $q(x) \geq 2$ for all nonzero $x \in M$.

Now let e be a minimal vector in M; that is, $q(e) = \min M$. Then e is primitive in M and hence primitive in L; so $e \notin 2L$. Moreover, e is *a fortiori* coset-minimal in M, so e is coset-minimal in L as well.

If v is any other coset-minimal vector in $e + 2L$, say $v = e + 2w$, write $w = w_1 + w_2$ with $w_1 \in M$ and $w_2 \in M^\perp$. Then

$$q(e) = q(v) = q(e + 2w_1) + q(2w_2).$$

But $e + 2w_1 \neq 0$, since $e \notin 2L$; therefore $q(e+2w_1) \geq q(e)$, since e is minimal in M. So in fact $w_2 = 0$, giving $v = e + 2w_1$. Therefore

$$q(e) = q(v) = q(e) + 4\big(B(e, w_1) + q(w_1)\big)$$

and hence $B(w_1, e + w_1) = 0$. This gives

$$q(e) = q(v) = q\big((e + w_1) + w_1\big) = q(e + w_1) + q(w_1).$$

But e is minimal in M, so we must have $w_1 = -e$ or $w_1 = 0$, and therefore $v = \pm e$. It follows that $\gamma(e) = 1$, contradicting the hypothesis. So our assumption that L has an indecomposable component M of rank ≥ 2 must have been wrong. □

9.7. Growth of Class Numbers with Rank

In the preceding section we used the Eichler–Kneser theorem to show that class numbers of positive definite even unimodular $\mathbb{Z}$-lattices increase in unbounded fashion as the rank increases. The reader has no doubt observed that these are very special lattices, and so the question arises as to whether *all* positive definite lattices of large rank have large class number. (A lattice L having large class number $h(L)$ is often taken as an indicator that classifying the lattices in $\operatorname{gen} L$ up to isometry is likely to be difficult.) The purpose of this section is to show that this is so and, in the process, to illustrate the interaction between local and global phenomena as we shuttle back and forth between the local and global settings. The key idea (developed in [**G1**]) builds on an approach of G. L. Watson [**W2**] that reduces the nonunimodular case to the even unimodular case already considered. One of the mathematical morals of this story is that sometimes an understanding of a special kind of object (in this case, even unimodular lattices) can be used to give information about a much more general class of objects.

Before proceeding, we note that none of this applies to *indefinite* lattices. In fact, a genus of lattices consists completely of odd lattices or completely of even lattices, since the 2-adic (in particular) localizations are isometric. Therefore from the earlier section on indefinite unimodular $\mathbb{Z}$-lattices it follows that all such lattices have class number 1, no matter what their rank. So whenever class numbers are discussed in the present section it should be understood that we are in the *definite* setting, whether or not it is stated explicitly.

First we need some results on localization. We use the notation of Chapter 6.

Lemma 9.38. *Let R be a principal ideal domain with quotient field F, let $A \in I(R)$, and let L and M be R-lattices on a quadratic F-space V. Finally, let p be a prime of R. Then*

(i) $(L \cap M)_p = L_p \cap M_p$;

(ii) $(L^A)_p = (L_p)^{A_p}$.

Proof. (i) First notice that if B and C are fractional ideals, say $B = (\prod p_i^{\beta_i})$ and $C = (\prod p_i^{\gamma_i})$, with the p_i nonassociated primes, then $B \cap C = (\prod p_i^{\max\{\beta_i,\gamma_i\}})$. And from this it follows that $(B \cap C)_p = B_p \cap C_p$. Statement (i) follows from this observation together with the Invariant Factor Theorem applied to L and M.

(ii) We have

$$(L^A)_p = (L \cap AL^\#)_p = L_p \cap (AL^\#)_p = L_p \cap A_p(L^\#)_p = L_p \cap A_p(L_p)^\# = L_p^{A_p}.$$

□

In words, the following result asserts that multiplying or scaling a $\mathbb{Z}$-lattice by a nonzero scalar does not change the class number of the lattice. The proof is left to the reader.

Lemma 9.39. *Let L be a $\mathbb{Z}$-lattice on a regular quadratic $\mathbb{Q}$-space, and let $\lambda \in \mathbb{Q}^*$. Then*

$$h(\lambda L) = h(L^\lambda) = h(L).$$

It follows that in proving that class numbers of all definite lattices grow with rank, it suffices to consider *integral* lattices (i.e., lattices with scale $\subseteq \mathbb{Z}$) whose Gram matrices have relatively prime entries. Given such a lattice L, every localization L_p has a Jordan splitting of the form

$$L_p = L_{p0} \perp L_{p1} \perp \cdots \perp L_{pt_p}$$

in which for $1 \le i \le t_p$ the Jordan component L_{pi} is (p^i)-modular or 0, and $L_{p0} \neq 0$. Equivalently, $L \cong A \in M_n(\mathbb{Z}) \cap GL_n(\mathbb{Q})$, where the matrix A has at least one 1 as an invariant factor. Such a lattice L is said to be **primitive**. Note that the definition of primitivity is independent of the particular Gram matrix chosen for L, since all L's Gram matrices have the same invariant factors.

For the rest of this chapter, all of our given lattices will be $\mathbb{Z}$-lattices. Accordingly, S will denote the set of prime numbers.

Definition 9.40. Let L be a $\mathbb{Z}$-lattice on a regular $\mathbb{Q}$-space, and let $p \in S$. Define

$$\mu_p L = L + p^{-1} L^{(p^2)}.$$

Proposition 9.41. *Let L be a primitive $\mathbb{Z}$-lattice, and let $p, q \in S$. Suppose there is a Jordan splitting $L_p = L_0 \perp \cdots \perp L_{t_p}$, with each L_i either (p^i)-modular or 0 and $L_{t_p} \neq 0$. Then*

$$(\mu_p L)_q = \begin{cases} L_q & \text{if } q \neq p, \\ L_0 \perp L_1 \perp p^{-1}(L_2 \perp \cdots \perp L_{t_p}) & \text{if } q = p. \end{cases}$$

Proof. We have $(\mu_p L)_q = (L + p^{-1}L^{(p^2)})_q = L_q + (p^{-1})_q L_q^{(p^2)_q}$.

(i) If $q \neq p$ then $|p|_q = 1$ and hence $(\mu_p L)_q = L_q$.

(ii) If $q = p$ then $(\mu_p L)_q = (\mu_p L)_p = L_p + p^{-1} L_p^{(p^2)}$, and here $(p^2) = p^2 \mathbb{Z}_p$. But

$$L_p^{(p^2)} = \perp_{0 \leq i \leq t_p} L_i^{(p^2)} = p^2 L_0 \perp p L_1 \perp L_2 \perp \cdots \perp L_{t_p}$$

and therefore $(\mu_p L)_p = L_0 \perp L_1 \perp p^{-1}(L_2 \perp \cdots \perp L_{t_p})$ as claimed. □

While the preceding proposition describes the action of the transformation μ_p on L in terms of the effect on given Jordan splittings for the local lattices L_q, the definition of $\mu_p L$ is "invariant" in the sense that it is independent of the particular choices of those Jordan splittings.

Notice that $\mu_p L \supseteq L$; and $\mu_p L \supset L$ if and only if L_p has a (p^ν)-modular Jordan component for some $\nu \geq 2$. Moreover, like L, the lattice $\mu_p L$ is integral, hence has volume $\subseteq \mathbb{Z}_p$. Therefore the inclusion chain

$$L \subseteq \mu_p L \subseteq \mu_p(\mu_p L) \subseteq \mu_p(\mu_p(\mu_p L)) \subseteq \cdots$$

eventually becomes constant, namely when the resulting lattice is either unimodular or is an orthogonal sum of a unimodular component and a p-modular component. Since a given $\mathbb{Z}$-lattice is unimodular at almost all $p \in S$, there are only finitely many primes p for which $\mu_p L \neq L$; and we let T denote that set. Thus T is the finite (possibly empty) set of primes $p \in S$ for which L_p has a (p^i)-modular Jordan component with $i \geq 2$; and we now define μL to be the lattice obtained from L by iterating the transformations

μ_p at all $p \in T$ until stability occurs. More precisely, define the $\mathbb{Z}$-lattice μL by

$$(\mu L)_p = \begin{cases} (\mu_p^{(\frac{t_p}{2})} L)_p & \text{if } t_p \text{ is even,} \\ (\mu_p^{(\frac{t_p-1}{2})} L)_p & \text{if } t_p \text{ is odd.} \end{cases}$$

Here the parenthetical exponent on μ_p indicates the number of times the map is applied. In particular, if $p \in S - T$ then $(\mu L)_p = L_p$.

Intuitively, μL is obtained from L by "pulling up" the lower Jordan components (i.e., the (p^i)-modular components with $i \geq 2$) until the resulting lattice's localizations at each $p \in S$ are either unimodular or the orthogonal sum of a unimodular and a (p)-modular component. Equivalently, the invariant factors of the Gram matrix of μL have no prime-power divisors p^i with $i \geq 2$.

Lemma 9.42. *Suppose L and M are $\mathbb{Z}$-lattices on the same quadratic $\mathbb{Q}$-space.*

(i) *If* $\operatorname{gen} L = \operatorname{gen} M$ *then* $\operatorname{gen} \mu L = \operatorname{gen} \mu M$.

(ii) *If* $L \cong M$ *then* $\mu L \cong \mu M$.

Proof. Part (i) is clear. And for (ii) it is enough to check that if $L \cong M$ then $\mu_p L \cong \mu_p M$ for all $p \in S$.

Suppose the map $\varphi \in O(V)$ satisfies $\varphi(L) = M$. Then

$$\varphi(\mu_p L) = \varphi(L + p^{-1}L^{(p^2)}) = \varphi(L) + p^{-1}\varphi(L^{(p^2)}) = M + p^{-1}M^{(p^2)} = \mu_p M.$$

□

Theorem 9.43. *Let L be a $\mathbb{Z}$-lattice on a definite $\mathbb{Q}$-space V. Then*

$$h(L) \geq h(\mu L).$$

Proof. Without loss of generality we will assume L is primitive. To prove the theorem it is enough to show that $h(L) \geq h(\mu_p L)$ whenever $p \in T = \{ p \in S \mid \mu_p L \neq L \}$. (And of course we can assume $T \neq \emptyset$; otherwise there is nothing to prove.)

Fix $p \in T$, let X and Y be the sets of classes in $\operatorname{gen} L$ and $\operatorname{gen} \mu_p L$, respectively, and define $\psi : X \to Y$ by $\psi(\operatorname{cls} M) = \operatorname{cls}(\mu_p M)$ for $M \in \operatorname{gen} L$. From the lemma, ψ is a well-defined function from X to Y, and we will be done if we can show ψ to be surjective. Equivalently, for $J \in \operatorname{gen} \mu_p L$ we must show there is a lattice $K \in \operatorname{gen} L$ such that $\mu_p K \cong J$. In fact we will find $K \in \operatorname{gen} L$ such that $\mu_p K = J$.

Let $L_p = L_0 \perp \cdots \perp L_{t_p}$ be a Jordan splitting, with L_i either (p^i)-modular or 0. Then there is a splitting (not necessarily a Jordan splitting)

$J_p = J_0 \perp \cdots \perp J_{t_p}$ with $J_i \cong L_i$ for $i = 1, 2$ and $J_i \cong p^{-1}L_i$ for $3 \leq i \leq t_p$. Now define the required $\mathbb{Z}$-lattice K by

$$K_q = \begin{cases} J_q & \text{if } q \in S - \{p\}, \\ J_0 \perp J_1 \perp p(J_2 \perp \cdots \perp J_{t_p}) & \text{if } q = p. \end{cases} \qquad \square$$

Corollary 9.44. *Let V be a positive definite quadratic $\mathbb{Q}$-space. Then there is a primitive lattice L on V whose class number $h(L)$ is minimal for $\mathbb{Z}$-lattices on V and whose Gram matrices have only square-free invariant factors.*

Lemma 9.45. *Let M be a unimodular $\mathbb{Z}_p$-lattice on a quadratic $\mathbb{Q}_p$-space of dimension $n \geq 5$. Then $M \cong \begin{pmatrix} 0 & 1 \\ 1 & 0 \end{pmatrix} \perp \cdots$.*

Proof. The result is clear if $p \neq 2$; so we can assume $p = 2$.

From the beginning of Chapter 8, M is an orthogonal sum $\perp_{i=1}^{t} M_i$ of indecomposable sublattices of rank 1 and/or 2; and the M_i of rank 2 in such a splitting are even (i.e., have norm ideal (2)). In view of the last part of Chapter 8, we will be done if we can show that M has a splitting $M_1' \perp M_2' \perp \cdots$ in which M_1' and M_2' are even and of rank 2.

If no two of $M_1, \ldots, M_t$ are even, then $M \cong \langle \varepsilon_1, \varepsilon_2, \varepsilon_3 \rangle \perp \cdots$ in some basis $\{v_1, v_2, v_3, \ldots\}$, with $\varepsilon_1, \varepsilon_2, \varepsilon_3$ 2-adic units. And since $\{1, 3, 5, 7\}$ is a representative set for the square classes of 2-adic units (from the end of Chapter 3), we can assume $\varepsilon_i \in \{1, 3, 5, 7\}$ for $i = 1, 2, 3$. The sublattice

$$M_1' = \mathbb{Z}_2(v_1 + v_2) + \mathbb{Z}_2(v_2 + v_3)$$

has a Gram matrix of the form $\begin{pmatrix} 2\lambda & \varepsilon_2 \\ \varepsilon_2 & 2\mu \end{pmatrix}$, with $\lambda, \mu \in \mathbb{Z}_2$, hence it is unimodular (and even). Therefore $M = M_1' \perp \cdots$. Now repeat this argument on the orthogonal complement of M_1' in M to get the desired M_2'. $\square$

Now we come to the main theorem of this section.

Theorem 9.46. *Let L be a $\mathbb{Z}$-lattice on an n-dimensional definite quadratic $\mathbb{Q}$-space V. Then*

$$h(L) \geq p\left(\left[\frac{n-5}{16}\right]\right),$$

where $[\ \ ]$ is the greatest integer function and $p(\ \ )$ is the partition function.

Proof. Without loss of generality we can assume that V is *positive* definite; moreover, we can safely assume that $n \geq 37$, since otherwise there is nothing to prove. So there is certainly an integer $t \geq 2$ such that $n \geq 16t + 5$. Moreover, by scaling, we can assume that L is primitive. Finally, since

$h(L) \geq h(\mu L)$, by replacing L with μL if necessary we can suppose that at each $p \in S$ the localized lattice L_p has a splitting $L_p = L_{p,1} \perp L_{p,2}$ in which $L_{p,1}$ is either unimodular or p-modular and of rank $\geq 8t + 3$.

By repeated applications of the lemma, there is a splitting $L_{p,1} = L'_{p,1} \perp L''_{p,1}$ in which $\operatorname{rank} L'_{p,1} = 8t$ and

$$L'_{p,1} \cong \begin{cases} \begin{pmatrix} 0 & 1 \\ 1 & 0 \end{pmatrix} \perp \cdots \perp \begin{pmatrix} 0 & 1 \\ 1 & 0 \end{pmatrix} & \text{if } L_{p,1} \text{ is unimodular,} \\ \begin{pmatrix} 0 & p \\ p & 0 \end{pmatrix} \perp \cdots \perp \begin{pmatrix} 0 & p \\ p & 0 \end{pmatrix} & \text{if } L_{p,1} \text{ is } p\text{-modular.} \end{cases}$$

Consider an $8t$-dimensional $\mathbb{Q}$-space $W_1 \cong \langle 1, \ldots, 1 \rangle$. There is an isometry $W_{1p} \cong \mathbb{Q}_p L'_{p,1}$ at each $p \in S$, by the classification theory for $\mathbb{Q}_p$-spaces, so by the Hasse–Minkowski theorem there is a representation $W_1 \to V$. Therefore, without loss of generality we can suppose that W_1 is a subspace of V; hence there is a splitting $V = W_1 \perp W_2$ with $W_{2p} \cong \mathbb{Q}_p(L''_{p,1} \perp L_{p,2})$ for all $p \in S$. It follows that there is a lattice M on V having a splitting $M = M_1 \perp M_2$ satisfying $M_{1p} \cong L'_{p,1}$ and $M_{2p} \cong L''_{p,1} \perp L_{p,2}$ for all $p \in S$. Thus $M \in \operatorname{gen} L$. We have $h(L) = h(M) \geq h(M_1)$. But M_1 can be obtained by scaling a positive definite even unimodular $\mathbb{Z}$-lattice of rank $8t$. (In fact, if $p_1, \ldots, p_k$ are the primes at which the localizations L'_{p_1} are not unimodular, then scaling M_1 by $(p_1 \cdots p_k)^{-1}$ yields a unimodular $\mathbb{Z}$-lattice; conversely, scaling that unimodular lattice by $(p_1 \cdots p_k)$ gives M_1.) Therefore $h(M_1) \geq p(t)$, by Example 9.26, and this proves the theorem. □

We will have some further comments on the growth rate of lattice class numbers at the end of the next section.

9.8. Introduction to Neighbor Lattices

We have seen that the genus of a $\mathbb{Z}$-lattice L—the family of lattices locally isometric to L and on the same space—can contain a large number of isometry classes, and it is sometimes useful to consider a family of lattices intermediate between the class and the genus. The spinor genus is such a family. This concept uses the spinor norm and spinorial kernel, defined in Chapter 2. Roughly speaking, two lattices are in the same spinor genus if they are locally isometric via rotations of spinor norm 1. This is close to the actual definition, but we need to adjust this rough definition in order to guarantee that the spinor genus of a lattice contains the lattice's isometry class.

Definition 9.47. Let L and M be $\mathbb{Z}$-lattices on the regular quadratic $\mathbb{Q}$-space V. We say that M is in the same **spinor genus** as L, and we write

$M \in \operatorname{spn} L$, if there is a lattice $M' \in \operatorname{cls} M$ with the property that for every prime number p there is a local isometry $\sigma_p \in O'(V_p)$ such that $\sigma_p L_p = M'_p$.

We leave it as an exercise for the reader to show that the relation of being in the same spinor genus is an equivalence relation on the set of lattices on V. From this it follows that the spinor genera partition the genus of a lattice, and a spinor genus is itself partitioned into isometry classes. A remarkable 1957 paper [**K2**] by Martin Kneser used the theory of the spinor genus to prove that if there is a unimodular $\mathbb{Z}$-lattice M on a given positive definite n-dimensional quadratic $\mathbb{Q}$-space V, then all classes of unimodular lattices on V can be obtained by making 2-adic changes in M. That is, if J is any other unimodular lattice on V, then there is a lattice L isometric to J with the property that $L_p = M_p$ for all $p \neq 2$. Therefore the invariant factors of L in M must be of the form $(2^{\nu_1}), \ldots, (2^{\nu_n})$; and here $\sum_i \nu_i = 0$ because L and M have the same volume. Kneser called L and M **neighbors** if the change from L to M is as small as possible; more precisely, if the invariant factors of L in M are $\left(\frac{1}{2}\right), \mathbb{Z}, \ldots, \mathbb{Z}, (2)$. And then Kneser proved the following result.

Theorem 9.48 (Kneser). *If L and M are unimodular $\mathbb{Z}$-lattices on the positive definite quadratic $\mathbb{Q}$-space V, then there is a sequence $L = L_1, \ldots, L_t = M$ of unimodular lattices on V such that L_i is a neighbor of L_{i+1} for $1 \leq i \leq t-1$.*

Back in Chapter 7 we used Hermite's inequality on the minimum of a lattice to prove that every positive definite $\mathbb{Z}$-lattice of rank $n \leq 5$ has an orthonormal basis. Now let us use Kneser's neighbor lattice concept to extend this result. We follow the basic argument in [**O'M1**].

Theorem 9.49. *Every positive definite unimodular $\mathbb{Z}$-lattice of rank $n \leq 7$ has an orthonormal basis and hence is isometric to $\mathbb{Z}^n$.*

Proof. By Hermite's inequality, we can suppose n is 6 or 7. Let V be an n-dimensional positive definite $\mathbb{Q}$-space supporting a unimodular $\mathbb{Z}$-lattice. From Theorem 9.18, V has an orthonormal basis $\{v_1, \ldots, v_n\}$; and now define $M = \mathbb{Z}v_1 \perp \cdots \perp \mathbb{Z}v_n$. By Kneser's Theorem it suffices to show that every neighbor of M is isometric to M.

Suppose L is such a neighbor. So there is a basis $\{w_1, \ldots, w_n\}$ for M such that

$$L = \mathbb{Z}\left(\frac{1}{2}w_1\right) + \mathbb{Z}w_2 + \cdots + \mathbb{Z}w_{n-1} + \mathbb{Z}(2w_n)$$

and hence $2M \subseteq L \subseteq \frac{1}{2}M$. Define $y = \frac{1}{2}w_1$. Then, after re-subscripting the

v_i as needed, we have

$$y = \frac{1}{2}(a_1v_1 + \cdots + a_rv_r) + a_{r+1}v_{r+1} + \cdots + a_nv_n,$$

for some $a_i \in \mathbb{Z}$, with a_i odd for $i \leq r$ and a_i even for $i \geq r+1$. Here $r \geq 1$ because $y \notin M$; so in fact we must have $r = 4$, since $q(y) \in \mathbb{Z}$.

Now define

$$w = \frac{1}{2}(v_1 + v_2 + v_3 + v_4) \quad \text{and} \quad w' = \frac{1}{2}(v_1 + v_2 + v_3 - v_4).$$

We will show that one of the symmetries $\tau_w, \tau_{w'}$ carries L onto M.

We have

$$y = w + z = w' + z'$$

with $z, z' \in M$ and $q(w) = q(w') = 1$.

Notice that either $B(y,w) \in \frac{1}{2}\mathbb{Z} - \mathbb{Z}$ or $B(y,w') \in \frac{1}{2}\mathbb{Z} - \mathbb{Z}$. Suppose the former is the case; say $B(y,w) = \dfrac{m}{2}$, with m odd. We have $\tau_w x = x - 2B(x,w)w$ for all $x \in V$, and so

$$\tau_w y = y - mw = z + (1-m)w \in M$$

Moreover, if $x \in L \cap M$ then $B(w,x) \in \mathbb{Z}$ and hence $\tau_w x = x - B(w,x)\cdot 2w \in M$. In particular,

$$\tau_w\big(\mathbb{Z}w_2 + \cdots + \mathbb{Z}w_{n-1} + \mathbb{Z}(2w_n)\big) \subset M$$

Therefore $\tau_w L \subseteq M$, and hence $\tau_w L = M$, since L and M have the same volume.

The argument if $B(y,w') \in \frac{1}{2}\mathbb{Z} - \mathbb{Z}$ is similar. □

The indecomposability of the lattice K_8 (also known in the literature as E_8 and as Γ_8) tells us that the preceding theorem does not extend to unimodular lattices on spaces of dimension $n \geq 8$.

To use Kneser's method effectively to find all the classes of unimodular lattices on a given positive definite space V of dimension $n \geq 8$, one needs to develop procedures for constructing all the neighbors of an initial unimodular lattice on V, determining which of these are in new isometry classes, finding the new isometry classes of *their* neighbors, and so on. The search for new classes ends when all neighbors of lattices already in hand have been shown to be in previously obtained classes. Incidentally, when $n = 8$ the neighbor-lattice method shows that the classes of $\mathbb{Z}^8$ and K_8 are the only classes of positive definite unimodulars. For details see Kneser's original paper [**K2**] or [**O'M1**], §106. Kneser carried the process through $n = 16$.

Other mathematicians—notably Niemeier, Borcherds, and Venkov—have extended the difficult work of determining isometry classes of positive definite unimodular $\mathbb{Z}$-lattices through dimension 25. These post-Kneser authors make heavy use of the **Mass Formula** of C. L. Siegel—a deep result from the analytic theory of quadratic forms, formulated in the 1930s—to determine that all the isometry classes have been found (if in fact they have) by whatever procedures have already been carried out. If L is a positive definite $\mathbb{Z}$-lattice, the **mass** of $\operatorname{gen} L$ is the sum

$$\sum_{M \in \operatorname{gen} L} \frac{1}{\#O(M)}$$

where the set of lattices involved in the sum should be understood to consist of one lattice from each isometry class in the genus of L. Note that the orthogonal group $O(M)$ of a positive definite $\mathbb{Z}$-lattice M is finite, since it is a discrete subgroup of the orthogonal group $O(\mathbb{R}^n)$ of the $\mathbb{R}$-space it spans; and $O(\mathbb{R}^n)$ is compact, being a closed and bounded subset of $\mathbb{R}^{n^2}$. (Identify each isometry with its matrix.) The cardinalities $\#O(M)$ involved in the mass can, in these low-dimensional lattices, be determined by careful analysis of the action of isometries on short vectors. Meanwhile, Siegel's Mass Formula computes the actual value of the mass from the local structure of the lattices in $\operatorname{gen} L$. Therefore, when enough lattice classes in a genus have been found so that the resulting sum $\sum \frac{1}{\#O(M)}$ is equal to the mass, one can conclude that representatives of all the classes in the genus have been found, and the search for more classes can be stopped. See the book by Conway and Sloane ([**CS**], Chapters 16–18) for more details and references.

In the preceding section we gave a lower bound (Theorem 9.46) for the number of isometry classes of positive definite $\mathbb{Z}$-lattices in terms of the partition function $p(t)$. Some back-of-the-envelope calculation suggests that $p(t)$ grows rapidly with t; for example, $p(3) = 3$, $p(6) = 11$, and $p(11) = 56$. (And it is a fact that $p(200) = 3,972,999,029,388$.) Indeed, in 1918 Hardy and Ramanujan showed that $p(t)$ is asymptotic to

$$\frac{1}{4t\sqrt{3}} \exp\left(\pi\sqrt{\frac{2t}{3}}\right)$$

as $t \to \infty$. But in fact the Mass Formula has been used to show that as the dimension n grows the lattice isometry classes begin to proliferate much sooner and much more rapidly than our partition-based formula indicates. Each orthogonal group $O(M)$ in the mass expression has at least two elements: the identity and its negative; therefore the mass of $\operatorname{gen} L$ is bounded from above by $\frac{1}{2}h(L)$, where $h(L)$ is the number of classes in the genus. So $h(L)$ is at least twice the mass. In 1937 Magnus [**Mag**] use Siegel's work to show that the mass of lattices grows very rapidly with n, hence class numbers do as well. In fact, it is known that there are more than 10^7 isometry classes of even unimodular lattices of dimension 32, and more than 10^{51}

classes in dimension 40. See Milnor and Husemoller ([**MH**], Chapter 2) for further exposition on this topic.

Exercises

(1) Assume L is an odd unimodular $\mathbb{Z}$-lattice of rank n. For which subsets $T \subseteq \{1, \ldots, n\}$ is there a basis $\{v_1, \ldots, v_n\}$ for L with the property that for all $i \in \{1, \ldots, n\}$,

$$q(v_i) \equiv 1 \pmod 2 \quad \text{if and only if} \quad i \in T \ ?$$

(2) Consider a matrix in $M_n(\mathbb{Z})$ of the form

$$A = \begin{pmatrix} 8 & 3 & \spadesuit \\ 3 & 4 & \spadesuit \\ \spadesuit & \spadesuit & 10 \end{pmatrix}.$$

Unfortunately the matrix entries in A at the locations marked with "♠" have been lost at sea.
(i) Supply entries that make A unimodular.
(ii) Is it possible that A is both *symmetric* and unimodular? Either prove it is impossible or supply entries that make it so. [Answer this part of the exercise in two ways: (i) using the material in this chapter; (ii) using results from the local theory in Chapter 8.]

(3) Show that every $\mathbb{Z}$-lattice with a Gram matrix of the form $\begin{pmatrix} 2a & b \\ b & 2c \end{pmatrix}$, with $a, b, c \in \mathbb{Z}$ and b odd, is indecomposable.

(4) Let L be a unimodular $\mathbb{Z}$-lattice with basis $\{v_1, \ldots, v_n\}$. Prove that a vector $v \in L$ is characteristic if and only if $B(v, v_i) \equiv q(v_i) \pmod 2$ for all i, $1 \leq i \leq n$.

(5) Suppose $L \cong \begin{pmatrix} 25 & 18 & 7 \\ 18 & 13 & 5 \\ 7 & 5 & 3 \end{pmatrix}$ in the basis $\{v_1, v_2, v_3\}$. Construct a characteristic vector for L. [Remark: By Theorem 9.29 this can be done by first constructing the dual of the given basis. But perhaps you can avoid going through all that.]

(6) Show that the relation of being in the same spinor genus is an equivalence relation on the set of lattices on a regular quadratic $\mathbb{Q}$-space.

(7) Let L be a positive definite integral $\mathbb{Z}$-lattice. Show that if there is a nontrivial splitting $L = M \perp K$, and $e \in M$, then e is coset-minimal in L if and only if e is coset-minimal in M.

(8) Prove that a lattice and its dual have the same class number.

(9) Show that among all the positive definite $\mathbb{Z}$-lattices of rank n, there is a lattice L whose class number is minimal and having the property that for each prime p the $\mathbb{Z}_p$-lattice L_p has a modular component of rank at least $\frac{n}{2}$.

(10) Show that a positive definite unimodular $\mathbb{Z}$-lattice has at most a finite number of neighbors.

(11) Show that the indecomposable even unimodular lattice K_8 (see Example 7.8) can be constructed as a neighbor of the completely decomposable odd unimodular lattice $\mathbb{Z}^8$. (Here the expression "completely decomposable" means "having an orthogonal basis.") [Remark: This exercise actually repeats an exercise from an earlier chapter, though the wording has changed.]

Chapter 10

Lattices over $\mathbb{F}_q[x]$

The classification of $\mathbb{Z}$-lattices up to isometry is a major open problem. It is generally impossible to explain exactly *why* we don't understand the things we don't understand; but as we reflect on the mysteries of $\mathbb{Z}$-lattices the number 2 presents itself—for both local and global reasons—as a major source of the difficulties. If p is an odd prime, we saw in Chapter 8 that classification of lattices over the p-adic integers is relatively straightforward: the ranks, modularities, and discriminants of the Jordan components tell the story. But classification over the 2-adic integers is much more complicated (and in fact we considered only special cases), in part due to the fact that the norm and scale ideals for lattices over $\mathbb{Z}_2$ need not coincide. Globally, the central role of 2 in the classification morass is best illustrated by Kneser's Theorem (9.48): lattices in all the classes of unimodular $\mathbb{Z}$-lattices on a given positive definite $\mathbb{Q}$-space can be produced from a given unimodular lattice on that space just by making 2-adic changes. Moreover, we know from Example 9.26 and Theorem 9.46 that the number of classes of definite $\mathbb{Z}$-lattices—and in particular of definite unimodular $\mathbb{Z}$-lattices—increases exponentially with the dimension. By way of contrast, we have already seen in Example 7.16 that on a given quadratic $\mathbb{F}_q(x)$-space—no matter what the dimension—there can be at most one isometry class of unimodular $\mathbb{F}_q[x]$-lattices.

A second source of difficulties over $\mathbb{Z}$—for instance in the representation problem—resides in the archimedean property of our measurement of "size": the ordinary absolute value. That is, the sum of many small numbers can be large.

In this chapter we will explore classification and representation issues for $\mathbb{F}_q[x]$-lattices on quadratic $\mathbb{F}_q(x)$-spaces, where $\mathbb{F}_q$ is a finite field of odd order q. The rings $\mathbb{F}_q[x]$ have much in common with $\mathbb{Z}$: they are Euclidean domains, and modulo the maximal ideals the residue class fields are finite. But neither of the difficulties described above for the theory over $\mathbb{Z}$ is a concern over $\mathbb{F}_q[x]$. That is, 2 is a unit in $\mathbb{F}_q[x]$, so all the spots on $\mathbb{F}_q(x)$ corresponding to prime ideals in $\mathbb{F}_q[x]$ are nondyadic. And the "infinite" spot on $\mathbb{F}_q(x)$—which measures "size" via the degree function—is nonarchimedean (and also nondyadic), so a sum of "small" elements in $\mathbb{F}_q(x)$ is small. In particular, the degree of a sum of polynomials is at most the largest degree of the summands.

As in the theory over $\mathbb{Z}$, the classification of lattices over $\mathbb{F}_q[x]$ splits into "definite" and "indefinite" cases, which we will define, and the simplifying features described above for $\mathbb{F}_q[x]$ will enable us to achieve (in §10.2) the chapter's main result: the complete classification of $\mathbb{F}_q[x]$-lattices in the definite case. Along the way we will get a start on the representation problem as well. For instance we will see that in the definite case Djokoviĉ's reduction algorithm (see 7.11) yields the minimum of a lattice, and we will obtain further representation results in the last section of the chapter.

To avoid confusion with the cardinality of the underlying finite field, we will use Q (instead of our usual q) for a generic quadratic form in this setting.

Before we get started on lattices, let us extend our observations on Legendre and Hilbert symbols over $\mathbb{Q}_p$, when p is an odd prime, to arbitrary nondyadic local fields. With $(F, \mathfrak{p})$ such a field, p a prime element of F at $\mathfrak{p}$, and $\alpha \in \mathfrak{u}(\mathfrak{p})$, define the **Legendre symbol**

$$\left(\frac{\alpha}{\mathfrak{p}}\right) = \left(\frac{\alpha}{p}\right) = \begin{cases} 1 & \text{if } \overline{\alpha} \in \overline{F}^2, \\ -1 & \text{otherwise.} \end{cases}$$

Notice that the particular choice of the prime p in F is irrelevant here. As in $\mathbb{Q}_p$ (with p an odd prime number), the value of the Legendre symbol $\left(\frac{\alpha}{\mathfrak{p}}\right)$ is completely determined by the leading term in a p-adic power series expansion (canonical or not) of α. And now it is easily checked that the basic properties of the Hilbert symbol in $\mathbb{Q}_p$, with p odd, carry over to this general nondyadic setting. In particular, the symbol is trivial on $\mathfrak{u} \times \mathfrak{u}$, the symbol is "bilinear" (e.g., $(\alpha\beta, \gamma)_\mathfrak{p} = (\alpha, \gamma)_\mathfrak{p}(\beta, \gamma)_\mathfrak{p}$), and if $\varepsilon \in \mathfrak{u}$ then

$$(p, \varepsilon)_\mathfrak{p} = \begin{cases} 1 & \text{if } \varepsilon \in \mathfrak{u}^2, \\ -1 & \text{if } \varepsilon \notin \mathfrak{u}^2. \end{cases}$$

10.1. An Initial Example

Example 10.1. Consider $\mathbb{F}_3[x]$-lattices L_1, L_2, L_3 satisfying

$$L_1 \cong C_1 = \begin{pmatrix} x & 1 \\ 1 & x \end{pmatrix}, \qquad L_2 \cong C_2 = \begin{pmatrix} 2x & 1 \\ 1 & 2x \end{pmatrix},$$

$$L_3 \cong C_3 = \begin{pmatrix} x^{13}+2x^6+x & x^{15}+2x^8+x^7+x^3+1 \\ x^{15}+2x^8+x^7+x^3+1 & x^{17}+2x^{10}+2x^9+x^5+2x^2+x \end{pmatrix}.$$

Are these lattices isometric? Equivalently, if $i \neq j$, is there a matrix $T \in GL_2(\mathbb{F}_3[x])$ such that $C_j = {}^tTC_iT$?

In fact, if

$$T = \begin{pmatrix} 1 & x^2 \\ x^6 & x^8+1 \end{pmatrix}$$

then ${}^tTC_1T = C_3$, and so $L_1 \cong L_3$. (Later on we will see how to produce such a matrix T.) As for whether $L_1 \cong L_2$, we approach the question in two ways, as follows.

(I) Certainly a necessary condition for an isometry $L_1 \cong L_2$ is that there be a representation $2x \to L_1$. But can this happen? Assume $L_1 \cong C_1$ in the basis $\{v_1, v_2\}$, and suppose $Q(v) = 2x$, with $v = av_1 + bv_2$ for some $a, b \in \mathbb{F}_3[x]$. We have

$$(*) \qquad 2x = Q(v) = (a^2+b^2)x + 2ab.$$

We claim that $\partial(a^2+b^2) = 2\max\{\partial a, \partial b\}$. This is clear if $\partial a \neq \partial b$. If $\partial a = \partial b$, let α, β be the leading coefficients of a, b, respectively; then the leading coefficient of a^2+b^2 is $\alpha^2+\beta^2$. This cannot be 0, since the space $\langle 1, 1\rangle$ is anisotropic over $\mathbb{F}_3$, and this proves the claim. Therefore $\partial\big((a^2+b^2)x\big) > \partial(2ab)$.

So, since the leading terms on both sides of (*) must be equal, and $2 \notin \mathbb{F}_3^2$, we must have $a, b \in \mathbb{F}_3^*$. Hence (*) is an impossibility!

(II) As our second approach to proving that L_1 and L_2 are not isometric, we show that their underlying quadratic $\mathbb{F}_3(x)$-spaces—call them V_1 and V_2, respectively—are not isometric. An isometry $V_1 \cong V_2$ would force local isometries $V_{1\mathfrak{p}} \cong V_{2\mathfrak{p}}$ at all spots $\mathfrak{p}$ on $\mathbb{F}_3(x)$, so for our purpose it will be enough to find one spot $\mathfrak{p}$ where no such isometry happens. Note that since $dV_1 = dV_2$, a failure of isometry at some $\mathfrak{p}$ would have to show up in inequality of the Hasse symbols at $\mathfrak{p}$.

From the Gram–Schmidt process we have

$$V_1 \cong \langle x, x(x^2-1)\rangle \qquad \text{and} \qquad V_2 \cong \langle 2x, 2x(x^2-1)\rangle.$$

Let $\mathfrak{p}$ be the $(x-1)$-adic spot. Since $x = 1 + (x-1)$, by the Local Square Theorem x is a square in $\mathbb{F}_3(x)_\mathfrak{p}$, and hence $S_\mathfrak{p}V_1 = \big(x, x(x^2-1)\big)_\mathfrak{p} = 1$. On the other hand, 2 is a nonsquare in $\mathbb{F}_3(x)_\mathfrak{p}$, since it is a nonsquare in the residue class field (namely $\mathbb{F}_3$) at $\mathfrak{p}$, and hence $2x$ is a nonsquare as well. But $2x(x^2-1) = 2x(x-1)(x+1)$ is a prime element of $\mathbb{F}_3(x)_{\mathfrak{p}}$, and so $S_{\mathfrak{p}}V_2 = \big(2x, 2x(x^2-1)\big)_{\mathfrak{p}} = -1$. It follows that $V_{1\mathfrak{p}} \not\cong V_{2\mathfrak{p}}$ and therefore $V_1 \not\cong V_2$ and $L_1 \not\cong L_2$.

10.2. Classification of Definite $\mathbb{F}_q[x]$-Lattices

Recall that a nonzero $\mathbb{Z}$-lattice L on a quadratic $\mathbb{Q}$-space V is said to be **indefinite** if it represents both positive and negative numbers; equivalently, the $\mathbb{Q}_\infty$-space V_∞ is isotropic. (Of course $\mathbb{Q}_\infty$ is just a fancy name for $\mathbb{R}$.) And L is **definite** if V_∞ is anisotropic. Similarly, we say that an $\mathbb{F}_q[x]$-lattice L on a quadratic $\mathbb{F}_q(x)$-space V is **definite** if the localized space $V_\infty = V \otimes \mathbb{F}_q(x)_\infty$ is anisotropic, and **indefinite** otherwise. The goal of this section is to classify definite $\mathbb{F}_q[x]$-lattices; that is, to produce a set of effectively computable criteria to determine when two given lattices are isometric. Since a necessary condition for lattice isometry is that the underlying spaces are isometric, there is no loss of generality in assuming that candidates for lattice isometry are on the same $\mathbb{F}_q(x)$-space V. If $\dim V = n \geq 5$, then we know from the local theory that V_∞ is isotropic, so we will necessarily be restricted to the situation in which $n \leq 4$. Our approach will be that of [**G5**].

Observe that the given Gram matrices for the lattices L_1 and L_2 in the previous section are in dominant diagonal form. (See Chapter 7.) Approach (I) for showing that $L_1 \not\cong L_2$ focused on the leading terms of the diagonal entries of those matrices. We start our work in this section by showing that this attention to the leading terms is the key to determining whether or not a given lattice is definite.

Lemma 10.2. *Let U be an n-dimensional quadratic space over a nondyadic local field F. Suppose $U \cong A = (a_{ij}) \in M_n(F)$, with $|a_{ii}| > |a_{ij}|$ for all $j \neq i$, $1 \leq i \leq n$. Then*

$$U \cong \langle a_{11}, \ldots, a_{nn} \rangle.$$

Proof. We have $a_{ii} \neq 0$ for all i, and

$$\begin{aligned} dU = |A| &= \prod_{i=1}^{n} a_{ii} + \sum_{\sigma \neq e} (\operatorname{sgn} \sigma) \prod_{i=1}^{n} a_{i\sigma(i)} \\ &= \prod_{1}^{n} a_{ii} \left(1 + \sum_{\sigma \neq e} (\operatorname{sgn} \sigma) \frac{\prod_{i=1}^{n} a_{i\sigma(i)}}{\prod_{i=1}^{n} a_{ii}} \right). \end{aligned}$$

But if $\sigma \neq e$ then

$$\left| \frac{\prod_{i=1}^{n} a_{i\sigma(i)}}{\prod_{i=1}^{n} a_{ii}} \right| < 1.$$

Therefore $dU = \prod_1^n a_{ii}$ by the Local Square Theorem, and from this the result follows by induction on n. $\square$

Remark 10.3. The assumption that F is nondyadic is essential in the preceding lemma. For instance, note that $\begin{pmatrix} 1 & 2 \\ 2 & 1 \end{pmatrix} \not\cong \langle 1, 1 \rangle$ over the dyadic local field $\mathbb{Q}_2$.

Lemma 10.4. *Let V be a quadratic $\mathbb{F}_q(x)$-space, and suppose $V \cong A = (a_{ij}) \in M_n(\mathbb{F}_q[x])$, where A has a dominant diagonal. For each i, suppose a_{ii} has leading coefficient λ_i and degree ν_i. Then*

$$V_\infty \cong \langle a_{11}, \ldots, a_{nn} \rangle \cong \langle \lambda_1 x^{\nu_1}, \ldots, \lambda_n x^{\nu_n} \rangle.$$

Proof. From the preceding lemma we have $V_\infty \cong \langle a_{11}, \ldots, a_{nn} \rangle$. Now for any $f(x) \in \mathbb{F}_q[x]$, say $f(x) = \sum_{i=0}^m \alpha_i x^i$, we can write

$$f = x^m \left(\alpha_m + \alpha_{m-1} x^{-1} + \cdots + \alpha_0 (x^{-1})^m \right).$$

But x^{-1} is a prime element in the local field $\mathbb{F}_q(x)_\infty$, and so f is in the square class of $\alpha_m x^m$. Applying this observation to each a_{ii} gives the result. $\square$

Recall from Chapter 7 that if k is a field and L is a $k[x]$-lattice on a quadratic $k(x)$-space, the **minimum** of L, denoted $\min L$, is the smallest degree of a nonzero element represented by L. Also recall from that chapter that a Gram matrix (and an associated lattice basis) over $k[x]$ is said to be **reduced** if its diagonal entries are linearly ordered—from smallest to largest, moving down the diagonal—by degree and it has a dominant diagonal: each diagonal entry's degree is strictly greater than the degrees of the other entries in its row or column. The following theorem begins the demonstration of the central role to be played by reduction in the classification of definite $k[x]$-lattices. In particular it shows that the degrees of the diagonal entries of a reduced Gram matrix for a $k[x]$-lattice constitute an invariant of the lattice, and the degree of the leading term in that sequence is the minimum of the lattice.

Theorem 10.5. *Suppose L is an $\mathbb{F}_q[x]$-lattice on the definite quadratic $\mathbb{F}_q(x)$-space V, and suppose further that $L \cong A = (a_{ij}) \in M_n(\mathbb{F}_q[x])$ in the reduced basis $\{v_1, \ldots, v_n\}$.*

(i) *Let $0 \neq v = \sum_{i=1}^n \alpha_i v_i \in L$, with $\alpha_i \in \mathbb{F}_q[x]$. Then the leading term of the polynomial $Q(v)$ is the leading term of $\sum_{i=1}^n \alpha_i^2 a_{ii}$, and $\partial Q(v) = \max_i \partial(\alpha_i^2 a_{ii})$.*

(ii) $\partial a_{11} = \min L$; *and for $i > 1$,*

$$\partial a_{ii} = \min\{\partial Q(x) \mid x \in L \text{ and linearly independent of } v_1, \ldots, v_{i-1}\}.$$

(iii) *Each element represented by L is represented only a finite number of times, and hence the orthogonal group $O(L)$ is finite.*

(iv) *The degree sequence $(\partial a_{11}, \ldots, \partial a_{nn})$ is an invariant of L. That is, if also $L \cong C = (c_{ij})$, with C reduced, then $(\partial c_{11}, \ldots, \partial c_{nn}) = (\partial a_{11}, \ldots, \partial a_{nn})$.*

Proof. (i) We have $Q(v) = \sum_{i=1}^{n} \alpha_i^2 a_{ii} + 2\sum_{1\le i<j\le n} \alpha_i\alpha_j a_{ij}$. Let $M = \max_i\{\partial(\alpha_i^2 a_{ii})\}$, and consider the expression $\sum_{i=1}^{n} \alpha_i^2 a_{ii}$; we claim that its degree is M. This is clear if only one term in the sum has degree M. In fact, at most two terms have degree M; for example, suppose $\partial\alpha_i^2 a_{ii} = M$ for $i = 1, 2, 3$. Then $\partial a_{ii} = M \pmod 2$, and, with λ_i the leading coefficient of a_{ii}, by Lemma 10.4 we would have

$$\langle a_{11}, a_{22}, a_{33}\rangle \cong \langle x^M\lambda_i, x^M\lambda_2, x^M\lambda_3\rangle \quad \text{over} \quad \mathbb{F}_q(x)_\infty.$$

But this would be isotropic, since $\langle\lambda_1, \lambda_2, \lambda_3\rangle$ is isotropic over $\mathbb{F}_q$, contradicting the fact that V is definite. Therefore, to finish the proof of the claim, we suppose (without loss of generality) that $\partial\alpha_1^2 a_{11} = \partial\alpha_2^2 a_{22} = M$, and let $\varepsilon_1, \varepsilon_2$ be the leading coefficients of α_1, α_2. Then the sum of the leading terms of $\alpha_1^2 a_{11}$ and $\alpha_2^2 a_{22}$ is $(\varepsilon_1^2\lambda_1 + \varepsilon_2^2\lambda_2)x^M$. If this were 0 then $\langle\lambda_1, \lambda_2\rangle$ would be a hyperbolic $\mathbb{F}_q$-plane, and hence, by Lemma 10.4, $\langle a_{11}, a_{22}\rangle$ would be hyperbolic over $\mathbb{F}_q(x)_\infty$, contradicting definiteness. Therefore $\partial\left(\sum_{i=1}^{n} \alpha_i^2 a_{ii}\right) = M$, as claimed. To finish the proof of statement (i), it suffices to check that $\partial\alpha_i\alpha_j a_{ij} < M$ whenever $i \neq j$. Say $\partial\alpha_i \le \partial\alpha_j$. Then, since A has dominant diagonal, we have

$$\partial\alpha_i\alpha_j a_{ij} \le \partial\alpha_j^2 a_{ij} < \partial\alpha_j^2 a_{jj} \le M.$$

Statements (ii) and (iii) now follow immediately from statement (i).

(iv) Write $\{\partial a_{11}, \ldots, \partial a_{nn}\} = \{m_1, \ldots, m_t\}$, with $m_1 < \cdots < m_t$ (so $1 \le t \le n$). Let $L_0 = \{0\}$, and if $1 \le j \le t$, let L_j be the sublattice of L generated by $\{v \in L \mid \partial Q(v) \le m_j\}$. Then, by statement (i), L_j is in fact generated by $\{v_i \mid \partial a_{ii} \le m_j\}$. Therefore, exactly rank L_j–rank L_{j-1} diagonal entries a_{ii} have degree m_j. □

Another proof of finiteness of $O(L)$ was given in [**O'M2**] using local methods.

Corollary 10.6. *If in Theorem* 10.5 *the inequality $\partial a_{11} < \partial a_{22}$ holds, then $v \in L$ is minimal (that is, $\partial Q(v) = \min L$) if and only if $v = \alpha v_1$ for some $\alpha \in \mathbb{F}_q^*$.*

Definition 10.7. In the notation of Theorem 10.5, the sequence $(m_1, \ldots, m_t)$ defined in the proof of statement (iv) is the sequence of **successive minima** of L.

The preceding definition follows the corresponding terminology for lattices over the ring $\mathbb{Z}$ of integers used in [**G4**], Section 2. (Some may prefer to call $(\partial a_{11}, \ldots, \partial a_{nn})$ the sequence of successive minima. E.g., see [**C2**], Chapter 12, Section 2, for this usage over $\mathbb{Z}$.)

Example 10.8. The purpose of this example is to show that the condition that L is definite is essential in Theorem 10.5. Suppose L is a binary lattice over $\mathbb{F}_5[x]$, with

$$L \cong A = \begin{pmatrix} x^4 + x^3 & 1 \\ 1 & x^4 \end{pmatrix} \quad \text{in} \quad \{v_1, v_2\}.$$

Then L is indefinite by Lemma 10.4 and the fact that $\langle 1, 1\rangle$ is isotropic over $\mathbb{Z}_5$. It is easily checked that

$$L \cong C = \begin{pmatrix} x^3 + 4 & 2x + 1 \\ 2x + 1 & x^5 + x^4 + x^2 + x \end{pmatrix}$$

in the basis $\{v_1 + 2v_2, -2xv_1 + (-4x + 1)v_2\}$. Thus

$$C = {}^tTAT \quad \text{with} \quad T = \begin{pmatrix} 1 & -2x \\ 2 & -4x + 1 \end{pmatrix}.$$

So different reduced Gram matrices for a given indefinite lattice may have different degree sequences for their diagonal entries.

Recall that for an $\mathbb{F}_q[x]$-lattice L on an $\mathbb{F}_q(x)$-space V the *dual lattice* is the lattice $L^\sharp$ on V defined by

$$L^\sharp = \{v \in V \mid B(v, L) \subseteq \mathbb{F}_q[x]\}.$$

If L has basis $\mathbb{B} = \{v_1, \ldots, v_n\}$ then $L^\sharp$ has the dual basis of V, namely

$$\mathbb{B}^\sharp = \left\{v_1^\sharp, \ldots, v_n^\sharp\right\},$$

where $B(v_i, v_j^\sharp) = \delta_{ij}$. (See Proposition 6.23.) If $L \cong A \in M_n(\mathbb{F}_q(x))$ in $\mathbb{B}$ then $L^\sharp \cong A^{-1}$ in $\mathbb{B}^\sharp$. In the following lemma, for a given matrix A the symbol $A(i \mid j)$ denotes the matrix obtained by deleting row i and column j of A.

Lemma 10.9. *If $\{v_1, \ldots, v_n\}$ is a reduced basis for the $\mathbb{F}_q[x]$-lattice L, then the reversed dual basis $\{v_n^\sharp, \ldots, v_1^\sharp\}$ is a reduced basis for the dual lattice $L^\sharp$; in particular, for $2 \le i \le n$ the inequality $\partial Q\left(v_i^\sharp\right) \le \partial Q\left(v_{i-1}^\sharp\right)$ holds. Moreover,*

$$\partial Q\left(v_i^\sharp\right) < \partial Q\left(v_{i-1}^\sharp\right) \quad \textit{if and only if} \quad \partial Q(v_{i-1}) < \partial Q(v_i).$$

Proof. Suppose $L \cong A = (a_{ij}) \in M_n(\mathbb{F}_q(x))$ in the basis $\{v_1, \ldots, v_n\}$. Then $L^\sharp \cong A^{-1}$ in $\left\{v_1^\sharp, \ldots, v_n^\sharp\right\}$. [Note: The reader who prefers to work with polynomials instead of with rational functions may temporarily scale the form by the discriminant dL, getting $\left(L^\sharp\right)^{dL} \cong \operatorname{adj} A$ in $\left\{v_i^\sharp\right\}_{1\le i\le n}$.]

Then

$$\partial Q\left(v_i^\sharp\right) = \partial \det A(i \mid i) = \sum_{j\neq i} \partial a_{jj} = -\partial a_{ii} + \sum_{j=1}^{n} \partial a_{jj}.$$

Therefore for $2 \le i \le n$ we have

$$\partial Q\left(v_i^\sharp\right) - \partial Q\left(v_{i-1}^\sharp\right) = \partial a_{i-1,i-1} - \partial a_{ii} \le 0,$$

so $\partial Q\left(v_i^\sharp\right) \le \partial Q\left(v_{i-1}^\sharp\right)$, and the statement on strict inequality is now clear. It remains to check the dominant diagonal property; that is, that $\partial \det A(i \mid j) < \partial \det A(i \mid j)$ when $i \neq j$.

Because A has dominant diagonal, the term in $\det A(i \mid i)$ of strictly largest degree is $\prod_{r\neq i} a_{rr}$; while when $i \neq j$ every term in $\det A(i \mid j)$ has the form $\prod_{r\neq i} a_{rj_r}$, with $j_r \neq r$ at least once (e.g., when $r = j$). So, again by the dominant diagonal, we have $\partial \det A(i \mid j) < \partial \det A(i \mid i)$; that is, $\partial B\left(v_i^\sharp, v_j^\sharp\right) < \partial Q\left(v_i^\sharp\right)$ when $i \neq j$. Thus adj A has dominant diagonal, hence so does A^{-1}. So the Gram matrix of $L^\sharp$ in $\left\{v_n^\sharp, \ldots, v_1^\sharp\right\}$ is reduced. □

Now for the main result of this chapter.

Theorem 10.10 (Classification of Definite $\mathbb{F}_q[x]$-Lattices). *Let L and M be $\mathbb{F}_q[x]$-lattices on a definite quadratic $\mathbb{F}_q(x)$-space V of dimension n, and suppose L and M have respective Gram matrices $A, C \in M_n(\mathbb{F}_q[x])$. Suppose further that A and C are reduced. Then*

$$L \cong M \quad \textit{if and only if} \quad C = {}^tTAT \quad \textit{for some} \quad T \in GL_n(\mathbb{F}_q).$$

Moreover, if for $1 \le i \le t$ the successive minimum m_i occurs with multiplicity n_i, then T has the form

$$\begin{pmatrix} B_1 & & \\ & \ddots & \\ & & B_t \end{pmatrix} \quad \textit{with} \quad B_i \in GL_{n_i}(\mathbb{F}_q).$$

Proof. Only the necessity requires proof, and for this we can assume without loss of generality that $L = M$. Because V is definite, we have $n \le 4$. Suppose $A = (a_{ij})$ and $C = (c_{ij})$ are the Gram matrices for L associated with reduced bases $\mathbb{B}_1 = \{v_1, \ldots, v_n\}$ and $\mathbb{B}_2 = \{w_1, \ldots, w_n\}$, respectively. We have $w_j = \sum_i t_{ij} v_i$ for $j = 1, \ldots, n$, with $T = (t_{ij}) \in GL_n(\mathbb{F}_q[x])$; and so $C = {}^tTAT$. We will be done if we can show that this T has the stated form.

(I) We first show that if T has the form

$$T = \begin{pmatrix} 1 & & & t_{1n} \\ & \ddots & & \vdots \\ & & 1 & t_{n-1,n} \\ & & & t_{n,n} \end{pmatrix} \in GL_n(\mathbb{F}_q[x])$$

then $t_{in} = 0$ for $1 \le i \le n-1$.

If $n = 2$ then $c_{12} = B(v_1, w_2) = t_{12}a_{11} + t_{22}a_{12}$. If $t_{12} \neq 0$ then $\partial c_{12} \geq \partial a_{11} = \partial c_{11}$ since $t_{22} \in \mathbb{F}_q^*$, contradicting the fact that C is reduced.

Now suppose $n = 3$. If $t_{13} = 0$ then the case $n = 2$ applied to the sublattice spanned by $\{v_2, v_3\}$ shows that $t_{23} = 0$, and conversely; so, without loss of generality, we can assume that $t_{13}t_{23} \neq 0$. We have

$$\begin{aligned} c_{13} &= B(v_1, w_3) = t_{13}a_{11} + t_{23}a_{12} + t_{33}a_{13} \quad \text{and} \\ c_{23} &= B(v_2, w_3) = t_{13}a_{12} + t_{23}a_{22} + t_{33}a_{23}. \end{aligned}$$

Since A and C are reduced and $t_{33} \in \mathbb{F}_q^*$, it follows that

$$\partial(t_{13}a_{11} + t_{23}a_{12}) < \partial a_{11} \quad \text{and} \quad \partial(t_{13}a_{12} + t_{23}a_{22}) < \partial a_{22};$$

therefore

$$\partial t_{13} + \partial a_{11} = \partial t_{23} + \partial a_{12} \quad \text{and} \quad \partial t_{13} + \partial a_{12} = \partial t_{23} + \partial a_{22},$$

which in turn yields the contradictory inequalities

$$\partial t_{23} = \partial t_{13} + (\partial a_{11} - \partial a_{12}) > \partial t_{13} \quad \text{and} \quad \partial t_{13} = \partial t_{23} + (\partial a_{22} - \partial a_{12}) > \partial t_{23},$$

finishing the proof of (I) in the case $n = 3$.

Finally, suppose $n = 4$. As in the preceding case, if $t_{14}t_{24}t_{34} = 0$ we are reduced to lower dimensions and hence finished; so there remains only the situation in which $t_{14}t_{24}t_{34} \neq 0$. Then for $i = 1, 2, 3$ we have $c_{i4} = B(v_i, w_4) = \sum_{j=1}^{4} t_{j4}a_{ij}$ and hence

$$\partial(t_{14}a_{i1} + t_{24}a_{i2} + t_{34}a_{i3}) < \partial a_{ii} \quad \text{for} \quad i = 1, 2, 3. \tag{10.1}$$

Then from (10.1) we have

$$\begin{aligned} \partial(t_{24}a_{12} + t_{34}a_{13}) &= \partial t_{14}a_{11}, \\ \partial(t_{14}a_{21} + t_{34}a_{23}) &= \partial t_{24}a_{22}, \\ \partial(t_{14}a_{31} + t_{24}a_{32}) &= \partial t_{34}a_{33}. \end{aligned}$$

The first of these three equations, together with the fact that A is reduced, gives that either $\partial t_{24} > \partial t_{14}$ or $\partial t_{34} > \partial t_{14}$. So $\partial t_{14} \neq \max\{\partial t_{14}, \partial t_{24}, \partial t_{34}\}$. Similarly, the other two equations show that neither ∂t_{24} nor ∂t_{34} is equal to

$\max\{\partial t_{14}, \partial t_{24}, \partial t_{34}\}$, an absurdity. Therefore the case in which $t_{14}t_{24}t_{34} \neq 0$ cannot occur. This completes the proof of part (I).

(II) Now suppose T is as in the statement of the theorem but otherwise unrestricted in $GL(\mathbb{F}_q[x])$. We will argue inductively. The case $n = 1$ is trivial, so we may assume that $1 < n \leq 4$ and that the theorem has been proved for lattices of rank $\leq n - 1$.

First suppose $\partial a_{nn} > \partial a_{n-1,n-1}$. Then, by Theorem 10.5, the matrix T has the form

$$\begin{pmatrix} & & & t_{1n} \\ & T_0 & & t_{2n} \\ & & & \vdots \\ 0 & 0 & 0 & t_{nn} \end{pmatrix}$$

with $T_0 \in GL_{n-1}(\mathbb{F}_q[x])$. In fact, from the induction hypothesis we have $T_0 \in GL_{n-1}(\mathbb{F}_q)$ (and having the appropriate block decomposition), and hence $\{v_1, \ldots, v_{n-1}, w_n\}$ is also a reduced basis for L. By part (I) of the proof, it then follows that $t_{1n} = \cdots = t_{n-1,n} = 0$, and we are done.

If $\partial a_{11} < \partial a_{22}$, the argument reduces to the preceding case by "dualizing" as follows. From Lemma 10.9, $\left\{v_n^\sharp, \ldots, v_1^\sharp\right\}$ and $\left\{w_n^\sharp, \ldots, w_1^\sharp\right\}$ are reduced bases for $L^\sharp$, with $\partial Q\left(v_n^\sharp\right) \leq \cdots \leq \partial Q(\left(v_2^\sharp\right) < \partial Q\left(v_1^\sharp\right)$. We have

$$L^\sharp \cong PA^{-1}P \text{ in } \left\{v_n^\sharp, \ldots, v_1^\sharp\right\} \text{ and } L^\sharp \cong PC^{-1}P \text{ in } \left\{w_n^\sharp, \ldots, w_1^\sharp\right\},$$

with

$$P = \begin{pmatrix} & & 1 \\ & \cdot^{\cdot^{\cdot}} & \\ 1 & & \end{pmatrix}.$$

Thus the preceding argument applies to $L^\sharp$, giving a matrix $S \in GL_n(\mathbb{F}_q)$ such that $PC^{-1}P = {}^tS(PA^{-1}P)S$. Upon taking inverses and setting $T = P\left({}^tS^{-1}\right)P$ we get $C = {}^tTAT$, with $T \in GL_n(\mathbb{F}_q)$ of the desired form.

It remains to consider the case $\partial a_{11} = \partial a_{22} < \partial a_{33} = \partial a_{44}$. In this situation we have

$$T = \begin{pmatrix} B_1 & * \\ 0 & B_2 \end{pmatrix} \quad \text{with } B_1, B_2 \in GL_2(\mathbb{F}_q).$$

Then

$$L \cong D = {}^tSAS = \begin{pmatrix} a_{11} & a_{12} & * & * \\ a_{12} & a_{22} & * & * \\ * & * & c_{33} & c_{34} \\ * & * & c_{34} & c_{44} \end{pmatrix} \quad \text{with } S = \begin{pmatrix} 1 & & t_{13} & t_{14} \\ & 1 & t_{23} & t_{24} \\ & & \multicolumn{2}{c}{B_2} \end{pmatrix},$$

in $\{v_1, v_2, w_3, w_4\}$. The matrix D is reduced, because v_1 and v_2 are $\mathbb{F}_q$-linear combinations of w_1 and w_2. Therefore, without loss of generality we can assume $v_1 = w_1$ and $v_2 = w_2$. We want to show that

$$\begin{pmatrix} t_{13} & t_{14} \\ t_{23} & t_{24} \end{pmatrix} = 0.$$

By symmetry it suffices to show that $t_{13} = 0$.

We have

$$\partial a_{11} > \partial B(v_1, w_3) = \partial(t_{13}a_{11} + t_{23}a_{12} + t_{33}a_{13} + t_{43}a_{14}),$$

with $t_{33}, t_{43} \in \mathbb{F}_q$. So if $t_{13} \neq 0$ then $\partial(t_{13}a_{11}) = \partial(t_{23}a_{12})$, from which it follows that $\partial t_{23} > \partial t_{13}$. But then since $\partial a_{11} = \partial a_{22} > \partial B(v_2, w_3)$, we could also deduce the inequality $\partial t_{13} > \partial t_{23}$, a contradiction. So we must have $t_{13} = 0$, as desired. $\square$

Remark 10.11. In light of Lemma 10.4, the hypothesis in Theorem 10.10 that V is definite guarantees that each n_i is equal to 1 or 2.

Corollary 10.12. *In the notation of Theorem* 10.10, *suppose further that* $\partial a_{11} < \cdots < \partial a_{nn}$. *Then there is essentially only one reduced basis for* L. *More precisely, if* $\{v_1, \ldots, v_n\}$ *and* $\{w_1, \ldots, w_n\}$ *are reduced bases for* L, *then* $w_i = \alpha_i v_i$, *with* $\alpha_i \in \mathbb{F}_q^*$, *for* $1 \leq i \leq n$. *Hence* $L \cong M$ *if and only if there is a diagonal matrix* $T \in GL_n(\mathbb{F}_q)$ *such that* $C = TAT$.

Example 10.13. Strict inequality is essential in the hypothesis of the preceding corollary. For suppose -1 is a nonsquare in $\mathbb{F}_q$, let

$$A = \begin{pmatrix} x^3 + x + 1 & x \\ x & x^3 \end{pmatrix} \quad \text{and} \quad C = \begin{pmatrix} 2x^3 + 3x + 1 & x + 1 \\ x + 1 & 2x^3 - x + 1 \end{pmatrix},$$

and take $\mathbb{F}_q[x]$-lattices L and M with respective Gram matrices A and C. The condition on -1 guarantees that L and M are definite. Then with

$$T = \begin{pmatrix} 1 & 1 \\ 1 & -1 \end{pmatrix} \in GL_2(\mathbb{F}_q)$$

we have $C = {}^tTAT$; and clearly no choice of T as a diagonal matrix in $GL_2(\mathbb{F}_q)$ would achieve this result.

Remark 10.14. For a lattice L as in the preceding corollary, the orthogonal group $O(L)$ satisfies the inclusion

$$O(L) \subseteq \left\{ \begin{pmatrix} \pm 1 & & \\ & \ddots & \\ & & \pm 1 \end{pmatrix} \right\}$$

(here we have identified an isometry with its matrix with respect to a reduced basis $\mathbb{B}$). Consider a graph G with $\mathbb{B}$ as vertex set and an edge v_iv_j if and only if $B(v_i, v_j) \neq 0$. Then $|O(L)| = 2^\nu$, where ν is the number of connected components of G. For example, if $B(v_1, v_{i+1}) \neq 0$ for all $i, 1 \leq i \leq n-1$, then the orthogonal group is trivial: $O(L) = \{\pm I\}$; in which case L is indecomposable.

Example 10.15. Over $\mathbb{F}_3[x]$, if

$$L \cong \begin{pmatrix} x^2 & 1 & & \\ 1 & x^3 & 2 & \\ & 2 & x^4 & 1 \\ & & 1 & x^5 \end{pmatrix}$$

then L is indecomposable.

10.3. On the Hasse–Minkowski Theorem over $\mathbb{F}_q(x)$

Our work in Chapter 5 leading to the Hasse–Minkowski Theorem over $\mathbb{Q}$ began by using Gauss's Law of Quadratic Reciprocity to prove Hilbert's Reciprocity Law. Then, with a lemma involving Strong Approximation and Dirichlet's Theorem on primes in an arithmetic progression, we went on to prove Hasse–Minkowski by a pair of inductions: one induction on the dimension of the spaces involved, another induction—in a fixed dimension—on the sizes of integers involved in a diagonalization of a locally isotropic space. One can retrace the whole argument from Chapter 5, modified appropriately (for instance, using polynomials instead of integers and degrees instead of absolute values) to obtain the Hasse–Minkowski Theorem and its several corollaries for $\mathbb{F}_q(x)$-spaces. [Note that both the Quadratic Reciprocity Law and Dirichlet's theorem have analogues over $\mathbb{F}_q[x]$. (Proofs can be found in [**R**].)] Or one can use a unified approach to simultaneously prove the Hasse–Minkowski Theorem over all global fields, though that requires extensive explorations into algebraic number theory for a complete treatment. See [**O'M1**], for example. Here we will be content to just give a taste of the approach to the $\mathbb{F}_q(x)$-theory that imitates the development over $\mathbb{Q}$, by showing how the Quadratic Reciprocity Law over $\mathbb{F}_q[x]$ leads to Hilbert's Reciprocity Law over $\mathbb{F}_q(x)$.

Theorem 10.16 (Quadratic Reciprocity Law for Polynomials). *Let p_1 and p_2 be distinct prime polynomials in $\mathbb{F}_q[x]$. Then*

$$\left(\frac{p_1}{p_2}\right)\left(\frac{p_2}{p_1}\right) = (-1)^{\partial p_1 \cdot \partial p_2 \cdot \frac{q-1}{2}}.$$

In the following, the expression "all primes" and all unrestricted products over primes should be understood to include the prime ∞.

Theorem 10.17 (Hilbert's Reciprocity Law over $\mathbb{F}_q(x)$). *Suppose $a, b \in \mathbb{F}_q(x)^*$. Then $(a,b)_p = 1$ for almost all p, and*

$$\prod_p (a,b)_p = 1.$$

Proof. As usual with Hilbert symbols, we can freely multiply or cancel square factors from elements involved in the symbols. When these manipulations are used in conjunction with the bilinearity of Hilbert symbols, we see that without loss of generality we can assume that

$$a, b \in \mathbb{F}_q^* \cup \{p \in \mathbb{F}_q[x] \mid p \text{ is a prime polynomial}\}.$$

(i) The theorem's first statement is immediate from the fact that $(\cdot,\cdot)_p$ is trivial on p-adic units. In particular this proves the theorem in the case in which $a, b \in \mathbb{F}_q^*$. [An alternative proof of the latter statement: The $\mathbb{F}_q$-space $\langle a, b \rangle$ is universal, hence represents 1, and therefore $(a,b)_p = 1$ for all p.]

(ii) Suppose $a \in \mathbb{F}_q^*$ and $b = p_0$, a prime. Then $(a, p_0)_p = 1$ for all primes $p \notin \{p_0, \infty\}$. So it remains only to show that $(a, p_0)_{p_0} \cdot (a, p_0)_\infty = 1$ in the event that a is not a square in $\mathbb{F}_q^*$. If ∂p_0 is even then p_0 is a square at ∞, so $(a, p_0)_\infty = 1$. And, since the residue class field at the p_0-adic spot is an even-degree extension of $\mathbb{F}_q$, every element of $\mathbb{F}_q$ is a square in that residue class field (by the Galois theory of finite fields) and hence a square in the p_0-adic completion of $\mathbb{F}_q(x)$. Therefore $(a, p_0)_{p_0} = 1$, so we are done in this case.

Now suppose ∂p_0 is odd. Then $(a, p_0)_\infty = (a, x^{-1})_\infty = -1$, since a is a nonsquare in $\mathbb{F}_q$, which is the residue class field at ∞, and x^{-1} is prime at ∞. Moreover, $(a, p_0)_{p_0} = -1$ as well, since a is a nonsquare in the p_0-adic residue class field, which is an odd-degree extension of $\mathbb{F}_q$. Thus $(a, p_0)_{p_0} \cdot (a, p_0)_\infty = 1$, as desired.

(iii) It remains to check that $\prod_p (p_1, p_2)_p = 1$ if p_1, p_2 are prime polynomials; and we can assume $p_1 \neq p_2$, since otherwise $(p_1, p_2)_p = (p_1, -1)_p$ for all p, and this was covered in the preceding case. We must show that

$$(p_1, p_2)_{p_1} \cdot (p_1, p_2)_{p_2} \cdot (p_1, p_2)_\infty = 1.$$

We have $(p_1, p_2)_\infty = 1$ if either ∂p_1 or ∂p_2 is even; and if $\partial p_1 \equiv \partial p_2 \equiv 1 \pmod 2$ then $(p_1, p_2)_\infty = (x^{-1}, -1)_\infty$. Because $\overline{\mathbb{F}_q(x)_\infty} = \mathbb{F}_q$, we know that $(x^{-1}, -1)_\infty = 1$ if and only if $-1 \in \mathbb{F}_q^{*2}$. In summary, the only situation in which $(p_1, p_2)_\infty = -1$ is this: ∂p_1 and ∂p_2 are odd and $q \equiv 3 \pmod 4$. This gives

$$(p_1, p_2)_\infty = (-1)^{\partial p_1 \cdot \partial p_2 \cdot \frac{q-1}{2}}.$$

Moreover,

$$(p_1, p_2)_{p_1} = \left(\frac{p_2}{p_1}\right) \quad \text{and} \quad (p_1, p_2)_{p_2} = \left(\frac{p_1}{p_2}\right).$$

Therefore our proof of Hilbert's Reciprocity Law over $\mathbb{F}_q(x)$ has been reduced to showing that

$$\left(\frac{p_1}{p_2}\right)\left(\frac{p_2}{p_1}\right) = (-1)^{\partial p_1 \cdot \partial p_2 \cdot \frac{q-1}{2}}.$$

But this is exactly the statement of the Quadratic Reciprocity Law for Polynomials! □

The statements of the Hasse–Minkowski Theorem over $\mathbb{F}_q(x)$ and its corollaries read exactly as they did over $\mathbb{Q}$, except that now $\mathbb{F}_q(x)$ replaces $\mathbb{Q}$.

10.4. Representation by $\mathbb{F}_q[x]$-Lattices

The fact that 2 is a unit in $\mathbb{F}_q[x]$ provides a major simplification in the transition from quadratic forms over $\mathbb{Z}$ to forms over $\mathbb{F}_q[x]$. We have already witnessed an important instance of this in Chapter 7, when we saw that unimodular lattices over a polynomial ring $k[x]$—with k any field of characteristic not 2—have an orthogonal basis, which in turn told us that a given quadratic $k(x)$-space could support at most one isometry class of unimodular $k[x]$-lattices. This is dramatically different from the unimodular theory over $\mathbb{Z}$, in which we saw (in Section 9.6) that class numbers of unimodular $\mathbb{Z}$-lattices can be arbitrarily large.

Let us turn now to representations by $\mathbb{F}_q[x]$-lattices; and we will restrict ourselves to **integral** lattices, meaning lattices that represent only elements of $\mathbb{F}_q[x]$. (Equivalently, $B(v, w) \in \mathbb{F}_q[x]$ for all v, w in the lattice.) Our approach will be that of [**G6**].

We saw in Example 9.11 (as a consequence of the Cassels–Pfister Theorem) that *every element of* $\mathbb{F}_q[x]$ *is a sum of three polynomial squares.* Now we consider sums of two squares. We will say that an $\mathbb{F}_q[x]$-lattice is **universal** if it represents every element of $\mathbb{F}_q[x]$.

Theorem 10.18. *Let L be a unimodular $\mathbb{F}_q[x]$-lattice of rank 2. If L is isotropic, then L is universal. If L is anisotropic, then L represents a polynomial $f \in \mathbb{F}_q[x]$ if and only if $\partial p \equiv 0 \pmod 2$ for all primes p occurring to an odd power in the canonical factorization of f.*

Proof. By Harder's Theorem and the theory of $\mathbb{F}_q$-spaces, we have $L \cong \langle 1, \lambda \rangle$ for some $\lambda \in \mathbb{F}_q^*$. By the Cassels–Pfister Theorem there is a representation $f \to L$ if and only if f is represented by the underlying quadratic space. Therefore without loss of generality we can assume that f is square-free. With the help of the Hasse–Minkowski Theorem over $\mathbb{F}_q(x)$, we have

$$\begin{aligned} f \to L \ &\Leftrightarrow\ \langle 1, \lambda, -f\rangle \cong \langle 1, -1, \lambda f\rangle \text{ over } \mathbb{F}_q(x) \\ &\Leftrightarrow\ S_p\langle 1, \lambda, -f\rangle = S_p\langle 1, -1, \lambda f\rangle \text{ for all primes } p \\ &\Leftrightarrow\ (\lambda, -f)_p = (-1, \lambda f)_p \text{ for all } p \\ &\Leftrightarrow\ (f, -\lambda)_p = 1 \text{ for all } p|f \text{ and } p = \infty \\ &\Leftrightarrow \text{ either } \lambda \in -\mathbb{F}_q^{*2}, \text{or } \lambda \notin -\mathbb{F}_q^{*2} \text{ and } \partial p \equiv 0 \pmod 2 \text{ for all } p|f. \ \square \end{aligned}$$

Now we restate the preceding theorem in less technical language. The reader should compare this statement to Fermat's Two-Square Theorem.

Corollary 10.19 (Leahey [Le]). *If $q \equiv 1 \pmod 4$ then every polynomial in $\mathbb{F}_q[x]$ is a sum of two squares in $\mathbb{F}_q[x]$. If $q \equiv 3 \pmod 4$ then $f \in \mathbb{F}_q[x]$ is a sum of two squares if and only if $\operatorname{ord}_p f \equiv 0 \pmod 2$ for all odd-degree prime divisors p of f.*

We have seen that the unimodular $\mathbb{F}_q[x]$-lattice $\langle 1,1,1\rangle$ is universal, but this is perhaps not surprising in view of the fact that the $\mathbb{F}_q$-space $\langle 1,1,1\rangle$ is isotropic. (Indeed, Leahey's Theorem shows that when $q \equiv 1 \pmod 4$, the space $\langle 1,1\rangle$ is already isotropic.) Of course it is not just isotropy of the lattice that causes universality; *it is isotropy of $\langle 1,1,1\rangle$ over $\mathbb{F}_q$ that does the job*, because $GL_n(\mathbb{F}_q) \subseteq GL_n(\mathbb{F}_q[x])$, hence there is a lattice splitting

$$\langle 1,1,1\rangle \cong \begin{pmatrix} 0 & 1 \\ 1 & 0 \end{pmatrix} \perp \langle -1\rangle$$

and here the hyperbolic component is clearly universal, since 2 is a unit in $GL_n(\mathbb{F}_q[x])$.

Now let us drop the unimodularity restriction and leave isotropy behind and consider representation by definite $\mathbb{F}_q[x]$-lattices, and in particular the property of being universal. So here definiteness requires that the underlying space V has the property that V_∞ is anisotropic. Clearly if a lattice on V is universal then V is universal as well; that is, V represents all of $\mathbb{F}_q(x)$, and hence at all spots $\mathfrak{p}$ the localizations $V_\mathfrak{p}$ represent all of $\mathbb{F}_q(x)_\mathfrak{p}$. So in order for a lattice on a definite space V to be universal, it is necessary that $\dim V = 4$ with V_∞ having the canonical structure of a 4-dimensional anisotropic space over a local field. In detail: $V_\infty \cong \langle 1, -\delta, x, -\delta x\rangle$, with δ a nonsquare in $\mathbb{F}_q$. This observation leads us to consider the following result.

Theorem 10.20. *Let δ be a nonsquare in $\mathbb{F}_q$, and suppose the $\mathbb{F}_q[x]$-lattice L satisfies*

$$L \cong \langle 1, -\delta, x, -\delta x\rangle.$$

Then L is universal.

Proof. We remark—though it is not necessary for the proof—that since every regular binary $\mathbb{F}_q$-space is universal, it is clear that L represents every polynomial of degree ≤ 1. But we need to go beyond this and show that L represents *every* polynomial.

We claim that L is $\mathbb{F}_q[x]$-maximal. Clearly for all prime polynomials $p \neq x$ the localization L_p is unimodular and hence $(\mathbb{F}_q[x])_p$-maximal, so only the x-adic localization needs a close look. In fact, from invariant factor theory, an $(\mathbb{F}_q[x])_x$- lattice J_x properly containing L_x would either fail to be integral (and so would not be $(\mathbb{F}_q[x])_x$-maximal) or would be unimodular. But if J_x were unimodular then the lattice L' on the underlying $\mathbb{F}_q(x)$-space—call it V—satisfying

$$L'_p = \begin{cases} L_p \text{ if } p \neq x, \\ J_x \text{ if } p = x, \end{cases}$$

would be unimodular, hence diagonalizable by Harder's Theorem, and hence isotropic, contradicting the anisotropy of V. Therefore L is $\mathbb{F}_q[x]$-maximal, as claimed.

Since $\dim V = 4$, by the Hasse–Minkowski Theorem V is universal. Therefore for every $f \in \mathbb{F}_q[x]$ there is an $\mathbb{F}_q[x]$-maximal lattice on V that represents f. But the $\mathbb{F}_q[x]$-maximal lattices on V constitute a single genus, namely the genus of L, so we will be done if we can show that L has class number 1; that is, that the class and genus of L coincide.

To this end, suppose $M \in \operatorname{gen} L$. Then $\partial(dM) = \partial(dL) = 2$; hence by the Lattice Reduction Theorem (7.11) we have

$$M \cong \langle 1, -\delta \rangle \perp M_1$$

with M_1 binary and $\partial(dM) = 2$. Moreover, since V is anisotropic, M_1 represents no elements of $\mathbb{F}_q^*$, so

$$M_1 \cong \begin{pmatrix} \alpha x + \beta & \gamma \\ \gamma & \varepsilon x + \eta \end{pmatrix}$$

with respect to some basis $\{w_1, w_2\}$, with $\alpha, \varepsilon \in \mathbb{F}_q^*$ and $\beta, \gamma, \eta \in \mathbb{F}_q$. Since $dL = dM$ we must have $dM_1 = -\delta x^2$. Therefore $\det \begin{pmatrix} \beta & \gamma \\ \gamma & \eta \end{pmatrix} = 0$, so there is a matrix $T = \begin{pmatrix} a & b \\ c & d \end{pmatrix} \in GL_2(\mathbb{F}_q)$ such that

$${}^tT \begin{pmatrix} \beta & \gamma \\ \gamma & \eta \end{pmatrix} T = \begin{pmatrix} 0 & 0 \\ 0 & * \end{pmatrix}.$$

Now use T as the transition matrix for a basis change for M_1. That is, define a new basis $\{w'_1, w'_2\}$ for M_1 by $w'_1 = aw_1 + cw_2$ and so on. This gives a new

Gram matrix for M_1 of the form

$$\begin{pmatrix} \alpha x & \beta x \\ \beta x & \varepsilon x + \gamma \end{pmatrix}$$

with respect to $\{w_1', w_2'\}$ for some $\alpha, \beta, \gamma, \varepsilon \in \mathbb{F}_q$. Replacing w_2' by $w_2' - \frac{\beta}{\alpha} w_1'$ and relabeling yields $M_1 \cong \langle \alpha x, \varepsilon x \rangle$ with $\alpha, \varepsilon \in \mathbb{F}_q^*$. But here $\alpha\varepsilon \in -\delta \mathbb{F}_q^{*2}$ and so $\langle \alpha, \varepsilon \rangle \cong \langle 1, -\delta \rangle$ over $\mathbb{F}_q$, and hence $\langle \alpha x, \varepsilon x \rangle \cong \langle x, -\delta x \rangle$. Thus $M \cong L$, as claimed. □

In 1993 Conway and Schneeberger announced the following result: *Every positive definite integral $\mathbb{Z}$-lattice that represents each of the integers* $1, 2, 3, 5, 6, 7, 10, 14, 15$ *represents every positive integer.* This is their celebrated **Fifteen Theorem**. It appeared in Schneeberger's 1997 Princeton dissertation [**Schn**], although the details of the underlying computations were not published, and it was subsequently given a shorter proof by Bhargava [**Bh**] in 2000. (Also see the background discussion in [**Con3**].) In considering definite $\mathbb{F}_q[x]$-lattices, it is natural to seek an analogue of the Fifteen Theorem. Based on Theorem 10.20 and other results, it was conjectured in [**G6**] that every definite integral $\mathbb{F}_q[x]$-lattice that represents each of $1, -\delta, x, -\delta x$ actually represents every element of $\mathbb{F}_q[x]$. (This was the **Four Conjecture**.) This conjecture was subsequently proved, first by Kim, Wang, and Xu [**KWX**] in 2004; then by Chan and Daniels [**ChD**] in 2005, using the theory of **regular** lattices: lattices that represent every element represented by some lattice in their genus; and, finally, in a relatively brief argument by Bureau in his 2006 dissertation [**Bu**]. Incidentally, while it is clear that every $\mathbb{F}_q[x]$-lattice of class number one is regular, Bureau goes on to prove that *a definite $\mathbb{F}_q[x]$-lattice is regular if and only if it has class number one*, though the corresponding result is known not to hold for definite $\mathbb{Z}$-lattices.

Exercises

(1) Consider the quadratic spaces V_1 and V_2 underlying the lattices L_1, L_2 in the chapter's opening example. For which spots $\mathfrak{p}$ on $\mathbb{F}_3(x)$ is there an isometry $V_{1\mathfrak{p}} \cong V_{2\mathfrak{p}}$?

(2) Evaluate the Legendre symbol $\left(\dfrac{x^2+1}{x^3+2x+1} \right)$.

(3) Show that if in Example 10.1 the coefficient ring were $\mathbb{F}_7[x]$ instead of $\mathbb{F}_3[x]$, then the lattice L_1 in that example would have a Gram matrix

of the form $\begin{pmatrix} 2x & 1 \\ 1 & * \end{pmatrix}$, but that nevertheless there is *no* value of q such that $\begin{pmatrix} x & 1 \\ 1 & x \end{pmatrix} \cong \begin{pmatrix} 2x & 1 \\ 1 & 2x \end{pmatrix}$ over $\mathbb{F}_q[x]$.

(4) Let L be an $\mathbb{F}_q[x]$-lattice on the regular $\mathbb{F}_q(x)$-space V. Suppose $L \cong A \in M_n(\mathbb{F}_q(x))$, and suppose all of A's invariant factors are squares. What can be concluded about the Hasse symbol $S_\infty V$?

(5) Consider a lattice

$$L \cong \begin{pmatrix} 2x^2 + 3x + 1 & 2 \\ 2 & 3x^2 + 2x + 5 \end{pmatrix}.$$

Show that $\min L = 2$ if the coefficient ring is $\mathbb{F}_{13}[x]$, but not if the coefficient ring is $\mathbb{F}_7[x]$.

(6) Suppose L is an $\mathbb{F}_7[x]$-lattice satisfying

$$L \cong \begin{pmatrix} 2x & 3x^2 & 3x^2 + 3x \\ 3x^2 & x^3 + 1 & x^3 + x^2 + x \\ 3x^2 + 3x & x^3 + x^2 + x & x^3 + 3x^2 + 5x \end{pmatrix}.$$

Use Djokoviĉ's Lattice Reduction Theorem (7.11) to find a reduced Gram matrix for L. Then answer this question: Is there a representation $6 \to L$?

(7) Suppose $L \cong \begin{pmatrix} 4x^3 + 3 & x^2 + 2x + 1 \\ x^2 + 2x + 1 & x^7 + x^2 + 1 \end{pmatrix}$ over the polynomial ring $\mathbb{F}_{19}[x]$. Show that no polynomial with leading term x^5 or $8x^3$ is represented by L.

Chapter 11

Applications to Cryptography

11.1. A Brief Sketch of the Cryptographic Setting

Where there is language and a need for secrecy, there is probably some form of **cryptography**: the art of disguising ("encrypting" or "enciphering") messages so that an unauthorized recipient is unable to learn their meaning in time to use the information they contain. Of course there is no point in encrypting a message unless the intended recipient knows how to "decrypt" or "decipher" it. And where there is cryptography there is probably also **cryptanalysis**: the art of decrypting when one is *not* the intended recipient. If a message has first been put into numerical form, then the methods of encryption and decryption very likely involve number theory. Our purpose in this section is to informally introduce the basic terminology of the subject and give the essence of some examples. In later sections we will examine a fundamental reduction process for $\mathbb{Z}$-lattices in $\mathbb{R}^n$, and then go on to sketch some ways in which $\mathbb{Z}$-lattices have been involved in both cryptography and cryptanalysis. What we will *not* do is give a formal presentation of cryptography or of computational complexity.

The message to be sent is the **plaintext**, the encrypted message is the **cyphertext**, and an effort by an unintended recipient of the cyphertext to construct the plaintext is an **attack**. A **cryptosystem** is a quintuple $(\mathbb{P}, \mathbb{C}, \mathbb{K}, \mathbb{E}, \mathbb{D})$ in which $\mathbb{P}$ and $\mathbb{C}$ are finite sets ($\mathbb{P}$ is the set of **plaintext symbols**, $\mathbb{C}$ is the set of **cyphertext symbols**); $\mathbb{K}$ is a set called the set of **keys**, where each $K \in \mathbb{K}$ is itself a set of mathematical objects; $\mathbb{E} = \{e_K\}_{K\in\mathbb{K}}$ is the set of **encryption functions** from $\mathbb{P}$ to $\mathbb{C}$; and, finally,

$\mathbb{D} = \{d_K\}_{K \in \mathbb{K}}$ is the set of **decryption functions** from $\mathbb{C}$ to $\mathbb{P}$, with $d_K \circ e_K = id_{\mathbb{P}}$. Incidentally, in the literature usually the sender of the message is **Alice**, the intended recipient is **Bob**, and the attacker is **Oscar**.

Example 11.1. Among the most classical cryptosystems is **affine encryption**, in which $\mathbb{P} = \mathbb{C} = S = \{0, 1, 2, \ldots, 25\}$ (corresponding to the letters of the alphabet),

$$\mathbb{K} = \{(a, b) \in S^2 \mid \gcd(a, 26) = 1\},$$

and for $K = (a, b) \in \mathbb{K}$, the associated encryption is given by

$$e_K(x) = ax + b \pmod{26}$$

for $x \in S$, and decryption by

$$d_K(x) = a^{-1}(x - b) \pmod{26}.$$

Of course the numbers in this example can be increased to allow for a wider range of possibilities.

For our purposes, the most important feature to notice about affine encryption—and this is typical of classical encryption systems—is that someone knowing what system is being used and knowing the encryption function will be able to construct the decryption function. (Of course, techniques for constructing fiendishly complicated encryption functions were developed over the centuries, so that "knowing the encryption function" could be an extraordinarily difficult thing for an attacker to achieve.) Alice and Bob will have had to agree in advance on an encryption key, their **secret key**, in order for Bob to be able to readily decrypt Alice's message. If Bob were to receive encrypted messages from many different individuals, for the sake of security he would need to agree on a different secret key with each sender. And if, for example, these communications took place in a highly competitive corporate environment, or perhaps in a sensitive diplomatic or military setting, and a security lapse were to occur—revealing the key used by one of the senders—then a successful attack on at least some of the communications sent to Bob would be the immediate outcome.

It is clear that cryptosystems based on secret keys were inadequate for the communications explosion that occurred in the final third of the Twentieth Century. And so, to make a long story short, in the 1970s **public key cryptography** was developed; and with that, number theory was instantly transformed from pure mathematics into applied mathematics. (For the history, see [**Y**] or the popular literature, e.g. [**Sin**].) In a public key cryptosystem, each key is a union of two parts: a **public key** and a **private key**. Bob announces the public part of the key K to the world; this is the part that Alice and everyone else will use to construct the encryption function e_K for sending their messages to him. But only Bob knows the private part, and that is the part needed to construct d_K. The security of the system requires that the construction of d_K from e_K be a computationally intractable task.

Example 11.2. (i) In the **RSA** system, created in 1978, Bob takes distinct massive primes p and q and computes their product $N = pq$. This number N is the heart of the public key and is used to construct e_K. Construction of d_K requires the computation of $\varphi(N)$, where φ is Euler's function; and this in turn is computationally equivalent in difficulty to factoring N. The factors of N are in Bob's secret key. Thus the security of RSA is based on the difficulty of the problem of factoring a properly prepared large integer. By the way, when we say that two mathematical procedures—in the present case, factoring N and computing $\varphi(N)$—are "computationally equivalent," we mean, roughly speaking, that someone with an algorithm to accomplish either one of these tasks will be then able to accomplish the other with relative ease. (A more precise definition requires deeper exploration into the theory of computational complexity.)

(ii) The **El Gamal** cryptosystem, dating from 1985, is based on the difficulty of the **discrete logarithm problem**, which goes as follows. If q is prime, the multiplicative group $\mathbb{F}_q^*$ is cyclic; a generator r is a **primitive root** for the group. Problem: Given $a \in \mathbb{F}_q^*$, find an exponent k such that $r^k = a$; this k is the **discrete logarithm of** a **to the base** r. (In the old number theory literature, k is the **index** of a with respect to r.) In El Gamal encryption, the function e_K involves "masking" a given piece of plaintext using a multiplier obtained by raising a given primitive root to a secret power (which is found in the secret key). An attacker will need to determine that secret power in order to construct d_K, but no effective algorithm is known for finding that power; that is, the discrete logarithm problem is believed to be computationally intractable.

Of course in both the RSA and El Gamal cryptosystems, the difficulties in successfully attacking encrypted messages will arise only if the numbers involved are very large. Details of both cryptosystems can be found in most of the elementary number theory books that have appeared in the last decade. Now let us focus on $\mathbb{Z}$-lattices in $\mathbb{R}^n$, the setting in which they are most useful for public key cryptography.

11.2. Lattices in $\mathbb{R}^n$

The dot product on $\mathbb{R}^n$ makes $\mathbb{R}^n$ into a regular quadratic $\mathbb{R}$-space, with the standard orthonormal basis $\{e_1, \ldots, e_n\}$; thus $\mathbb{R}^n \cong I_n$ with respect to that basis. In terms of our usual notation: $B(v, w) = v \cdot w$ and $q(v) = \|v\|^2$, where $\| \quad \|$ is the Euclidean length function. (So to say that $q(v)$ is large is equivalent to saying that $\|v\|$ is large.) Now let $\mathbb{B} = \{v_1, \ldots, v_n\}$ be another basis, and consider the $\mathbb{Z}$-lattice $L = \mathbb{Z}v_1 + \cdots + \mathbb{Z}v_n$. If $v_j = \sum_i t_{ij}e_i = {}^t(t_{1j}, \ldots, t_{nj})$, with $(t_{ij}) = T$, then L has Gram matrix ${}^tTT = A \in GL_n(\mathbb{R})$ with respect to $\mathbb{B}$. The matrix T, whose columns are the basis vectors v_j, is said to be a **generator matrix** for L. (Note: Some authors call the transpose ${}^tT$—whose *rows* are the basis vectors—a generator matrix instead.) Extending our discussion in Section 6.2, the half-open parallelepiped

$$\{\sum_{i=1}^{n} \alpha_i v_i \mid 0 \leq \alpha < 1\} \subset \mathbb{R}^n$$

is the **fundamental region** determined by the basis. Its n-dimensional volume is the absolute value of the determinant of the generator matrix T (see [**MB**]), which in turn is equal to $(dL)^{\frac{1}{2}}$.

As we will see, most cryptographic applications of lattices involve the search for short vectors. Hermite's inequality in Theorem 7.5 gave us an upper bound on the minimum of a lattice, but Hermite's proof does not yield an explicit construction of a minimal vector. [By way of contrast, Theorem 10.5 shows that application of reduction theory to a definite $\mathbb{F}_q[x]$-lattice actually produces a minimal vector .] Our main goal in this section and the next is to explore results that move in that direction. We will be guided in part by the treatment in Cohen [**Coh**].

Theorem 11.3 (Hadamard's Inequality). *Let $\mathbb{B} = \{v_1, \ldots, v_n\}$ be a basis for $\mathbb{R}^n$, and let L be the $\mathbb{Z}$-lattice spanned by $\mathbb{B}$. Then*

$$dL \leq \prod_{i=1}^{n} q(v_i).$$

Proof. As in 2.13, construct an orthogonal basis $\{v_1^*, \ldots, v_n^*\}$ for $\mathbb{R}^n$ from $\mathbb{B}$ by first putting $v_1^* = v_1$, and, having defined $v_1^*, \ldots, v_{i-1}^*$, put

$$v_i^* = v_i - \sum_{j=1}^{i-1} \mu_{ij} v_j^*$$

with $\mu_{ij} = \dfrac{B(v_i, v_j^*)}{q(v_j^*)}$. (So the sets $\{v_1, \ldots, v_i\}$ and $\{v_1^*, \ldots, v_i^*\}$ span the same subspace.) Then, since the transition matrix for the basis change has determinant 1, we have

$$dL = \prod_{i=1}^{n} q(v_i^*) \leq \prod_{i=1}^{n} q(v_i). \qquad \square$$

Upon taking square roots, we have the following geometric interpretation of Hadamard's inequality: *The volume of a parallelepiped determined by a basis of $\mathbb{R}^n$ is less than or equal to the product of the lengths of the basis vectors.*

Remarks 11.4. (i) The equation $v_j = v_j^* + \sum_{i=1}^{j-1} \mu_{ij} v_i^*$ shows that v_j^* is the projection of v_j onto the orthogonal complement of the space spanned by $v_1, \ldots, v_{j-1}$.

(ii) Suppose $0 \neq v \in L$, say $v = \sum_{j=1}^{k} \alpha_j v_j$, with all $\alpha_j \in \mathbb{Z}, \alpha_k \neq 0$, and $k \leq n$. Then

$$v = \sum_{j=1}^{k} \alpha_j \Big(v_j^* + \sum_{i=1}^{j-1} \mu_{ij} v_i^* \Big) = \alpha_k v_k^* + \sum_{i=1}^{j-1} \beta_j v_j^*$$

with $\beta_j \in \mathbb{Q}$. Therefore $q(v) \geq q(\alpha_k v_k^*) \geq q(v_k^*)$, and hence

$$\min L \geq \min_{1 \leq j \leq n} q(v_j^*).$$

The intuition behind the inequalities $q(v_j) \geq q(v_j^*)$ is this: For the parallelepipeds with a given base B and a given volume, the edge not in B will be shortest if it is orthogonal to B. For this reason, the transition from the vectors in the initial basis for L to a lattice basis closer to $\{v_1^*, \ldots, v_n^*\}$ is known as **size reduction**. (This is part of the reduction process to be discussed shortly.)

In human calculations real numbers are truncated after finitely many decimal places; so we will be particularly interested in examples of lattices that have Gram matrices in $GL_n(\mathbb{Q})$. Indeed, by scaling such lattices, there will be no loss of generality in focusing on $\mathbb{Z}$-lattices that are not only positive definite (equivalently, span $\mathbb{R}^n$ for some n), but integral: $B(v, w) = v \cdot w \in \mathbb{Z}$ for all v, w in the lattice. (But we will not need to restrict ourselves to such lattices when we consider lattice reduction in the next section.)

Examples 11.5. (i) Consider the lattice

$$A_n \cong \begin{pmatrix} 2 & 1 & & & \\ 1 & 2 & 1 & & \\ & \ddots & \ddots & \ddots & \\ & & \ddots & \ddots & 1 \\ & & & 1 & 2 \end{pmatrix}$$

in the basis $\{v_1, \ldots, v_n\}$. A short induction argument shows that

$$q(v_k^*) = \frac{k+1}{k}$$

for $1 \leq k \leq n$. From this it follows that $\min A_n > 1$, and hence $\min A_n = 2$. (Of course we have known this for a long time by other arguments.)

(ii) More generally, consider a $\mathbb{Z}$-lattice of the form

$$L \cong A = \begin{pmatrix} a_1 & 1 & & & \\ 1 & a_2 & 1 & & \\ & \ddots & \ddots & \ddots & \\ & & \ddots & \ddots & 1 \\ & & & 1 & a_n \end{pmatrix}$$

in the basis $\{v_1, \ldots, v_n\}$, with integers $a_i \geq 2$ for all i. We claim that $\min L = \min_i a_i$. One way to see this is to observe that for all $v \in L$ we have $q(v) = q_0(v) + q_1(v)$, where $(L, q_0) \cong A_n$ and $(L, q_1) \cong \langle a_1 - 2, \ldots, a_n - 2\rangle$. (For this argument, the a_i need not be integers.)

Now let us use the results from the present chapter instead. By Remark 11.4(ii) it suffices to check that $q(v_j^*) > a_j - 1$ for all j. Trivially $q(v_1^*) = a_1$; and, having checked the result for $j = k$, we have

$$q(v_{k+1}^*) = q\left(v_{k+1} - \frac{1}{q(v_k^*)} v_k^*\right) = a_{k+1} - \frac{1}{q(v_k^*)} > a_{k+1} - \frac{1}{a_k - 1} \geq a_{k+1} - 1,$$

which proves the claim.

11.3. LLL-Reduction

Now we consider an approach to reduction for $\mathbb{Z}$-lattices in $\mathbb{R}^n$ that has important cryptographic applications. [This reduction process had its birth in the pioneering paper [**LLL**] by A. Lenstra, H. Lenstra, and L. Lovasz, whose abbreviated names provide the label for the process.] We continue to use the notation already introduced in this chapter.

Definition 11.6. Let $\mathbb{B} = \{v_1, \ldots, v_n\}$ be a basis for a lattice L. We say that $\mathbb{B}$ is an **LLL-reduced** (or $\mathbf{L}^3$**-reduced**) basis if the following two conditions are satisfied:

(i) $|\mu_{i,j}| \leq \frac{1}{2}$ whenever $1 \leq j < i \leq n$;

(ii) $q(v_i^* + \mu_{i,i-1} v_{i-1}^*) \geq \frac{3}{4} q(v_{i-1}^*)$ for all $i \geq 2$. [Note: condition (ii) is equivalent to this inequality: $q(v_i^*) \geq \left(\frac{3}{4} - \mu_{i,i-1}^2\right) q(v_{i-1}^*)$.]

Notice that

$$\mu_{i,j} = \frac{B(v_i, v_j^*)}{q(v_j^*)} = \frac{v_i \cdot v_j^*}{\|v_j^*\|^2} = \frac{\|v_i\| \cos\theta}{\|v_j^*\|},$$

where θ is the angle between v_i and v_j^*. So condition (i) tells us that the projection of v_i on v_j^* is at most half the length of v_j^* if $1 \leq j < i$. The intuition: v_i is somewhat close to being orthogonal to the space spanned by $\{v_1, \ldots, v_{i-1}\}$.

Now for reduction condition (ii). The line $\mathbb{R}v_i^*$ is orthogonal to the space spanned by $\{v_1, \ldots, v_{i-1}\}$. The vector v_i^* itself is the projection of v_i onto that line, and $\mu_{i,i-1}v_{i-1}^*$ is the projection of v_i onto $\mathbb{R}v_{i-1}^*$; so $v_i^* + \mu_{i,i-1}v_{i-1}^*$ is the projection of v_i onto the plane $\mathbb{R}v_{i-1}^* \perp \mathbb{R}v_i^*$. Therefore condition (ii) tells us that the length of the projection of v_i onto $\mathbb{R}v_{i-1}^* \perp \mathbb{R}v_i^*$ must be at least $\frac{\sqrt{3}}{2}$ times the length of the projection of v_{i-1} onto that plane.

The LLL-algorithm for achieving such a basis consists of an oscillation between two processes: size reduction to achieve property (i), and vector interchanges to achieve property (ii). It is a fact—but not at all an *obvious* fact—that the process will terminate, and it will do so having achieved an LLL-reduced basis. To get the flavor of the algorithm, let us first have a detailed look at its application to a binary lattice (to keep the computations short) and pretend that we have not already seen the reduction of binary lattices in Chapter 7.

Example 11.7. Suppose $L \cong A = \begin{pmatrix} 5 & 6 \\ 6 & 8 \end{pmatrix}$ in $\{v_1, v_2\}$. Here $v_2^* = v_2 - \frac{6}{5}v_1$. While changing v_2 to v_2^* gives a basis for the underlying space, $\{v_1, v_2^*\}$ is not a basis for L since the transition matrix $\begin{pmatrix} 1 & -\frac{6}{5} \\ 0 & 1 \end{pmatrix}$ is not unimodular. So we *approximate* that matrix with the unimodular matrix $T = \begin{pmatrix} 1 & \lfloor \frac{-6}{5} \rceil \\ 0 & 1 \end{pmatrix} = \begin{pmatrix} 1 & -1 \\ 0 & 1 \end{pmatrix}$. (Here $\lfloor \alpha \rceil$ denotes the **nearest integer** to α.) Then $L \cong {}^tTAT = \begin{pmatrix} 5 & 1 \\ 1 & 1 \end{pmatrix}$ in the new basis, which we relabel as $\{v_1, v_2\}$. Now

$$\mu_{2,1} = \frac{B(v_2, v_1^*)}{q(v_1^*)} = \frac{1}{5},$$

so property (i) in the definition has been achieved. But $q(v_2^*) = \frac{4}{5}$, while $\left(\frac{3}{4} - \mu_{2,1}^2\right) q(v_1^*) = \frac{71}{20}$, hence we do not yet have property (ii).

Now interchange v_1 and v_2 and relabel, getting $L \cong \begin{pmatrix} 1 & 1 \\ 1 & 5 \end{pmatrix}$ in the current $\{v_1, v_2\}$. Then $v_2^* = v_2 - v_1 \in L$, and now the transition matrix from $\{v_1, v_2\}$ to $\{v_1, v_2^*\}$ is unimodular. We have $L \cong \begin{pmatrix} 1 & 0 \\ 0 & 4 \end{pmatrix}$ in this new basis, which is clearly a reduced basis.

The reduction process typically is much more turbulent in higher dimensions, in the sense that making an adjustment in the basis with the goal of achieving one of the properties in the definition can cause a setback with respect to the other property, and one (more honestly, one's computer) must shuttle back and forth to achieve both properties simultaneously. Now let us look more closely at the algorithm.

The LLL-algorithm. Given a lattice L and a basis $\mathbb{B} = \{v_1, \ldots, v_n\}$ for L that is not reduced, the goal is to produce a reduced basis. Set $\mathbb{B}_i = \{v_1, \ldots, v_i\}$, for $1 \leq i \leq n$. As in the proof of Hadamard's Inequality (11.3), apply the Gram–Schmidt process to $\mathbb{B}$ to obtain the orthogonal basis $\{v_1^*, \ldots, v_n^*\}$ and the values

$$\mu_{ij} = \frac{B(v_i, v_j^*)}{q(v_j^*)}.$$

Trivially $\{v_1\}$ is a reduced basis for the sublattice $\mathbb{Z}v_1$. Now suppose that $k \geq 2$ and k is the largest value for which $\mathbb{B}_{k-1} = \{v_1, \ldots, v_{k-1}\}$ is a reduced basis for $\bigoplus_{i=1}^{k-1} \mathbb{Z}v_i$. So $\mathbb{B}_k = \{v_1, \ldots, v_k\}$ is not a reduced basis of $\bigoplus_{i=1}^{k} \mathbb{Z}v_i$, and the blame for this failure must somehow fall on v_k.

If $\mathbb{B}_k$ fails to be reduced because it fails condition (i) in Definition 11.6, choose l maximal such that

$$|\mu_{k,j}| \leq \frac{1}{2} \text{ for } l < j < k, \text{ while } |\mu_{k,l}| > \frac{1}{2}.$$

In this event, replace v_k by $v_k' = v_k - \lfloor \mu_{k,l} \rceil v_l$, and upon noticing that

$$q(v_l^*) = B\Big(v_l - \sum_{i<l} \mu_{l,i} v_i^*, v_l^*\Big) = B(v_l, v_l^*)$$

we see that

$$\left| \frac{B(v_k', v_l^*)}{q(v_l^*)} \right| = \left| \mu_{k,l} - \lfloor \mu_{k,l} \rceil \frac{B(v_l, v_l^*)}{q(v_l^*)} \right| = \Big| \mu_{k,l} - \lfloor \mu_{k,l} \rceil \Big| \leq \frac{1}{2}.$$

Now relabel v_k' as v_k (so we now have $|\mu_{k,l}| \leq \frac{1}{2}$), and recalculate the values $\mu_{k,j}$ for $j \leq l-1$. (Namely, replace $\mu_{k,j}$ with $\mu_{k,j} - \lfloor \mu_{k,l} \rceil \mu_{l,j}$.) Repeat this ritual as needed until we have "size reduced" v_k; that is, $|\mu_{k,j}| \leq \frac{1}{2}$ for all j satisfying $1 \leq j \leq k-1$.

If the new $\mathbb{B}_k$ still fails to be reduced, it must be due to noncompliance with condition 11.6(ii), with v_k once again the villain. (Remember: $\mathbb{B}_{k-1}$ is reduced.) That is, it must be the case that

$$q(v_k^* + \mu_{k,k-1} v_{k-1}^*) < \frac{3}{4} q(v_{k-1}^*);$$

equivalently, $q(v_k^*) < \left(\frac{3}{4} - \mu_{k,k-1}^2\right) q(v_{k-1}^*)$. In this event, interchange v_k and v_{k-1} and relabel them. Having done this, without further computation we know only that $\mathbb{B}_{k-2}$ is reduced, and we start the whole process over again, with k replaced by $k-1$ (an apparent regression).

On the other hand, if $\mathbb{B}_k$ is now reduced, either we have achieved our desired reduced basis or we return to the start of the algorithm, but with k replaced by some $k+t$, with $\mathbb{B}_{k+t}$ not reduced.

In either event, we continue the process until a reduced basis is achieved.

For a presentation of the LLL-algorithm in a pseudocode format, proof that it eventually terminates, discussion of running time, and variations of the algorithm, we refer the reader to the extensive treatment in [**Coh**], §2.6. Also see Micciancio–Goldwasser ([**MG**], Chapter 2), Pohst–Zassenhaus ([**PZ**], Chapter 3), or the original paper by Lenstra, Lenstra, and Lovasz [**LLL**] in which the algorithm appeared.

Example 11.8. Suppose

$$L \cong \begin{pmatrix} 2 & 1 & 0 \\ 1 & 2 & 1 \\ 0 & 1 & 2 \end{pmatrix} \text{ in the basis } \mathbb{B} = \{v_1, v_2, v_3\}.$$

We have

$$v_1^* = v_1, \quad v_2^* = v_2 - \frac{1}{2}v_1, \quad q(v_2^*) = \frac{3}{2}, \text{ and } \mu_{2,1} = \frac{1}{2}.$$

Therefore $q(v_2^* + \mu_{2,1}v_1^*) = q(v_2) = 2$. So $\mathbb{B}_2 = \{v_1, v_2\}$ is reduced.

Clearly $\mu_{3,1} = 0$, but $\mu_{3,2} = \frac{2}{3}$; so v_3 requires size reduction. Replace v_3 by $v_3' = v_3 - \lfloor \mu_{3,2} \rceil v_2 = v_3 - v_2$, and relabel this as v_3. Having made this change, we obtain

$$L \cong \begin{pmatrix} 2 & 1 & -1 \\ 1 & 2 & -1 \\ -1 & -1 & 2 \end{pmatrix} \text{ in the basis } \{v_1, v_2, v_3\},$$

and now $\mu_{3,1} = -\frac{1}{2}$ and $\mu_{3,2} = -\frac{1}{3}$. So condition 11.6(i) is satisfied.

As for condition 11.6(ii), we have

$$v_3^* = v_3 - \mu_{3,1}v_1^* - \mu_{32}v_2^* = \frac{1}{3}v_1 + \frac{1}{3}v_2 + v_3,$$

which gives $q(v_3^*) = \frac{4}{3}$, and hence

$$q(v_3^* + \mu_{3,2}v_2^*) = \frac{3}{2} = q(v_2^*).$$

Therefore the new $\mathbb{B}$ is reduced. Incidentally, if we replace v_3 by $-v_3$, we get

$$L \cong \begin{pmatrix} 2 & 1 & 1 \\ 1 & 2 & 1 \\ 1 & 1 & 2 \end{pmatrix}$$

which is still LLL-reduced. (Only the signs of $\mu_{3,1}$ and $\mu_{3,2}$ have changed.)

The following theorem gives the most important consequence of LLL-reduction; namely, that in constructing an LLL-reduced basis $\mathbb{B}$ for a lattice, we may well have found a vector short enough to be useful in many applications. The key to the proof lies in bounding the vectors v_i in $\mathbb{B}$ by suitable constants times the lengths of the corresponding vectors v_i^* in the orthogonal basis for $\mathbb{R}^n$ constructed from $\mathbb{B}$ by the Gram–Schmidt process.

Theorem 11.9. *Let* $\{v_1, \dots, v_n\}$ *be an LLL-reduced basis for a* $\mathbb{Z}$*-lattice* L *on* $\mathbb{R}^n$*. Then* (i) $q(v_1) \le 2^{n-1} \min L$.

(ii) *More generally, if* $x_1, \dots, x_t$ *are linearly independent elements of* L*, then*

$$q(v_j) \le 2^{n-1} \max\{q(x_1), \dots, q(x_t)\} \text{ if } 1 \le j \le t.$$

Proof. (i) From Definition 11.6, for $i \ge 2$ we have

$$q(v_i^*) \ge \left(\frac{3}{4} - \mu_{i,i-1}^2\right) q(v_{i-1}^*) \ge \frac{1}{2} q(v_{i-1}^*)$$

and therefore

$$q(v_j^*) \le 2^{i-j} q(v_i^*) \quad \text{if} \quad j \le i.$$

And from the fact that $v_i^* = v_i - \sum_{j=1}^{i-1} \mu_{ij} v_j^*$ we then have

$$\begin{aligned} q(v_i) &= q(v_i^*) + \sum_{j=1}^{i-1} \mu_{ij}^2 q(v_j^*) \\ &\le q(v_i^*) + \frac{1}{4} q(v_i^*) \sum_{j=1}^{i-1} 2^{i-j} \\ &= q(v_i^*) \left(\frac{2^{i-1}+1}{2}\right). \end{aligned}$$

Upon combining these inequalities when $1 \le j \le i \le n$ we obtain

$$q(v_j) \le q(v_j^*) \left(\frac{2^{j-1}+1}{2}\right) \le \left(2^{i-2} + 2^{i-j-1}\right) q(v_i^*) \le 2^{i-1} q(v_i^*).$$

Now suppose v is a minimal vector in L. Say $v = \sum_{j=1}^{i} a_j v_j$, with $a_j \in \mathbb{Z}$ for all j, and $a_i \ne 0$. Then we have $v = \sum_{j=1}^{i} b_j v_j^*$, with $b_j \in \mathbb{Q}$ for all j and with $b_i = a_i$; and from this we have

$$q(v) \ge b_i^2 q(v_i^*) = a_i^2 q(v_i^*) \ge q(v_i^*) \ge 2^{1-i} q(v_1) \ge 2^{1-n} q(v_1)$$

and therefore $q(v_1) \le 2^{n-1} q(v)$, which finishes the proof of part (i).

(ii) Write $x_j = \sum_{i=1}^{i(j)} r_{ij} v_i$, with all $r_{ij} \in \mathbb{Z}$ and $r_{i(j),j} \ne 0$. Then

$$x_j = r_{i(j),j} v_{i(j)}^* + \sum_{i=1}^{i(j)-1} \lambda_i v_i^*$$

for some $\lambda_i \in \mathbb{Q}$, and hence $q(x_j) \geq q(v^*_{i(j)})$. Now order the vectors $x_1, \ldots, x_t$ so that $i(1) \leq \cdots \leq i(t)$. Since the vectors $x_1, \ldots, x_j$ are linearly independent, they cannot all be linear combinations of $v_1, \ldots, v_{j-1}$. Therefore $i(j) \geq j$ for $1 \leq j \leq t$, and we have

$$\begin{aligned} q(v_j) &\leq 2^{i(j)-1} q(v^*_{i(j)}) \text{ from the proof of part (i)} \\ &\leq 2^{n-1} q(v^*_{i(j)}) \\ &\leq 2^{n-1} q(x_j). \end{aligned}$$

□

Corollary 11.10. *If $\{v_1, \ldots, v_n\}$ is an LLL-reduced basis for L, then*

$$\min L \leq q(v_1) \leq 2^{\frac{n-1}{2}} (dL)^{\frac{1}{n}}.$$

Proof. From the proof of part (i) of the theorem we have the n inequalities

$$q(v_1) \leq 2^{i-1} q(v^*_i) \quad \text{for } 1 \leq i \leq n.$$

Upon taking the product of these inequalities we obtain

$$q(v_1)^n \leq 2^{\sum_i^n (i-1)} \prod_{i=1}^{n} q(v^*_i) = 2^{\frac{n(n-1)}{2}} dL,$$

from which the result follows. □

Remark 11.11. The inequality achieved in Corollary 11.10 is weaker than the inequality

$$\min L \leq \left(\frac{4}{3}\right)^{\frac{n-1}{2}} (dL)^{\frac{1}{n}}$$

of Hermite's inequality (7.5), but the LLL-algorithm actually produces a vector satisfying the inequality, whereas Hermite's result is not constructive. In fact, it is a consequence of Minkowski's Convex Body Theorem in the geometry of numbers that a $\mathbb{Z}$-lattice L on $\mathbb{R}^n$ satisfies an inequality of the form $\min L \leq c_n (dL)^{\frac{1}{n}}$, where c_n is asymptotic to $\dfrac{2n}{\pi e}$ as $n \to \infty$. See [**MH**], Chapter 2, for more details.

11.4. Lattice Attacks on Knapsack Cryptosystems

The **knapsack** or **subset sum** problem is the following: Given a sequence $W = \{a_1, \ldots, a_n\}$ of positive integers (the **weights**) and a positive integer S such that $S = \sum_{i=1}^n x_i a_i$ for some $x_1, \ldots, x_n \in \{0, 1\}$, *find* values $x_1, \ldots, x_n$ that do the job.

Example 11.12. There are some choices for W for which the knapsack problem has an easy solution. For example, if $S < 2^n$ and $a_i = 2^{i-1}$ for $1 \leq i \leq n$, then the x_i are just the digits of S (in reverse order) when written in binary. More generally, if for all $j > 1$ we have $a_j > \sum_{i=1}^{j-1} a_i$ (such a sequence $\{a_i\}$ is said to be **superincreasing**), just choose the largest subscript j for which $a_j < S$; then $x_j = 1$, while $x_i = 0$ for all $i > j$. Now repeat this argument on $S - a_j$ to get $x_{j-1}, \ldots, x_1$.

If the set W of weights is chosen in such a way that there is no useful pattern or coherence among the weights, then solving the knapsack problem can be very difficult, and this difficulty is the foundation for knapsack cryptosystems.

Remark 11.13. The knapsack problem is among the problems labeled **NP-complete** by computer scientists. This is the class of problems for which the following are true:

1. Checking whether a conjectured solution (that is, a possible solution produced "**nondeterministically**") is an actual solution is relatively straightforward; more precisely, it can be carried out in a time bounded by a polynomial function of the input length (i.e., in **polynomial time**). If this is the case, the problem is in the class **NP**.

2. No algorithm is known that actually leads to a solution in polynomial time to all instances of the problem. (Were such an algorithm known, the problem would be said to be in the class **P**.)

3. The problem is among the most difficult of those problems satisfying properties 1 and 2, in the following sense. If someone could exhibit an algorithm that solved all instances of the problem in polynomial time—thereby putting it into the class **P**—it would follow that every other problem in the class **NP** could also be solved in polynomial time; in other words, we would have the class equality **P=NP**. Given the efforts that have gone—without success—into the search for polynomial time algorithms for these problems, this appears highly unlikely. In the terminology of the subject: Many believe, though no proof has been given, that **P≠NP**. See Garey and Johnson [**GJ**] for more details.

In a **knapsack cryptosystem**, Bob publicly announces a set $W = \{a_1, \ldots, a_n\}$ of weights to be used to encrypt messages to him; that is, W is the public key. Then Alice encrypts her message $x = (x_1, \ldots, x_n)$, with $x_i \in \{0, 1\}$ for all i, by sending Bob

$$e_K(x) = \sum_{i=1}^{n} x_i a_i = S.$$

So here

$$\mathbb{P} = \mathbb{Z}_2^n, \quad \mathbb{C} = \{0, 1, \ldots, \sum_{i=1}^{n} a_i\}, \quad \text{and} \quad K = W = \{a_1, \ldots, a_n\}.$$

Bob hopes that his W is chosen cleverly enough to make solution of the knapsack problem—and hence construction of the decryption function d_K—inaccessible to attacker Oscar, while at the same time easy for Bob himself to decrypt with the help of the appropriate secret key.

The original ideas for knapsack cryptosystems were first set forth by Merkle and Hellman [**MeH**] in 1978. To design such a cryptosystem, one starts with a set $W' = \{b_1, \ldots, b_n\}$ for which decryption is easy (e.g., a superincreasing weight set, as in Example 11.12) and transforms it into the system's more challenging publicly announced weight system W, for instance by choosing relatively prime positive integers m and t, with $t > \sum_{i=1}^n b_i$, and making the substitutions

$$b_i \longmapsto a_i = mb_i \pmod{t}$$

for $1 \leq i \leq n$. (Here m and t are part of Bob's secret key.) Bob, knowing m and t, can compute $k = m^{-1} \pmod{t}$ and recast the decryption problem in terms of his secret "easy" weight set W'. Thus, given

$$x \overset{e_K}{\longmapsto} e_K(x) = \sum_{i=1}^n x_i a_i = S,$$

we then have

$$\begin{aligned} kS = k\sum_{i=1}^n x_i a_i &\equiv \sum_{i=1}^n x_i(kmb_i) \pmod{t} \\ &\equiv \sum_{i=1}^n x_i b_i \pmod{t}. \end{aligned}$$

Upon setting $S' = kS \pmod{t}$, Bob can readily solve the knapsack problem $S' = \sum_{i=1}^n x_i b_i$ for $x = (x_1, \ldots, x_n)$.

There are many variations and elaborations on the theme of changing "easy" weight systems to "hard" weight systems in such a way that encryption via the "hard" system can be reformulated in terms of the "easy" system with the help of the secret key. See Lagarias–Odlyzko [**LO**] and Odlyzko [**Od**] for details.

As it turns out, many knapsack cryptosystems can be successfully attacked with lattices. Here is a method due to Lagarias and Odlyzko in 1985. Suppose that by using the weights in the public key $W = \{a_1, \ldots, a_n\}$ Alice sends Bob the encrypted message

$$e_K(x) = \sum_{i=1}^n x_i a_i = S.$$

Oscar then considers the lattice $L = \mathbb{Z}v_1 + \cdots + \mathbb{Z}v_{n+1} \subset \mathbb{R}^{n+1}$ with upper-triangular generator matrix

$$\begin{pmatrix} 1 & & & & \\ 0 & 1 & & & \\ \vdots & \vdots & \ddots & & \\ 0 & 0 & & 1 & \\ -a_1 & -a_2 & \cdots & -a_n & S \end{pmatrix}.$$

(So v_j is the jth column.) Notice that if the weights a_j and S are large (as will typically be the case in practice), then these basis vectors are correspondingly large.

On the other hand, consider the vector

$$w = \left(\sum_{i=1}^{n} x_i v_i\right) + v_{n+1} = \begin{pmatrix} x_1 \\ x_2 \\ \vdots \\ x_n \\ 0 \end{pmatrix} \in L.$$

Since here $x_i \in \{0, 1\}$ for all i, the vector w is probably among the shortest vectors in L, and (it turns out) in fact it is often the shortest one. Therefore if Oscar has an algorithm for finding short vectors in a lattice, he is very likely to be able to solve the knapsack problem and decrypt Alice's message. From Theorem 11.9 we know that LLL-reduction produces a supply of short vectors to try as candidates for the knapsack solution.

The **density** of a weight set $W = \{a_1, \ldots, a_n\}$ is the ratio

$$d = \frac{n}{\max_i\{\log_2 a_i\}}.$$

Thus W has high density if all a_i are small, and low density otherwise. Lattice attacks on knapsack cryptosystems using LLL-reduction are generally most effective when W has low density. (Intuitively this is reasonable, since the presence of large weights corresponds to big vectors in the basis of Oscar's lattice, and this is likely to make short vectors more exceptional; see [**Od**] for the technical details.) But there are other forms of lattice reduction and lattice attack that are more effective for high density systems. For example, in 1994 Schnorr and Euchner [**SE**] used a variant of **Korkine–Zolotareff (KZ) lattice reduction** for this purpose.

Definition 11.14. A basis $\mathbb{B} = \{v_1, \ldots, v_n\}$ for a lattice L is **KZ-reduced** if

(i) $\mathbb{B}$ is size reduced; that is, condition 11.6(i) is satisfied; and

(ii) For each i, $1 \leq i \leq n$, the projection $\pi_i(L)$ of L on the subspace spanned by $\{v_i^*, \ldots, v_n^*\}$ satisfies the condition $\min \pi_i(L) = q(v_i^*)$.

See [**SE**] for a presentation of reduction algorithms and a comparative analysis of the effectiveness of attacks on knapsack cryptosystems via several forms of lattice reduction.

11.5. Remarks on Lattice-Based Cryptosystems

The previous section described an application of the structure theory of lattices to an attack on a cryptosystem whose design had no apparent connection with lattices. Over roughly the last decade there have been efforts in the opposite direction; that is, to design cryptosystems whose encryption schemes use lattices and whose security rests on the difficulty of lattice problems. The details involve technical issues far removed from the number-theoretic focus of this book, and our purpose here is only to give a brief indication of two such cryptosystems and cite some relevant literature. In both cases the basic idea is that a question about a $\mathbb{Z}$-lattice L on $\mathbb{R}^n$ that may be readily solvable if the lattice is given in terms of one basis may be much more difficult to solve if some other basis is given instead. The "difficult" basis will be part of the system's public key, while the "easy" basis is in the private key.

We have given a great deal of attention—starting with Hermite's inequality and continuing through LLL-reduction—to the search for short vectors in a lattice. In the cryptographic literature this is referred to as the **shortest vector problem** (**SVP**). The more general problem for lattices in $\mathbb{R}^n$ is the **closest vector problem (CVP)**: Given a point $P \in \mathbb{R}^n$ and a $\mathbb{Z}$-lattice L in $\mathbb{R}^n$, find a vector $v \in L$ whose distance from P is the smallest. Were we to choose P sufficiently close to 0, then such a vector v would be a minimal vector in L; so CVP has SVP as a special case, and we can therefore feel confident that CVP is a difficult problem.

Now imagine that L is the lattice generated by the standard orthonormal basis $\{e_1, \dots, e_n\}$ of $\mathbb{R}^n$. If we know this to be the case, and a vector $P = (\alpha_1, \dots, \alpha_n) \in \mathbb{R}^n$ is given to us, then a nearest lattice vector is $v = (\lfloor \alpha_1 \rceil, \dots, \lfloor \alpha_n \rceil)$. That is, we obtain v by rounding each coordinate of P to the nearest integer. But how would we solve this instance of CVP if some other basis were given for L and we did not know how to reconstitute a basis for L that would make the problem tractable? The difficulty of this problem is essential to the security of the **GGH cryptosystem**, proposed by Goldreich, Goldwasser, and Halevi [**GGH**] in 1997. Here the public key includes a basis $\{v_1, \dots, v_n\}$ for a lattice L on $\mathbb{R}^n$, along with a short "error" or "perturbation" vector e. Encryption of an input vector $x = (x_1, \dots, x_n) \in \mathbb{Z}^n$ is carried out by

$$e_K(x) = \sum_{i=1}^{n} x_i v_i + e$$

and decryption requires finding the closest lattice vector to $e_K(x)$. Security of the system relies on careful selection of the vectors v_i and the error vector e, and on the exact way in which the message being sent is embedded in the vector x. And underlying it all is an analysis of how approachable CVP is with respect to a given basis.

For a second lattice-based cryptosystem we turn to a system proposed in 1996 by Ajtai and Dwork [**AjD**], which they base on what they call **hidden hyperplanes**, and which requires constructing a special sort of lattice L on $\mathbb{R}^n$. For this, start with real numbers M and d, with $0 < M < d$, and in $\mathbb{R}^n$ construct a set $\mathbb{B}_0 = \{v_1, \dots, v_{n-1}\}$ of linearly independent vectors of length at most M. This starts off a basis for the lattice L under construction; write L' for the $\mathbb{Z}$-lattice generated by $\mathbb{B}_0$. Then use Gram–Schmidt to produce a vector $w \in \mathbb{R}^n$ orthogonal to L' and with $\|w\| \geq d$. Finally, take as a last basis vector for L any vector v_n of the form $w + \sum_{i=1}^{n-1} \alpha_i v_i$ with $\alpha_i \in \mathbb{R}$. Then it is easy to check that L'—and the hyperplane H that it spans—is generated by all the vectors in L that have length $\leq M$; all other vectors in L have length $\geq d$. The hyperplane H and the cosets of H of the form $av_n + H$, with $a \in \mathbb{Z}$, are the "hidden hyperplanes" of the system. Part of the private key is the basis $\mathbb{B}_0$ for L'. The input for the encryption is a binary vector, and the encryption is done bit by bit with the help of a public basis for L that reveals nothing about the hidden hyperplanes. For $x \in \{0, 1\}$, the encrypted bit $e_K(x)$ is a point in the vicinity of one of the hidden hyperplanes. If the point is sufficiently close to that hyperplane, then $e_K(x)$ is decrypted as 0; otherwise $e_K(x)$ is decrypted as 1.

For details on these and other lattice-based cryptosystems, see the cited papers, the book of Micciancio and Goldwasser [**MG**], and the collection edited by Silverman [**Si**], with particular attention to the paper by Nguyen and Stern [**NS**].

Appendix: Further Reading

No book on a subject as rich as the theory of quadratic forms can fully explore—or even touch upon—all aspects of the subject. It is my hope that the reader of this book will be interested in going beyond what has been presented here. My purpose in this appendix is to give a guide to the quadratic forms literature by listing some of the other books in which quadratic forms are the central topic, with ultra-brief comments on each one. *I will restrict my list to books that are accessible after or alongside this one without requiring a substantial amount of additional background.* Some have already been cited in this book—some quite frequently—while others have received no mention up to now. Even when the subject matter is familiar there is something to be gained by seeing other perspectives and examples. The books will be listed in approximate alphabetical order by author's name. Needless to say (but I will say it anyway), each of these books has its own bibliography, and the union of all these bibliographies constitutes much of the work on quadratic forms over the last hundred and fifty years.

Geometric Algebra, by Emil Artin [**A**]. The geometry of vector spaces. The second half of the book is almost entirely about quadratic forms over fields.

Binary Quadratic Forms, by Johannes Buchmann and Ulrich Vollmer [**BV**]. A thorough exploration of the special universe of binary quadratic forms, ranging from the pioneering work by Gauss and Euler through current applications to cryptography.

Rational Quadratic Forms, by J. W. S. Cassels [**C2**]. A broad look at quadratic forms over $\mathbb{Q}$ and over $\mathbb{Z}$, significantly overlapping with and significantly departing from the material in this book.

The Sensual (Quadratic) Form, by John H. Conway [**Con2**]. The author's personal insights into integral quadratic forms, with particular attention to binary forms and to representation issues, for instance to isospectral lattices—lattices representing the same numbers (and equally often)—over the integers.

Sphere Packings, Lattices and Groups, by John Conway and Neil Sloane [**CS**], and *Lattices and Codes*, by Wolfgang Ebeling [**Eb**]. These are the places to learn about the interaction of the theory of integral quadratic forms and coding theory. The two-part bibliography in [**CS**] is extraordinary.

Primes of the Form $x^2 + ny^2$, by David A. Cox [**Cox**]. This book uses the theory of binary quadratic forms as the launching pad for explorations in algebraic number theory and class field theory.

Contemporary Mathematics [**KHKS**], [**BLR**], [**BHJP**]. These three collections of papers presented at conferences in 1998, 1999, and 2002, offer a broad sampling of recent research in the algebraic and arithmetic theories of quadratic forms.

Quadratische Formen und Orthogonale Gruppen, by Martin Eichler [**Ei2**]. This wide-ranging book really deepened the modern local-global approach to the classification of lattices over the rings of number theory. For example, it includes the introduction of the spinor genus and its applications.

Representations of Integers as Sums of Squares, by Emil Grosswald [**Gr**]. A thorough examination of the title subject, largely via the application of analytic number theory.

Classical Groups and Geometric Algebra, by Larry C. Grove [**Gro**]. The classical groups are the interesting subgroups of the general linear group $GL_n(V)$, where V is a vector space, and many of these groups relate to quadratic or hermitian forms on V.

The Classical Groups and K-Theory, by Alexander J. Hahn and O. Timothy O'Meara [**HO**]. A thorough treatment of the topics in the title, with plenty of attention to the groups associated with quadratic and hermitian forms.

Differentiable Manifolds and Quadratic Forms, by Friedrich Hirzebruch, Walter D. Neumann, and S. S. Koh [**HNK**]. Applications of unimodular $\mathbb{Z}$-lattices to the classification of manifolds.

The Arithmetic Theory of Quadratic Forms, by Burton W. Jones [**Jo**]. A classical treatment of quadratic forms over $\mathbb{Q}, \mathbb{Z}$, and the p-adic integers, including the composition theory of binary forms over $\mathbb{Z}$.

Linear Algebra and Geometry, by Irving Kaplansky [**Kap**]. Quadratic forms, mostly over fields, and their interactions with and applications to investigations in linear algebra and geometry.

Arithmetic of Quadratic Forms, by Yoshiyuki Kitaoka [**Kit**]. The first third of Kitaoka's book has considerable overlap with this book, and then the author goes on to extensive discussions of the analytic theory of quadratic forms and his studies of tensor products of lattices.

Quadratische Formen, by Martin Kneser [**K3**]. A wide-ranging study of virtually all aspects of quadratic forms over $\mathbb{Q}$ and $\mathbb{Z}$ and over their localizations.

Quadratic and Hermitian Forms over Rings, by Max-Albert Knus [**Knu**]. An extensive study of the many algebraic structures associated with quadratic and hermitian forms over rings.

Introduction to Quadratic Forms over Fields, by T. Y. Lam [**Lam**]. Lam's book is the place to go for everything you want to know about the algebraic theory of quadratic forms.

Perfect Lattices in Euclidean Spaces, by Jacques Martinet [**Mar**]. A broad and detailed study of lattices in $\mathbb{R}^n$ and their applications, with plenty of explicit examples.

Complexity of Lattice Problems, by Daniele Micciancio and Shafi Goldwasser [**MG**], and *Cryptography and Lattices*, edited by Joseph Silverman [**Si**]. These two books are prime sources for the recent work on the interaction of lattices with public key cryptography. It is an unfortunate sociological fact that to date there has not been much interaction among number theorists interested in lattices because of their connections with quadratic forms and cryptographers interested in lattices for other purposes. These books may facilitate that interaction.

Symmetric Bilinear Forms, by John Milnor and Dale Husemoller [**MH**]. A selection of topics from the arithmetic, analytic, and algebraic sectors of the theory of quadratic forms, with applications to sphere packing and topology.

Integral Matrices, by Morris Newman [**N1**]. Explorations of assorted matrix relations, including congruence and matrix equivalence, so important in the theory of quadratic forms and in linear algebra, with applications to the classical groups.

Introduction to Quadratic Forms, by O. Timothy O'Meara [**O'M1**]. Essentially unchanged since its original publication in 1963, this book continues to provide the best background for studying quadratic forms over the interesting subrings of global fields and their completions.

Variations on a Theme of Euler, by Takashi Ono [**O**]. Connections of the theory of quadratic forms to topics in geometry, especially to the theory of elliptic curves.

Quadratic Forms with Applications to Algebraic Geometry and Topology, by Albrecht Pfister [**Pf2**]. The title gives a succinct summary of the contents. The book includes some of Pfister's pioneering work from the 1960s that launched the algebraic theory of quadratic forms.

Squares, by A. R. Rajwade [**Ra**]. Sums of squares in many settings, but especially over fields—particularly function fields—and polynomial rings.

Quadratic and Hermitian Forms, by Winfried Scharlau [**Scha**]. An interesting blend of the algebraic and arithmetic aspects of the theory of quadratic and Hermitian forms, along with the associated algebraic structures.

From Fermat to Minkowski: Lectures on the Theory of Numbers and Its Historical Development, by Winfried Scharlau and Hans Opalka [**SO**]. A fascinating history of work in number theory by the early masters of the subject. Much of that history involves the theory of quadratic forms.

A Course in Arithmetic, by Jean-Pierre Serre [**Se**]. A compact introduction to quadratic forms over $\mathbb{Q}$ and $\mathbb{Q}_p$, and to the unimodular theory over $\mathbb{Z}$, including applications of analytic number theory to the representation theory of positive definite unimodular $\mathbb{Z}$-lattices. Serre's construction of the p-adic numbers via inverse limits is quite different from the treatment in this book.

Bilinear Algebra, by Kazimierz Szymiczek [**Sz**]. Bilinear forms and the many structures associated with the algebraic theory of quadratic forms.

Integral Quadratic Forms, by G. L. Watson [**W1**]. Watson eschewed most of the algebraic machinery available by the second half of the Twentieth Century. His writing is highly idiosyncratic, and it can be slow going for the reader. Nevertheless, his book and his many papers constitute a remarkable body of work and are well worth studying.

Bibliography

[AjD] M. Ajtai and C. Dwork, *A public-key cryptosystem with worst-case/average case equivalence*, in Proceedings 29th Annual Assoc. Comput. Mach. Symposium on Theory of Computing, 284–293, ACM, 1997.

[A] E. Artin, *Geometric Algebra* (reprint of the 1957 original), John Wiley & Sons, New York, 1988.

[Ba] G. Bachman, *Introduction to p-adic Numbers and Valuation Theory*, Academic Press, New York-London 1964.

[BHJP] R. Baeza, J. S. Hsia, B. Jacob, A. Prestel (editors), *Algebraic and Arithmetic Theory of Quadratic Forms*, Contemporary Mathematics v. 344, American Mathematical Society, Providence, 2004.

[BLR] E. Bayer-Fluckiger, D. Lewis, A. Ranicki (editors), *Quadratic Forms and Their Applications*, Contemporary Mathematics v. 272, American Mathematical Society, Providence, 2000.

[Be] C. N. Beli, *Representations of integral quadratic forms over dyadic local fields*, Electron. Res. Announc. Amer. Math. Soc. 12 (2006), 100–112.

[Bh] M. Bhargava, *On the Conway-Schneeberger fifteen theorem*, in Quadratic Forms and Their Applications, 27–37, Contemp. Math. 272, Amer. Math. Soc., Providence, RI, 2000.

[Bl] F. van der Blij, *An invariant of quadratic forms mod* 8, Indag. Math. 21 (1959), 291–293.

[Bo] C. B. Boyer, *A History of Mathematics* (revised by U. C. Merzbach), John Wiley & Sons, 1991.

[BR] R. H. Bruck and H. J. Ryser, *The nonexistence of certain finite projective planes*, Canadian J. Math. 1 (1949), 88–93.

[BV] J. Buchmann and U. Vollmer, *Binary Quadratic Forms*, Springer; Berlin, Heidelberg, New York, 2007.

[Bu] J. E. Bureau, *Representation properties of definite lattices in function fields*, Ph.D. dissertation, Louisiana State University, December 2006.

[Ca] R. Casse, *Projective Geometry: An Introduction*, Oxford University Press, Oxford, 2006.

[C1] J. W. S. Cassels, *On the representation of rational functions as sums of squares*, Acta Arith. 9 (1964), 79–82.

[C2] J. W. S. Cassels, *Rational Quadratic Forms*, Academic Press, Inc. [Harcourt Brace Jovanovich, Publishers], London-New York, 1978.

[C3] J. W. S. Cassels, *Local Fields*, London Mathematical Society, Cambridge University Press, Cambridge, 1986.

[ChD] W. K. Chan and J. Daniels, *Definite regular quadratic forms over* $\mathbb{F}_q[T]$, Proc. Amer. Math. Soc. 133 (2005), 3121–3131.

[Coh] H. Cohen, *A Course in Computational Algebraic Number Theory*, Springer-Verlag, Berlin, 1993.

[Con1] J. H. Conway, *A characterisation of Leech's lattice*, Invent. Math. 7 (1969), 137–142.

[Con2] J. H. Conway, *The Sensual (Quadratic) Form*, Carus Math. Monographs 26, Mathematical Association of America, 1997.

[Con3] J. H. Conway, *Universal quadratic forms and the fifteen theorem*, in Quadratic Forms and Their Applications, 23–26, Contemp. Math. 272, Amer. Math. Soc., Providence, RI, 2000.

[CS] J. H. Conway and N. J. A. Sloane, *Sphere Packings, Lattices and Groups* (third edition), Springer-Verlag, New York, 1999.

[Cox] D. A. Cox, *Primes of the Form* $x^2 + ny^2$, John Wiley and Sons, New York, 1997.

[CR] C.W. Curtis and I. Reiner, *Representation Theory of Finite Groups and Associative Algebras* (reprint of the 1962 original), AMS Chelsea Publishing, Providence, RI, 2006.

[D] L. E. Dickson, *Lowest integers representing sides of a right triangle*, Amer. Math. Monthly 1 (1894), 6–11.

[Dj] D. Djoković, *Hermitian matrices over polynomial rings*, J. Algebra 43 (1976), 359–374.

[EN] A. G. Earnest and G. Nipp, *On the theta series of positive quaternary quadratic forms*, C. R. Math. Rep. Acad. Sci. Canada 13 (1991), 33–38.

[Eb] W. Ebeling, *Lattices and Codes* (second revised edition), Friedr. Vieweg & Sohn, Braunschweig, 2002.

[Ei1] M. Eichler, *Note zur Theorie der Kristallgitter*, Math. Ann. 125 (1952), 51–55.

[Ei2] M. Eichler, *Quadratische Formen und Orthogonale Gruppen*, Springer-Verlag, Berlin-Göttingen-Heidelberg, 1952

[El1] N. D. Elkies, *A characterization of the* Z^n *lattice*, Math. Res. Lett. 2 (1995), 321–326.

[El2] N. D. Elkies, *Lattices and codes with long shadows*, Math. Res. Lett. 2 (1995), 643–651.

[FS] R. Fintushel and R. J. Stern, *Definite 4-manifolds*, J. Differential Geom. 28 (1988), 133–141.

[GJ] M. R. Garey and D. S. Johnson, *Computers and Intractability*, W. H. Freeman and Company, 1979.

[Gar] D. A. Garbanati, *An algorithm for the representation of* 0 *by a quadratic form*, J. Pure and Applied Algebra 13 (1978), 57–63.

[Ga] M. Gaulter, *Lattices without short characteristic vectors*, Math. Res. Lett. 5 (1998), 353–362.

[G1] L. J. Gerstein, *The growth of class numbers of quadratic forms*, Amer. J. Math. 94 (1972), 221–236.

[G2] L. J. Gerstein, *A new proof of a theorem of Cassels and Pfister*, Proc. Amer. Math. Soc. 41 (1973), 327–328.

[G3] L. J. Gerstein, *A remark on the quadratic analogue of the Quillen-Suslin theorem*, J. Reine Angew. Math. 337 (1982), 166–170.

[G4] L. J. Gerstein, *Nearly unimodular quadratic forms*, Annals of Math. 142 (1995), 597–610.

[G5] L. J. Gerstein, *Definite quadratic forms over $\mathbb{F}_q[x]$*, J. Algebra 268 (2003), 252–263.

[G6] L. J. Gerstein, *On representation by quadratic $\mathbb{F}_q[x]$-lattices*, Algebraic and Arithmetic Theory of Quadratic Forms, Contemp. Math. 344 (2004), 129–134, Amer. Math. Soc., Providence, RI.

[G7] L. J. Gerstein, *Characteristic elements of unimodular $\mathbb{Z}$-lattices*, Linear and Multilinear Algebra 52 (2004), 381–383.

[GGH] O. Goldreich, S. Goldwasser, and S. Halevi, *Public-key cryptosystems from lattice reduction problems*, Advances in Cryptology—CRYPTO '97 (Santa Barbara, CA, 1997), 112–131, Lecture Notes in Comput. Sci. 1294, Springer, Berlin, 1997.

[Gr] E. Grosswald, *Representations of Integers as Sums of Squares*, Springer-Verlag, New York, 1985.

[Gro] L. C. Grove, *Classical Groups and Geometric Algebra*, American Mathematical Society, Providence, RI, 2002.

[HO] A. J. Hahn and O. T. O'Meara, *The Classical Groups and K-Theory*, Springer-Verlag, New York, 1989.

[H] M. Hall, *Combinatorial Theory* (reprint of the 1986 second ed.), John Wiley & Sons, Inc., New York, 1998.

[HNK] F. Hirzebruch, W. D. Neumann, and S. S. Koh, *Differentiable Manifolds and Quadratic Forms*, Marcel Dekker, New York, 1971.

[IR] K. Ireland and M. Rosen, *A Classical Introduction to Modern Number Theory*, Springer-Verlag, 1982.

[J] N. Jacobson, *A note on hermitian forms*, Bull. Amer. Math. Soc. 46 (1940), 264–268.

[Jo] B. W. Jones, *The Arithmetic Theory of Quadratic Forms*, by Burton W. Jones, Mathematical association of America, 1950.

[Kap] I. Kaplansky *Linear Algebra and Geometry*, Dover Publications, 2003. (Reprint of the 1974 Edition by Chelsea Publications.)

[KWX] M.-H. Kim, Y. Wang, and F. Xu, *Universal quadratic forms over $\mathbb{F}_q[T]$*, preprint, 2004.

[KHKS] M.-H. Kim, J. S. Hsia, Y. Kitaoka, R. Schulze-Pillot (editors), *Integral Quadratic Forms and Lattices*, Contemporary Mathematics v. 249, American Mathematical Society, Providence, 1999.

[Kit] Y. Kitaoka, *Arithmetic of Quadratic Forms*, Cambridge University Press, Cambridge, 1999.

[Kb] M. Knebusch, *Grothendieck- und Wittringe von nichtausgearteten symmetrischen Bilinearformen*, S.-B. Heidelberger Akad. Wiss. Math.-Natur. Kl. (1969/70), 93–157.

[K1] M. Kneser, *Zur Theorie der Kristallgitter*, Math. Ann. 127 (1954), 105–106.

[K2] M. Kneser, *Klassenzahlen definiter quadratischer Formen*, Arch. Math. 8 (1957), 241–250.

[K3] M. Kneser, *Quadratische Formen*, Springer-Verlag, Berlin-Heidelberg-New York, 2002.

[Knu] M.-A. Knus, *Quadratic and Hermitian Forms over Rings*, Springer-Verlag, Berlin-Heidelberg, 1991.

[LO] J. C. Lagarias and A. M. Odlyzko, *Solving low-density subset sum problems*, J. Assoc. Comput. Mach. 32 (1985), 229–246.

[Lam] T. Y. Lam, *Introduction to Quadratic Forms over Fields*, American Mathematical Society, Providence, RI, 2005.

[Le] W. Leahey, *Sums of squares of polynomials with coefficients in a finite field*, Amer. Math. Monthly 74 (1967), 816–819.

[Lp] D. Leep, *Editor's Endnotes*, Amer. Math. Monthly 112 (2005), 943–944; journal error corrected in vol. 113 (2006), 671.

[LLL] A. K. Lenstra, H. W. Lenstra, Jr., and L. Lovasz, *Factoring polynomials with rational coefficients*, Math. Ann. 261 (1982), 515–534.

[LN] R. Lidl and H. Niederrreiter, *Introduction to Finite Fields and Their Applications*, Cambridge University Press, 1986.

[MB] S. MacLane and G. Birkhoff, *Algebra* (third edition), Chelsea Publishing Co., New York, 1988.

[Mag] W. Magnus, *Über die Anzahl der in einem Geschlecht enthaltenen Klassen von positiv-definiten quadratischen Formen*, Math. Ann. 114 (1937), 465–475.

[Mar] J. Martinet, *Perfect Lattices in Euclidean Space*, Springer-Verlag, Berlin-Heidelberg, 2003.

[MeH] R. C. Merkle and M. Hellman, *Hiding information and signatures in trapdoor knapsacks*, IEEE Trans. Inform. Theory IT-24: 525–530, September, 1978.

[MG] D. Micciancio and S. Goldwasser, *Complexity of Lattice Problems*, Kluwer Academic Publishers, Boston, MA, 2002.

[MH] J. Milnor and D. Husemoller, *Symmetric Bilinear Forms*, Springer-Verlag, New York-Heidelberg, 1973.

[N1] M. Newman, *Integral Matrices*, Academic Press, New York-London, 1972.

[N2] M. Newman, *Tridiagonal matrices*, Linear Algebra Appl. 201 (1994), 51–55.

[NS] P. Q. Nguyen and J. Stern, *The two faces of lattices in cryptology*, in Cryptography and Lattices, 146–180, Lecture Notes in Comput. Sci. 2146, Springer, Berlin, 2001.

[Od] A. M. Odlyzko, *The rise and fall of knapsack cryptosystems*, in Cryptology and Computational Number Theory, 75–88, Proc. Sympos. Appl. Math., 42, Amer. Math. Soc., Providence, RI, 1990.

[O'M1] O. T. O'Meara, *Introduction to Quadratic Forms* (reprint of the 1973 edition), Springer-Verlag, Berlin, 2000.

[O'M2] O. T. O'Meara, *The automorphisms of the orthogonal groups and their congruence subgroups over arithmetic domains*, J. Reine Angew. Math. 238 (1969), 169–206.

[O] T. Ono, *Variations on a Theme of Euler*, Plenum Press, New York and London, 1994.

[P] S. Parimala, *Failure of a quadratic analogue of Serre's conjecture*, Bull. Amer. Math. Soc. 82 (1976), 962–964.

[Pf] A. Pfister, *Zur Darstellung von* -1 *als Summe von Quadraten in einem Körper*, J. London Math. Soc. 40 (1965), 159–165.

[Pf2] A. Pfister, *Quadratic Forms with Applications to Algebraic Geometry and Topology*, Cambridge University Press, 1995.

[PZ] M. Pohst and H. Zassenhaus, *Algorithmic Algebraic Number Theory*, Cambridge University Press, 1997.

[PD] H. Pollard and H. G. Diamond, *The Theory of Algebraic Numbers* (third edition), Dover Publications, Inc., Mineola, NY, 1998.

[Ra] A. R. Rajwade, *Squares*, Cambridge University Press, 1993.

[Ri] C. Riehm, *On the integral representation of quadratic forms over local fields*, Amer. J. Math. 86 (1964), 25–62.

[R] M. Rosen, *Number Theory in Function Fields*, Springer-Verlag, New York, 2002.

[Rog] C. A. Rogers, *Packing and Covering*, Cambridge University Press, 1964.

[SO] W. Scharlau and H. Opolka, *From Fermat to Minkowski: Lectures on the Theory of Numbers and Its Historical Development*, Springer-Verlag, New York, 1985.

[Scha] W. Scharlau, *Quadratic and Hermitian Forms*, Springer-Verlag, Berlin 1985.

[Schn] W. A. Schneeberger, *Arithmetic and Geometry of Integral Lattices*, Ph.D. dissertation, Princeton University, November 1997.

[SE] C.-P. Schnorr and M. Euchner, *Lattice basis reduction: improved practical algorithms and solving subset sum problems*, Math. Programming 66 (1994), no. 2, Ser. A, 181–199.

[Sen] M. Senechal, *Quasicrystals and Geometry*, Cambridge University Press, Cambridge, 1995.

[Se] J.-P. Serre, *A Course in Arithmetic*, Springer-Verlag, New York-Heidelberg, 1973.

[Si] J. Silverman (editor), *Cryptography and Lattices*, Lecture Notes in Computer Science 2146, Springer-Verlag, Berlin, 2001.

[Sin] S. Singh, *The Code Book*, Doubleday, New York, 1999.

[Sz] K. Szymiczek, *Bilinear Algebra*, Gordon and Breach, Amsterdam, 1997.

[W1] G. L. Watson, *Integral Quadratic Forms*, Cambridge University Press, 1960.

[W2] G. L. Watson, *The class-number of a positive quadratic form*, Proc. London Math. Soc. (3) 13 (1963), 549–576.

[We] A. Weil, *Number Theory, An Approach Through History, From Hammurapi to Legendre*, Birkhäuser Boston, Boston, MA, 1984.

[Y] S. Y. Yan, *Number Theory for Computing* (second edition), Springer-Verlag, New York-Heidelberg-Berlin, 2002.

Index

Titles in This Series

90 **Larry J. Gerstein,** Basic quadratic forms, 2008

89 **Anthony Bonato,** A course on the web graph, 2008

88 **Nathanial P. Brown and Narutaka Ozawa,** C^*-algebras and finite-dimensional approximations, 2008

87 **Srikanth B. Iyengar, Graham J. Leuschke, Anton Leykin, Claudia Miller, Ezra Miller, Anurag K. Singh, and Uli Walther,** Twenty-four hours of local cohomology, 2007

86 **Yulij Ilyashenko and Sergei Yakovenko,** Lectures on analytic differential equations, 2007

85 **John M. Alongi and Gail S. Nelson,** Recurrence and topology, 2007

84 **Charalambos D. Aliprantis and Rabee Tourky,** Cones and duality, 2007

83 **Wolfgang Ebeling,** Functions of several complex variables and their singularities (translated by Philip G. Spain), 2007

82 **Serge Alinhac and Patrick Gérard,** Pseudo-differential operators and the Nash–Moser theorem (translated by Stephen S. Wilson), 2007

81 **V. V. Prasolov,** Elements of homology theory, 2007

80 **Davar Khoshnevisan,** Probability, 2007

79 **William Stein,** Modular forms, a computational approach (with an appendix by Paul E. Gunnells), 2007

78 **Harry Dym,** Linear algebra in action, 2007

77 **Bennett Chow, Peng Lu, and Lei Ni,** Hamilton's Ricci flow, 2006

76 **Michael E. Taylor,** Measure theory and integration, 2006

75 **Peter D. Miller,** Applied asymptotic analysis, 2006

74 **V. V. Prasolov,** Elements of combinatorial and differential topology, 2006

73 **Louis Halle Rowen,** Graduate algebra: Commutative view, 2006

72 **R. J. Williams,** Introduction the the mathematics of finance, 2006

71 **S. P. Novikov and I. A. Taimanov,** Modern geometric structures and fields, 2006

70 **Seán Dineen,** Probability theory in finance, 2005

69 **Sebastián Montiel and Antonio Ros,** Curves and surfaces, 2005

68 **Luis Caffarelli and Sandro Salsa,** A geometric approach to free boundary problems, 2005

67 **T.Y. Lam,** Introduction to quadratic forms over fields, 2004

66 **Yuli Eidelman, Vitali Milman, and Antonis Tsolomitis,** Functional analysis, An introduction, 2004

65 **S. Ramanan,** Global calculus, 2004

64 **A. A. Kirillov,** Lectures on the orbit method, 2004

63 **Steven Dale Cutkosky,** Resolution of singularities, 2004

62 **T. W. Körner,** A companion to analysis: A second first and first second course in analysis, 2004

61 **Thomas A. Ivey and J. M. Landsberg,** Cartan for beginners: Differential geometry via moving frames and exterior differential systems, 2003

60 **Alberto Candel and Lawrence Conlon,** Foliations II, 2003

59 **Steven H. Weintraub,** Representation theory of finite groups: algebra and arithmetic, 2003

58 **Cédric Villani,** Topics in optimal transportation, 2003

57 **Robert Plato,** Concise numerical mathematics, 2003

56 **E. B. Vinberg,** A course in algebra, 2003

55 **C. Herbert Clemens,** A scrapbook of complex curve theory, second edition, 2003

54 **Alexander Barvinok,** A course in convexity, 2002
53 **Henryk Iwaniec,** Spectral methods of automorphic forms, 2002
52 **Ilka Agricola and Thomas Friedrich,** Global analysis: Differential forms in analysis, geometry and physics, 2002
51 **Y. A. Abramovich and C. D. Aliprantis,** Problems in operator theory, 2002
50 **Y. A. Abramovich and C. D. Aliprantis,** An invitation to operator theory, 2002
49 **John R. Harper,** Secondary cohomology operations, 2002
48 **Y. Eliashberg and N. Mishachev,** Introduction to the h-principle, 2002
47 **A. Yu. Kitaev, A. H. Shen, and M. N. Vyalyi,** Classical and quantum computation, 2002
46 **Joseph L. Taylor,** Several complex variables with connections to algebraic geometry and Lie groups, 2002
45 **Inder K. Rana,** An introduction to measure and integration, second edition, 2002
44 **Jim Agler and John E. McCarthy,** Pick interpolation and Hilbert function spaces, 2002
43 **N. V. Krylov,** Introduction to the theory of random processes, 2002
42 **Jin Hong and Seok-Jin Kang,** Introduction to quantum groups and crystal bases, 2002
41 **Georgi V. Smirnov,** Introduction to the theory of differential inclusions, 2002
40 **Robert E. Greene and Steven G. Krantz,** Function theory of one complex variable, third edition, 2006
39 **Larry C. Grove,** Classical groups and geometric algebra, 2002
38 **Elton P. Hsu,** Stochastic analysis on manifolds, 2002
37 **Hershel M. Farkas and Irwin Kra,** Theta constants, Riemann surfaces and the modular group, 2001
36 **Martin Schechter,** Principles of functional analysis, second edition, 2002
35 **James F. Davis and Paul Kirk,** Lecture notes in algebraic topology, 2001
34 **Sigurdur Helgason,** Differential geometry, Lie groups, and symmetric spaces, 2001
33 **Dmitri Burago, Yuri Burago, and Sergei Ivanov,** A course in metric geometry, 2001
32 **Robert G. Bartle,** A modern theory of integration, 2001
31 **Ralf Korn and Elke Korn,** Option pricing and portfolio optimization: Modern methods of financial mathematics, 2001
30 **J. C. McConnell and J. C. Robson,** Noncommutative Noetherian rings, 2001
29 **Javier Duoandikoetxea,** Fourier analysis, 2001
28 **Liviu I. Nicolaescu,** Notes on Seiberg-Witten theory, 2000
27 **Thierry Aubin,** A course in differential geometry, 2001
26 **Rolf Berndt,** An introduction to symplectic geometry, 2001
25 **Thomas Friedrich,** Dirac operators in Riemannian geometry, 2000
24 **Helmut Koch,** Number theory: Algebraic numbers and functions, 2000
23 **Alberto Candel and Lawrence Conlon,** Foliations I, 2000
22 **Günter R. Krause and Thomas H. Lenagan,** Growth of algebras and Gelfand-Kirillov dimension, 2000
21 **John B. Conway,** A course in operator theory, 2000
20 **Robert E. Gompf and András I. Stipsicz,** 4-manifolds and Kirby calculus, 1999
19 **Lawrence C. Evans,** Partial differential equations, 1998